Biological and Medical Physics, Biomedical Engineering

BIOLOGICAL AND MEDICAL PHYSICS, BIOMEDICAL ENGINEERING

This series is intended to be comprehensive, covering a broad range of topics important to the study of the physical, chemical and biological sciences. Its goal is to provide scientists and engineers with textbooks, monographs, and reference works to address the growing need for information. The fields of biological and medical physics and biomedical engineering are broad, multidisciplinary and dynamic. They lie at the crossroads of frontier research in physics, biology, chemistry, and medicine.

Books in the series emphasize established and emergent areas of science including molecular, membrane, and mathematical biophysics; photosynthetic energy harvesting and conversion; information processing; physical principles of genetics; sensory communications; automata networks, neural networks, and cellular automata. Equally important is coverage of applied aspects of biological and medical physics and biomedical engineering such as molecular electronic components and devices, biosensors, medicine, imaging, physical principles of renewable energy production, advanced prostheses, and environmental control and engineering.

More information about this series at http://www.springer.com/series/3740

Richard De La Rue · Hans Peter Herzig ·
Martina Gerken
Editors

Biomedical Optical Sensors

Differentiators for Winning Technologies

Springer

Editors
Richard De La Rue
Optoelectronics Research Group
University of Glasgow
School of Engineering
Glasgow, UK

Hans Peter Herzig
STI-DO
Ecole Polytechnique Federale Lausan
Neuchâtel, Switzerland

Martina Gerken
Faculty of Engineering
Kiel University
Kiel, Schleswig-Holstein, Germany

ISSN 1618-7210 ISSN 2197-5647 (electronic)
Biological and Medical Physics, Biomedical Engineering
ISBN 978-3-030-48385-2 ISBN 978-3-030-48387-6 (eBook)
https://doi.org/10.1007/978-3-030-48387-6

This Springer imprint is published by the registered company Springer Nature Switzerland AG
The registered company address is: Gewerbestrasse 11, 6330 Cham, Switzerland

Preface

Biomedical sensing is an increasingly important domain of research and development activity across many countries, with the potential to have widespread impact across the globe. Biomedicine is itself of central importance for the human race and the development of advanced civilisation, because it not only addresses the ever-growing need for reliable and sustainable healthcare of a world population that continues to expand, but also because it is likely to be required (i.e. have an essential role) when humanity faces the challenges that will inevitably arise in the future—and have already arisen to a large extent. The increasing importance of biomedical sensors is directly linked to the development of medical technologies—and medicine more generally. Biomedical sensors are already being used to increase the reliability and speed of medical diagnosis, both prior to and after medical events experienced by individual human beings and, potentially, by large numbers of human beings at the same time, as well as sequentially.

This book is intended to form a useful, but limited, introduction to the topic of biomedical sensing based on optical or photonic techniques. The light that is available for purposes such as biomedical sensing—and has been demonstrated as useful—covers a very wide (multi-octave) spectrum, including parts of the Ultra-Violet (UV) spectrum, all of the visible spectrum and much of the infrared spectrum—and frequencies as low as the TeraHertz (THz) region of the electromagnetic spectrum. The utility of light in biomedical sensing arises, in part, from the fact that the wavelength (and therefore the frequency and photon energy) of the light can be selected, with considerable precision, to match the characteristic resonant frequencies of the bio-molecules, e.g. proteins, of interest. In this context, we note that the bond vibration spectrum of many biological molecules lies in the Mid-Infrared Region (MIR) of the electromagnetic spectrum. This mid-infrared region is readily accessible using standard laboratory apparatus such as Fourier-Transform Infrared (FTIR) spectrometers, together with weak and spectrally broad light sources. Alternatively, coherent light sources such as diode lasers can be tuned to well-defined target wavelengths—and useful identification and measurement of biomaterial may be obtained using a specific single optical wavelength or a specific set of different wavelengths. Such identification may be obtained via the use of fluorescent labelling of characteristic bio-molecules or cells—and by means of label-free specific binding processes.

Our intention, in the following chapters, is to identify scientific and technological concepts that are both intellectually interesting and potentially applicable in biomedicine. Inevitably there will be some overlap between the material that is treated in any one chapter and material appearing in one or more other chapters.

Chapters 1–3 are concerned with optical/photonic sensors that are based on planar optical waveguide techniques. Chapter 4 is also concerned with biosensors that typically exploit planar fabrication techniques based on flat substrates, but the interaction of the light with the sensor is typically not through waveguiding action. The photonic crystal structures covered in Chap. 5 may be realised in various ways. In many cases, the photonic crystal structures are realised by planar fabrication techniques. But other fabrication techniques have also been used, e.g. the realisation of synthetic opal structures by colloidal deposition of approximately spherical silica micro-balls from solution. This last technique can be used in situations where the substrate for the photonic crystal structure is non-planar, e.g. the walls of a cylindrical tube.

Chapter 1 'Towards Refractive Index Sensing Using Nanoscale Slot Waveguide Cavities' identifies convincingly the merits of the slot waveguide configuration for bio-sensing applications.

Chapter 2 specifically addresses the topic of waveguide biosensors based on plasmonic techniques—exploiting, in particular, the advantages that result from the use of long-range plasmon propagation. The application of this approach to disease detection in complex fluids is also discussed.

Chapter 3 explores the merits of using spectroscopy based on the collective modal characteristics of multimode planar optical waveguides—where the hybrid modes of plasmonic–dielectric structures can be exploited.

Chapter 4 has the title: 'Refractive Index Sensing with Anisotropic Hyperbolic Metamaterials'—and it addresses the more general topic of refractive index (change) based sensing—but includes bio-sensing as a lead application.

Chapter 5 covers the topic of 'Photonic Crystal Biosensors' in essentially three distinct categories: (a) sensor surfaces based on 1D and 2D periodic slabs—and their application in Photonic Crystal Enhanced Microscopy (PCEM), (b) devices based on both 2D and 1D photonic crystal structures realised in planar waveguides and (c) primarily 3D photonic crystal structures, but also closely related 2D photonic crystal structures. In this third part, non-planar, modified or partially planar fabrication processes play a significant role. Notably, because of its distinct nature, the topic of 'photonic crystal fibers' is not addressed in this chapter. It is however covered, to some extent, in the following chapter.

Chapter 6 is concerned with 'Optical Fiber Sensors'. Optical fibers are the predominant optical waveguide technology that is in current use. Their use in modern telecommunications has become nearly universal—and they have manifold other applications, including various types of sensor.

Chapter 7 has the title: 'Planar Optofluidics for On-Chip Particle Manipulation'. Delivery of biomaterial via micro-fluidics and, possibly, nano-fluidics (locally) is arguably the norm for optics based bio-sensing. For cells, in particular, or small enough biomaterial-loaded beads, optical techniques—including optical tweezing and trapping—are of great interest. This chapter is particularly concerned with using the transfer of momentum from photons to bio-particles and its application in bio-sensing.

Chapter 8 'New Directions in Sensing Using Raman Analysis on Paper and Microfluidic Platforms'. Raman scattering provides a powerful technique for biomedical sensing. An important example, because of its potential for very large levels of sensitivity enhancement, is the *Surface-Enhanced* Raman Scattering (SERS) process.

Glasgow, UK
Neuchâtel, Switzerland
Kiel, Germany

Richard De La Rue
Hans Peter Herzig
Martina Gerken

Contents

Contributors

J. Stewart Aitchison Department of Electrical and Computer Engineering, University of Toronto, Toronto, ON, Canada

Farshid Bahrami Department of Electrical and Computer Engineering, University of Toronto, Toronto, ON, Canada

Elsie Barakat École Polytechnique Fédérale de Lausanne (EPFL), Neuchâtel, Switzerland

Pierre Berini School of Electrical Engineering and Computer Science, University of Ottawa, Ottawa, Canada;
Department of Physics, University of Ottawa, Ottawa, Canada;
Centre for Research in Photonics, University of Ottawa, Ottawa, Canada

Gilberto Brambilla Optoelectronics Research Centre, University of Southampton, Southampton, UK

Hong Cai School of Engineering, University of California, Santa Cruz, Santa Cruz, CA, USA

Derek Craig SUPA, School of Physics and Astronomy, University of St Andrews, St Andrews, Fife, UK

Richard De La Rue School of Engineering, University of Glasgow, Glasgow, UK

Kishan Dholakia SUPA, School of Physics and Astronomy, University of St Andrews, St Andrews, Fife, UK

Ming Ding School of Instrument Science and Opto-electronics Engineering, Beihang University, Beijing, People's Republic of China

Martina Gerken Faculty of Engineering, Kiel University, Kiel, Germany

Hans Peter Herzig École Polytechnique Fédérale de Lausanne (EPFL), Neuchâtel, Switzerland

Oleksiy Krupin Department of Biological and Chemical Engineering, University of Ottawa, Ottawa, ON, Canada

Kaelyn D. Leake School of Engineering, University of California, Santa Cruz, Santa Cruz, USA;
Margaret Jones Wyllie '45 Engineering Program, Sweet Briar College, Sweet Briar, VA, USA

Mo Mojahedi Department of Electrical and Computer Engineering, University of Toronto, Toronto, ON, Canada

Gaël Osowiecki École Polytechnique Fédérale de Lausanne (EPFL), Neuchâtel, Switzerland

Holger Schmidt School of Engineering, University of California, Santa Cruz, Santa Cruz, CA, USA

Nikolaos Vasilantonakis Institute of Electronic Structure and Laser Foundation for Research and Technology Hellas (FORTH), Heraklion, Crete, Greece

Gregory A. Wurtz Department of Physics, University of North Florida, Jacksonville, FL, USA

Anatoly V. Zayats Department of Physics and London Centre for Nanotechnology, King's College London, Strand, London, United Kingdom

Chapter 1
Towards Refractive Index Sensing Using Nanoscale Slot Waveguide Cavities

Elsie Barakat, Gaël Osowiecki, and Hans Peter Herzig

Abstract Optical techniques are nowadays broadly used for high precision diagnostics and process monitoring in many aspects of our lives, with applications ranging from security to health. Interferometry is one of the fundamental techniques used for such a class of diagnostics. It relies on changes in the refractive index of the medium arising from a variation of the ambient properties. The miniaturization of these analytical systems generally responds to multiple needs, such as in situ or in vivo detection, reducing cost, speed of analysis and allowing processing of samples of very small volume. This requires the development of powerful sensors, sometimes miniaturized down to the limits imposed by physics.

1.1 Introduction

Optical techniques are nowadays broadly used for high precision diagnostics and process monitoring in many aspects of our lives, with applications ranging from security to health. Interferometry is one of the fundamental techniques used for such a class of diagnostics. It relies on changes in the refractive index of the medium arising from a variation of the ambient properties. The miniaturization of these analytical systems generally responds to multiple needs, such as in situ or in vivo detection, reducing cost, speed of analysis and allowing processing of samples of very small volume. This requires the development of powerful sensors, sometimes miniaturized down to the limits imposed by physics.

Most studies have concentrated on the advantages of miniaturization. However, rare are the studies that show, in addition to the advantages, the drawbacks of miniaturization. Hence, the design as well as the geometry of the sensor are highly dependent on the application and, therefore, there are many types of sensing devices. Each device has certain applications for which it is well suited.

E. Barakat · G. Osowiecki · H. P. Herzig (✉)
École Polytechnique Fédérale de Lausanne (EPFL), Rue de la Maladière 71b, 2000 Neuchâtel, Switzerland
e-mail: hanspeter.herzig@epfl.ch

R. De La Rue et al. (eds.), *Biomedical Optical Sensors*,
Biological and Medical Physics, Biomedical Engineering,
https://doi.org/10.1007/978-3-030-48387-6_1

Research activities in the latter area aim to combine nanotechnology, nano-optical and biological sciences, in order to develop sensors with increased sensitivity and stability that are sufficiently well-suited for rapid action and continuity. This activity will allow monitoring and control of physico-chemical and biological parameters in all fields of human activity. In the context of increasing societal demands, we shall illustrate, with a few examples, the challenges in this area of research for the coming years with a particular emphasis on optical sensors that are easily miniaturized.

The chapter is divided into two sections. In the first section, we discuss the basic principles of sensing based on the measurement of refractive index. Discussion includes the principles of various interferometer types such as Michelson, Fabry–Perot and micro-ring interferometers. We shall define the design constraints in relation to the application. In the second section, we shall discuss integrated refractive index sensors that can be used to detect changes in refractive index in sub-attoliter volumes of sample fluid.

1.2 Existing Interferometers as Optical Sensors

An interferometer is typically formed by splitting a single beam into two beams. After propagation, the different beams are combined to generate interference. These interferences depend on the wavelength and the delay between the beams. The delay is also called Optical Path Difference (OPD). By changing the OPD between the beams, one can obtain spectral information or one can measure the change in refractive index. Interferometers are basic optical tools that are used to precisely measure wavelength, distance and index of refraction. Note that, in the case of a spectrometer, the refractive index of the medium is known (air in general)—and, in the case of a refractive index sensor, the wavelength is known.

In order to enable a comparison between interferometers and resonators for sensing applications, we shall first define different key parameters and figures of merit. The performance of resonators can be expressed in terms of resolution ($\delta\lambda$), free spectral range (FSR), finesse ($\mathscr{F}$), and quality factor (Q).

1.2.1 Michelson Interferometer

Using an optical interferometer, physical distances can be measured in terms of wavelengths by counting interference fringes that move when one or the other of two objects is displaced. In order to see interference, the beams must be coherent so that a definite phase relationship will exist between them. A schematic view of the Michelson interferometer is shown in Fig. 1.1. The light emitted from the source is split into two beams, which are then recombined after reflecting from the mirrors M_1 and M_2, respectively.

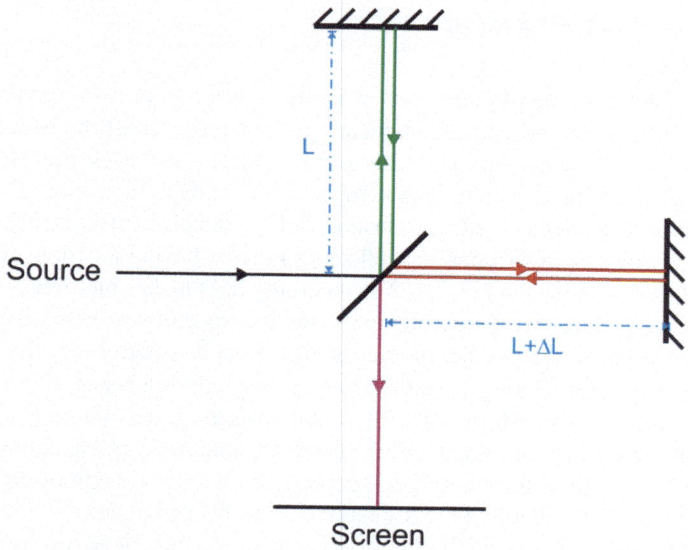

Fig. 1.1 Schematic of a Michelson interferometer. The collimated light from the source is split into two beams. One beam is reflected from a fixed mirror while the other one is reflected from a movable mirror. The beams are recombined and the intensity is detected

A key parameter of an interferometric system is the detectable wavelength separation in the spectral region of interest, for different conditions. This parameter is defined as the resolving power (R) of the interferometer and is defined through the following equation [1]:

$$R = \frac{\lambda}{\delta\lambda} = \frac{\text{OPD}}{\lambda} \tag{1.1}$$

λ is the central wavelength, $\delta\lambda$ is the resolution limit for a spectrometer and OPD is the effective optical path difference. Thus the maximum achievable resolution is directly related to the motion range of the moving mirror and the wavelength. In the case of a Michelson interferometer, the optical path difference is $2\Delta L$ between the rays reflected from M_1 and M_2, if M_1 is supposed to be movable. There are multiple alternative definitions that describe the resolution of two spectra and one of the most frequently used is the Rayleigh criterion [1] . Note that the resolution $\delta\lambda$ is not the smallest detectable quantity. The measurable wavelength shift, and with that also the detectable refractive index change, is much smaller when only one wavelength is used. In that case, the limitation on the smallest detectable quantity, i.e. the resolution, is defined by the signal-to-noise ratio.

The Michelson interferometer has been taken as a common example of an interferometer. However, the expression for the resolving power is valid for interferometers more generally. It is directly related to the ratio of the optical path length over the wavelength, i.e. to the optical path-length measured in wavelengths.

1.2.2 Fabry–Pérot Resonators

In order to decrease the physical size of interferometers, optical resonators show significant advantages for sensing applications. The presence of the optical cavity allows a longer effective optical path, due to the multiple round trips inside the cavity. One can therefore obtain a higher resolving power for small dimensions.

A common and indeed the simplest example of an optical cavity is the Fabry–Pérot resonator. It consists of two parallel reflecting mirrors having a certain separation distance (L) [2] as shown in Fig. 1.2. The incoming light makes multiple round trips within the cavity. Interference occurs due to the superposition of both reflected and transmitted beams at the parallel interfaces. Constructive interference occurs if the beams are in-phase, modulo 2π, leading to high-transmission peaks. If the reflected beams are out-of-phase, a destructive interference occurs and a minimum appears in the transmission spectrum. Whether the reflected beams are in-phase or not depends on the optical length of the cavity i.e. it depends on the wavelength of the incident light (λ), the physical length of the resonator (L), the refractive index of the medium between the interfaces (n), and the angle of incidence (θ). For plane waves, the phase difference for a single propagation round-trip between the mirrors of the Fabry–Perot interferometer is simply given by [3]:

$$\phi = \frac{2\pi}{\lambda} n2L = \frac{2\pi}{\lambda} \text{OPD} \tag{1.2}$$

where $2nL$ is defined as the effective optical path difference (OPD). The attenuation during the round trip is described by an amplitude attenuation factor $h = |r| \exp(i\phi)$ that results from losses due to the two mirror reflections and absorption in the medium. The corresponding intensity attenuation factor for a round trip is $|r|^2$ with $|r| < 1$. The number of round trips is then represented by the finesse $\mathscr{F}$ [4]:

$$\mathscr{F} = \frac{\pi \sqrt{|r|}}{1 - |r|} \tag{1.3}$$

The transmission function of the resonator is given by [4]:

$$T = \frac{I_{\text{max}}}{1 + (2\mathscr{F}/\pi)^2 \sin^2\left(\frac{\phi}{2}\right)} \tag{1.4}$$

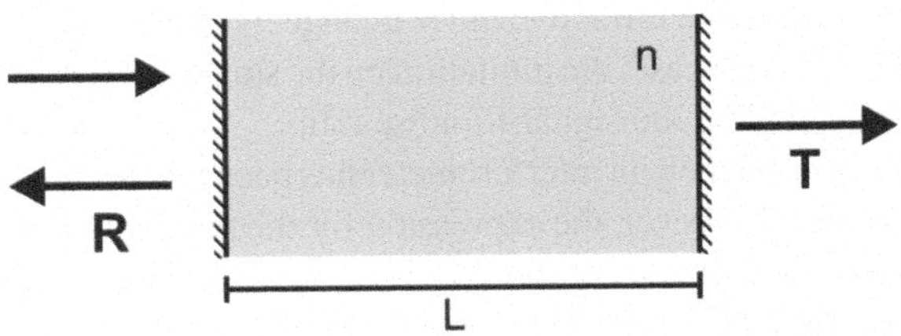

Fig. 1.2 Fabry–Perot interferometer, T defines the transmission, R the reflection of the system, L the distance between the mirrors and n the refractive index of the medium

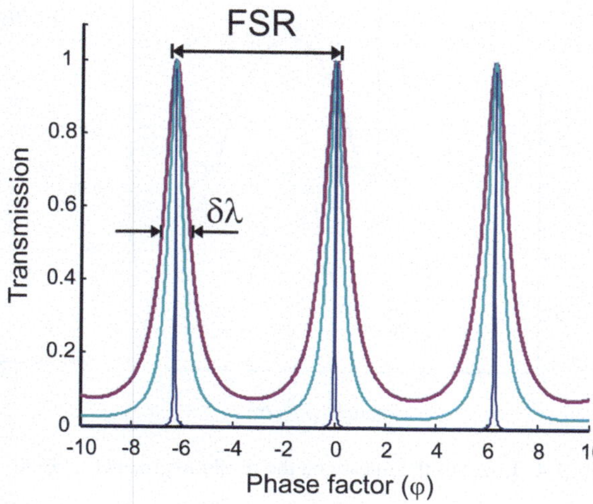

Fig. 1.3 Transmission response of a Fabry–Perot resonator for different values of the finesse, $\mathscr{F} = 11.75$, 37.53 and 9900. The highest finesse leads to the narrowest peak

With $I_{\max} = \frac{I_0}{(1-|r|)^2}$

Figure 1.3 presents the transmission response of a Fabry–Perot interferometer versus the wavelength, as described in (1.4). The higher the reflection coefficient of the mirrors, the narrower is the width $\delta\lambda$ of the resonance peak. This is a consequence of the increased number of interfering beams in the Fabry–Perot cavity. Note that the spectral width is related to the finesse by $\delta\lambda = \mathrm{FSR}/\mathscr{F}$ [4]. The free spectral range (FSR) is the distance between two subsequent resonance peaks.

The sensing application requires the study of either the absolute value or the relative change of the refractive index that is related to the ambient chemical properties. Since the two adjacent interference maxima have a phase difference of 2π, it is possible to theoretically calculate the absolute refractive index (n) or the absolute length (L) of the cavity if one of them is known. In many cases, only the relative refractive index change is of interest and the range of refractive index variation is small, so the phase shift is less than 2π. In this case, the phase ambiguity issue can be avoided. The relative refractive index change can be calculated on the basis of the spectral shift of the interferogram. The interference signal reaches its maximum ($I_{\max}$) when the phase of the sine term becomes an odd integer number of π. That is $I = I_{\max}$, when

$$\frac{2\pi Ln}{\lambda_v} + \varphi_0 = (2m+1)\pi \tag{1.5}$$

where m is an integer and λ is the center wavelength of the specific interference valley where the transmission value is highest.

As already mentioned, the free spectral range (FSR) is the wavelength separation between adjacent peaks. From (1.5) we get:

$$\mathrm{FSR} \approx \frac{\lambda^2}{2nL} = \frac{\lambda^2}{\mathrm{OPD}} \tag{1.6}$$

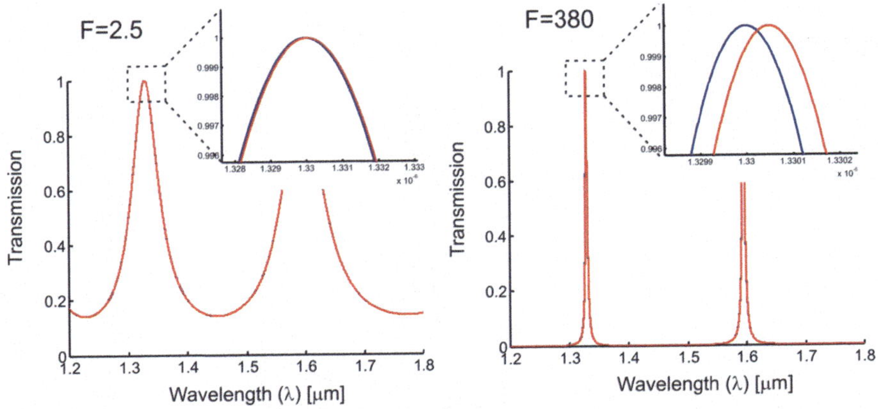

Fig. 1.4 Effect of the finesse on the resolution (detection limit) of the optical sensor

where OPD is the effective optical path difference for one round-trip. The FSR corresponds to the maximum wavelength range that can be measured without overlap. Equation (1.6) shows that the smaller the cavity length (L), the larger is the free spectral range, FSR.

The primary factor that affects the finesse is the reflectivity of the interfaces because it directly affects the number of oscillating beams in the cavity. An example of a FP response for two different reflection coefficient factors is given in Fig. 1.4. A higher reflection coefficient, which is equivalent to a higher finesse, indicates a high number of interfering beams, resulting in a higher resolving power. This resolution refers to the smallest peak that a sensor can distinguish. However, one should note that the detection limit also depends on the instrument in use, as well as the adjustable parameters such as the sampling frequency and the number of points per measurement.

In summary, the finesse $\mathscr{F}$ defines the number of round trips in a resonator. It depends on the losses by reflection and absorption. The quality factor (Q) can also be used as a parameter that characterizes the losses of an optical resonator. The quality factor (Q) has been proposed by Connes [5] as enabling a direct comparison of interferometers. He assumed that the quality factor will be larger for better instruments. Q measures the relationship between stored energy and lost energy per cycle. Q is also equal to the resolving power given by (1.1) [6, 7]. Consequently it depends on the optical wavelength and the total optical path difference. In the case of a Michelson interferometer the finesse is equal to one ($\mathscr{F} = 1$).

The quality factor (Q), the resolving power and the Finesse (F) are related by (1.7):

$$Q = \frac{\lambda}{\delta\lambda} = \frac{\text{OPD}}{\lambda}\mathscr{F} \tag{1.7}$$

The shorter OPD gives a larger FSR. Even though a large FSR gives a wide dynamic range in a sensor it gives, at the same time, a poor resolution due to blunt peak signals [8]. Therefore, it is important to design the OPD of the Fabry–Pérot interferometer, depending on the application, for satisfaction of both the dynamic range and the resolution required.

1.2.3 Microring Resonators

Another common resonator that is used for miniaturization is the microring resonator. Microring resonators can be used as optical sensors because their resonances can respond sensitively to the change in the effective index of the guided optical mode [9]. One of their advantages is the integrated aspect of the resonator and the increase in the signal-to-noise ratio. A high quality factor results in the highest resonance extinction ratio and thus leads to a maximum signal-to-noise ratio [10]. Microring resonators are widely used for sensing applications. Because of the generally high quality factor, and therefore a large number of round trips inside the ring, light interacts repeatedly with an analyte deposited on the surface. In general, a microring resonator consists of a ring waveguide and one or two bus waveguides. The ring plays the role of the resonant cavity, while the first bus waveguide is used to couple light into the ring. The light is then coupled back into the output ports. The resonance occurs when the round-trip phase shift in the ring is equal to 2π, which leads to constructive interference between propagating light of different cycles. As for the Fabry–Perot resonator, the round-trip phase-shift of a ring resonator depends strongly on the wavelength and the refractive index of the material. Therefore, the observed modification of the output signal due to the changes in the refractive index is widely used for sensing applications.

In the case of lossless ring resonators, the free spectral range can be deduced from the resonance condition [11]:

$$\mathrm{FSR} = -\frac{2\pi}{L_{\mathrm{cav}}}\left(\frac{\delta\beta}{\delta\lambda}\right)^{-1} \approx \frac{\lambda^2}{n_{\mathrm{eff}}L_{\mathrm{cav}}} = \frac{\lambda^2}{OPD} \tag{1.8}$$

where L_{cav} is the length of the cavity,β is the propagation constant, n_{eff} is the effective index of the medium in the cavity and c is the speed of light. As in the case of the Fabry–Pérot resonator, the width of the resonance peak $\delta\lambda$ depends on the losses, e.g. the fractional intensity loss (γ), the intensity coupling coefficient of the coupler (k) and/or the attenuation loss (α). The intensity is therefore proportional to the round trip loss via the following equation [12]: $A \propto e^{-\alpha L}$

For a sharp resonance, the finesse of the resonator can be approximated as [12]:

$$F \cong \frac{\pi\sqrt{A}}{1-A} \tag{1.9}$$

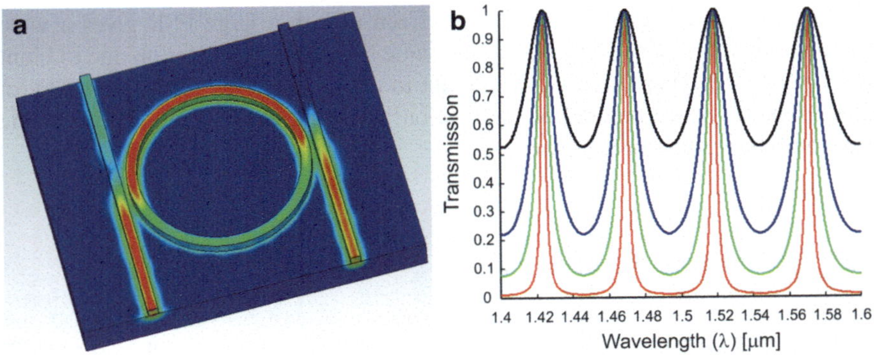

Fig. 1.5 **a** Electric field distribution of a ring resonator having a radius $R = 8\ \mu\text{m}$ and a refractive index of $n = 3.4$. **b** Transmitted spectra as function of the wavelength for different loss factor coefficients ($\alpha = 0.4, 0.6, 0.75$ and 0.95)

This expression shows a similarity with the Fabry–Pérot finesse (1.3), where A includes the round trip loss factor (α). Insertion loss, fractional coupling loss and other quantities can be included in A as well: $A \propto e^{-\alpha L}\sqrt{(1-\gamma)(1-k)(1-\eta)}$, where η is the insertion loss.

Figure 1.5a shows an 8 μm microring resonator evanescently side-coupled to a pair of signal waveguides. The ring supports a traveling wave that represents the total power through any cross-section of the ring. This traveling wave in the ring evanescently couples to the transmitted wave. The relationship that relates the travelling wave and the transmitted wave includes the fraction power (k^2) that is coupled out of the ring. By considering that input and output are equally coupled to the ring, and by only considering losses induced by the coupling factor k, we obtain [13]:

$$\delta\lambda \approx \frac{2k^2\lambda^2}{(2\pi)^2 R n_e} = \frac{k^2}{\pi}\frac{\lambda^2}{\text{OPD}} \tag{1.10}$$

k^2 is defined as the fraction of power coupled out of/into the ring over the interaction region. The resolution $\delta\lambda$ is proportional to the power coupled into the ring over the interaction region.

Figure 1.5b shows the transmitted spectra for different loss-factor coefficients. Higher losses induce lower half-power bandwidth and therefore lower quality factor. For bio-sensing applications, the sensors should be constructed from microresonators with a high extinction over a large dynamic range. In such a case, the normalized optical power at the output of the fiber can be expressed as a function of the round trip loss factor (α). α ranges between zero and one, where zero loss corresponds to $\alpha = 1$.

The quality factor, Q, of the resonator can be calculated by using its definition as the time average stored energy per optical cycle, divided by the power coupled into the two outgoing waves. Based on [13], Q can be expressed as:

Table 1.1 ddff

Free spectral range	$\mathrm{FSR} = \frac{\lambda^2}{\mathrm{OPD}}$
Quality factor	$Q = \frac{\mathrm{OPD}}{\lambda} F$
Sensitivity	$S = \frac{\mathrm{d(OPD)}}{\mathrm{d}n}$

$$Q \approx \frac{2\pi R n_e}{\lambda_0}\frac{\pi}{k^2} = \frac{\mathrm{OPD}}{\lambda}\mathscr{F} \tag{1.11}$$

In another form, the finesse can be related to the coupling factor via the following expression $\mathscr{F} = {}^{\pi}/_{k^2}$. Since the Finesse is inversely proportional to the square of the coupling coefficient, a high finesse in the optical cavity radically increases the effective optical path length of light. A key mechanism that affects the quality factor is the coupling coefficient, k, between the cavity and the adjacent waveguide. Q remains higher for weaker coupling because the coupling factor is directly proportional to the losses in the resonator. Besides coupling, k, can also describe losses due to, e.g., fabrication imperfections or absorption in the materials used. Note, that the coupling coefficient, k, is equivalent to the coupling by a mirror in a FP cavity. Thus we can write $1 - R_{\mathrm{mirror}} = k$.

As a conclusion, in optical resonators the quality factor is a function of the optical path difference and the Finesse, while the sensitivity is a function of the optical path difference. It can be controlled by the ring radius and external coupling.

Discussion

The objective in the above sections was to summarize the general formulae describing the performance of a sensor, independently of the type of device, as in Table 1.1. This summary is very helpful for the design and discussion of instruments. One can see in the previous sections that, for ring and Fabry–Pérot resonators, the free spectral range, FSR, is similarly expressed as the ratio of the squared wavelength over the optical path difference, OPD. Depending on geometry and structure, the values have to be replaced by their effective values, i.e. effective wavelength and effective OPD.

Q is increased by increasing the cavity length—and therefore it is relatively easy to realize a high Q value if the system is large. The finesse is independent of the cavity length. However, to obtain a high Q value with a miniaturized system, requires a large finesse. The problem is that obtaining a large finesse requires low losses and consequently also low coupling, which makes it difficult to couple more light into the cavity.

The sensitivity, which is used to measure the resonance wavelength shift in response to the index change of the medium, is independent of the resonance spectral profile. However, the detection limit (DL), which is defined as the minimum refractive index change in the sensing medium that can be detected by the sensor system, is proportional to the resonance linewidth ($\delta\lambda$) or inversely proportional to the resonance Q-factor [14]. As has been shown previously, the quality factor is defined by the resonator loss, as well as by the coupling loss.

The sensitivity in detecting the index of refraction change can be estimated from the size of the cavity and the effective refractive index. The detection limit however is defined as the minimum detectable change and is usually limited by the noise equivalent power of the optical detector. High sensitivity implies a large resonance wavelength shift for a given index change, but it is more difficult to detect a small refractive index change. On the contrary, a high Q-factor improves the detection limit but reduces the shift in the resonance wavelength.

As a conclusion, the resolving power is defined as the ratio of the OPD over the wavelength. Thus, the miniaturization consists in folding the optical path. The benefit of folding is limited by coupling losses, absorption and scattering.

1.3 Plasmonic Sensors

The general equations shown in Table 1.1 are also valid for other sensing concepts, such as surface plasmon resonance (SPR) sensors. Such sensors are of interest because the light can be confined near the surface, where a biological substance is deposited for analysis. The light distribution is therefore concentrated on the sensing area.

SPR sensors have been used since the first demonstration of gas sensing in the early 1980s [15]. They have been extensively explored and developed as one of the most powerful label-free sensing techniques available today. The principles, platforms, and applications of SPR sensors are described in excellent reviews [16–18]. A surface plasmon (SP) is a resonant coupling between the free electrons at the surface of a metal and the electromagnetic field, while most of the energy probes the medium in contact with the metal. Different excitation methods [18] including prism coupling, waveguide coupling, grating coupling, and optical fiber coupling are applied to fulfill the propagation-constant matching condition for TM polarized light. SPR sensors are based on the resonant excitation of the SP. A change in the refractive index of the medium modulates the optical properties of the light coupling to the SP. The properties of the light that are modulated may be the angle of incidence, the resonant wavelength, the intensity, or the phase.

Many SPR sensors utilize either the angular (θ) or the spectral (λ) dependence of the reflection to perform the refractive index measurement. The primary difference between angular and spectral measurements is that angular detection illuminates the surface with a collimated monochromatic light beam while spectral detection fixes the incident angle (θ) for a wide spectral range. Due to their definitions, the sensitivity of the two methods shows different wavelength dependences. For angular modulation, the sensitivity decreases with increasing wavelength, while it increases with wavelength for spectral modulation. Spectral modulation exhibits high sensitivity values, on the order 10^4 nm/RIU [19] for visible light, resulting from its dependence on the dispersion of the metal and the prism. The material dispersion results in a small value in the denominator of the sensitivity calculation [19]. The sensitivity for angular modulation is in the order of 100°/RIU in the visible range.

It can be shown [20] that the ultimate resolution of SPR sensors depends predominantly on the noise present in the light source and the detector, while the performance is independent of the coupler and the modulation methods. In addition to the improvement in resolution, research on SPR sensors is directed towards the multiplexing of channels. The SPR imaging (SPRI) technique [21] shows promise for high throughput detection, in which a sample array on the order of hundreds is imaged. Typical SPRI is based on the prism coupling of monochromatic light at one angle of incidence and the intensity of the reflected light embodies the RI variation. The reflected light from the entire array is collected by a camera.

Recently, new sensing technologies based on surface plasmons have appeared that offer considerable promise for the design of robust miniature sensors. Among the diverse configurations developed for refractive index (RI) detection, plasmonic devices using nanoscale patterned metals are particularly promising. Nano-structured metallic devices enable field confinement in a volume smaller than the diffraction limit. Moreover, they offer possibilities for device miniaturization and sensor multiplexing on the same substrate.

Considering the development of compact and portable sensing systems, optical devices have been promoted for the realization of complex and functional systems through device miniaturization, which follows the advances achieved in fabrication technologies. Nevertheless, the light control and manipulation are difficult, due to light scattering when the size of devices is reduced from the macro- to the nano-scale. As a result of the surface plasmon wave, light can be squeezed into far subwavelength dimensions and nano-structured metal exhibits previously unforeseen optical properties. Therefore, the field confinement at nanoscale and the high sensitivity to environmental change are interesting for local RI measurements. In this chapter, nano-structured metallic devices are chosen as a starting point for the design of optical sensing elements

1.3.1 Integrated Plasmonic Refractive Index Sensor

Optical sensors can also be miniaturized by using integrated optics. Such components are compact and can be integrated simultaneously on a single substrate. In contrast to fiber optics, integrated optics is not strongly limited by the material properties or the structure of the optical system, thereby giving additional degrees of freedom for optimized miniaturization. However, integrated sensors unfortunately have drawbacks such as the scattering effects that are caused by the inhomogeneity of chemical solutions or defects in fabrication. Therefore, for the development of any new integrated optical sensor, one should specify the characteristic properties of the sensor in order to match the application requirements with the strong points of the optical sensors.

Integrated optical sensors are mainly used for parallel sensing of different chemicals and for measuring very small volumes. Integrated optical sensors typically include microfluidic systems that help with transportation of the sample volumes to the sensing region.

The sensor presented in the next section will be required to detect the presence of proteins in liquids. In order to be able to miniaturize an optical cavity, it is essential to reduce, as far as possible, the wavelength of the light used. Logically one should therefore work with ultra-violet or even extreme ultra-violet light. However, since the nature of what we want to detect may be strongly modified by these high energy wavelengths, another approach needs to be used. Therefore, one might consider increasing the refractive index of the medium in which the light travels which would reduce the effective wavelength. Or, alternatively, one could use photonic crystal micro-cavities which offer a high quality factor and low effective cavity volume, but are limited in size to half of the effective wavelength. Additionally, one of the main drawbacks of the photonic crystal cavity approach is the difficulty of coupling light into the cavity. Nevertheless, adopting this strategy led us to investigate bounded surface waves on metallic surfaces, which are more commonly called surface plasmon polaritons. These surface plasmon resonators are promising candidates for strong confinement of the electromagnetic field in small volumes. Nanostructures then appear as the ideal solution for the integration of different optical sensors having high sensitivities.

With the idea of mounting a sensor integrated onto a chip, in order to achieve further low-cost mass production, we choose a waveguide-coupled approach for the excitation of the plasmonic cavities. The free space wavelength that will be used is in the telecom wavelength range, which is $\lambda = 1200$–1700 nm. This choice also gives access to the technology of photonic integrated circuits based on silicon. Many of the fabrication processes required have already been developed for microelectronics—and therefore it may be feasible to realise a CMOS compatible optical sensor.

1.4 Periodic Plasmonic Slots Excited with a Waveguide

Extraordinary optical transmission (EOT) through sub-wavelength hole arrays [22] has led to much previous discussion. It has been shown that EOT is caused by the coupling of propagating light to surface plasmons on the surface of the metal film, which is combined with a Fabry–Pérot resonance located in the holes of the metallic structure. Therefore, under EOT conditions, the wavelength peak of light transmitted through the metal layer is strongly dependent on the refractive index of the surrounding medium [23]. These research results have inspired us to implement EOT in an integrated chip for a simple implementation of biomolecular sensing. To transform the metallic hole array used for EOT into a two-dimensional structure, we have simply taken a cross-section. This structure, which is shown in Fig. 1.6 must be excited by an integrated waveguide on the chip. In order to simplify the fabrication process, we have opted for evanescent coupling to the plasmonic structure. The evanescent field extending from the waveguide surface into an upper medium with lower refractive index will excite the metallic slot array.

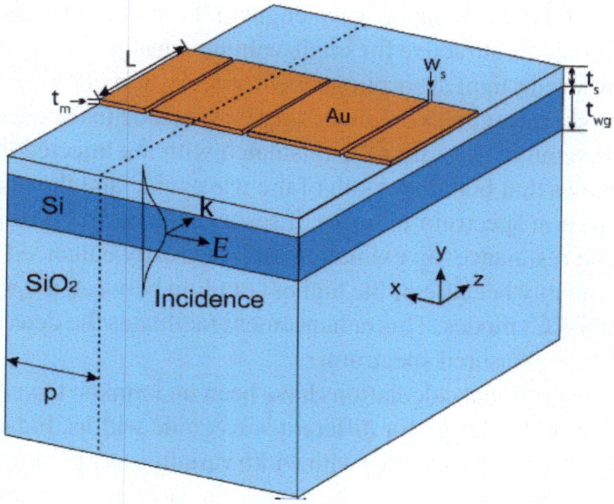

Fig. 1.6 Initial EOT inspired integrated sensor structure of the periodic slot waveguide cavity [24]

The ability of metallo-dielectric interfaces to sustain localized electromagnetic modes known as surface plasmon polaritons (SPP) allows confinement and guiding of the light in volumes that are smaller than the diffraction limit over a several micrometer wavelength region in the visible spectrum. Because of these properties, we propose a geometry that is based on a plasmonic slot waveguide cavity and also enables the significant improvement of highly multiplexed detection on an integrated silicon platform [25, 26]. Due to the large enhancement of the electromagnetic field by the SPP, the sensing volume is greatly reduced, which allows localized detection of molecules in sub-attoliter volumes. We expect our proposed device to be able to detect fluctuations of refractive index induced by as few as one or two molecules. A plasmonic slot waveguide (PSW) enables easy integration with wafer-level micro-fabrication, thanks to its planar geometry. In addition, it shows deep sub-wavelength localization and enhancement of EM fields. Our proposed hybrid plasmonic-dielectric sensor provides the possibility of bringing focused light to the desired location on a planar surface, in order to detect and analyze the binding of small molecules. The small volume of the resonating cavity enables refractive index sensing in sub-attoliter volumes, which is of great interest for bio-molecular sensing especially when it comes to breaking the concentration limit of optical single-molecule detection. The refractive index change can be obtained by measuring the shift of the resonant wavelength in the transmitted spectrum or by measuring the transmitted intensity change at a single wavelength.

1.4.1 From Multimode to Monomode Waveguide Coupling

There are two major classifications of waveguide systems: multimode and single-mode. First we started our study with multimode waveguides that have a thickness much greater than the wavelength of the excitation light. The localized field in such a tiny slot is attractive for small-volume sensing. The field confinement is accompanied by strong light absorption, due to energy transfer to local electrons. However, multi-mode waveguides exhibited large losses, due to the excitation of several modes in the waveguide as well as the mismatch with the injected light. Additionally, simulation shows that both the width of the waveguide and the number of slots affect the transmission spectrum in terms of resonance spectral position and transmitted intensity. The resonance dip extinction ratio is greatly enhanced if the slot number is reduced, probably because of the higher coupling between the waveguide and the PSW cavity (PSWC) modes. This enhancement facilitates the detection of the resonance position in the measured spectrum.

Additional calculations have been undertaken to verify the optical response variation of the device for different waveguide widths. Figure 1.7 shows the transmission spectra for waveguides with width variation as a parameter (width = period × number of slots). For the single slot device, the large transmission dip extinction ratio is apparently due to the field being squeezed into the center of the waveguide where the slot is located. Light is therefore more efficiently coupled into the slot cavity and the amplitude of the transmission dip is enhanced. When the number of slots increases, the transmission spectrum evolves to be close to that of the infinite number case.

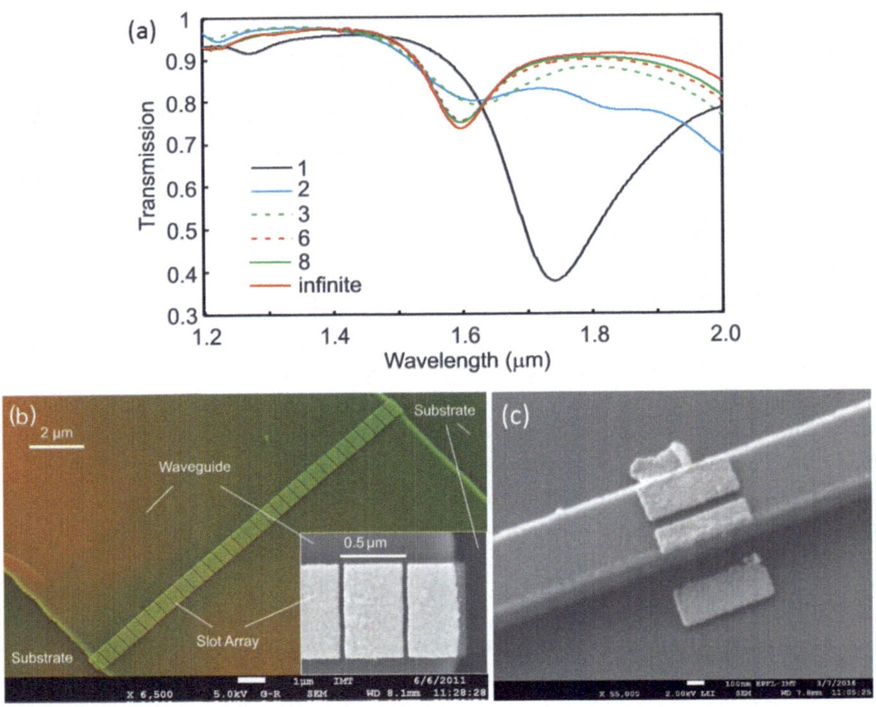

Fig. 1.7 **a** Transmission spectra for various waveguide widths (period × slot number). **b** SEM image of a periodic slot waveguide cavity [27]. **c** SEM image of a single slot waveguide cavity [28]

For a number of slots higher than 8, a difference of only 2% is observable in the transmission dip, compared to the case of the infinite structure.

Figure 1.7b shows a fabricated structure with 30 slots, that corresponds closely to an infinite structure, theoretically. The chosen width (15 μm) of the waveguide is adapted to the modal field diameter (about 10 μm) of SMF28 single mode fiber at a wavelength of 1.55 μm. With a width of 15 μm, the silicon waveguide supports multiple modes that will influence the transmission spectrum. In theory, the fundamental mode of the waveguide is required for characterization. We assume that the fundamental mode is excited predominantly because of its strong modal overlap integral with the mode of the single mode fiber. However, higher order modes may also be excited, because of waveguide imperfections. These higher order modes may cause modifications of the transmission dip. These higher order modes couple to the cavity in an unwanted manner. Accordingly, the desired transmission dip may be widened or multiple dips may appear. We therefore opt to work with single mode waveguides for the excitation of the single-slot cavity. Although the sensitivity is in theory reduced, the sensing volume is deceased by more than an order of magnitude and the detection limit is increased by the higher contrast of the resonance dip in the spectrum. The excited Fabry–Perot (F–P) mode resonates through the metallic slot layer, where it propagates very slowly (low group velocity) giving the opportunity to interact with a molecule. The increase in the coupling between the two inner metal dielectric interfaces gives strong optical confinement in the nano-slot.

The cavity can also serve as a conduit, allowing molecules to flow freely in and out, by using micro-fluidics. Our general objectives are to achieve a better detection limit with the plasmonic slot waveguide sensor, as well as higher localization of the sensing area—thereby enabling single cell analysis and single molecule detection.

1.4.2 Plasmonic Slot Waveguide Cavity (PSWC) on a Single Mode Waveguide

The integrated plasmonic cavity and its typical transmission response are shown in Fig. 1.8. It lies on top of a silicon oxide spacer layer and a silicon waveguide core. The silicon waveguide is designed to support only two modes in the near infrared wavelength range between 1200 and 1700 nm. Based on the field profiles, the fundamental quasi-TE mode of the silicon waveguide is the only mode that is able to excite the PSWC modes in contrast with the fundamental quasi-TM mode. The silicon waveguide core has a thickness of 220 nm and a width of 450 nm and is used to guide light from the source to the cavity, as well as from the cavity back into the waveguide and on to the detector.

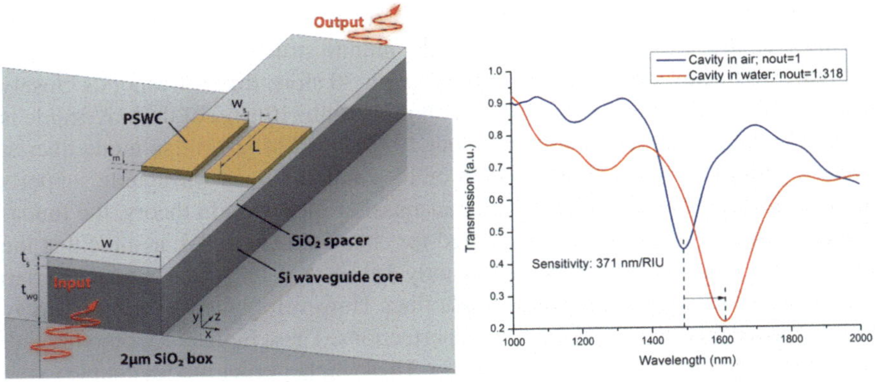

Fig. 1.8 (Left) Hybrid Photonic-Plasmonic structure. A plasmonic slot waveguide cavity is sitting on top of a single crystal silicon waveguide and is separated from it by a silicon dioxide film. (Right) Sensitivity of the PSWC obtained by FDTD simulation method

1.4.2.1 Parameter Study

The transmission spectrum is measured at the output of the silicon waveguide and depends on its geometrical parameters: the cavity length (L), the metal thickness (t_m), the spacer thickness (t_s) and the slot width (w_s), shown in Fig. 1.8a. In order to gain an idea of the influence of these parameters on the resonant optical properties of the device, we primarily calculate the transmission spectrum for different lengths of the PSWC. Then we investigate the impact of the changes in coupling on the resonance properties, by changing the oxide spacer thickness (t_s). We perform our numerical study with the commercial software CST Microwave using both frequency and time-domain calculations. A non-uniform three-dimensional hexahedral meshing is applied in the x, y, and z directions, in order to define the structure. The grid size varies continuously between 2 nm for the plasmonic cavity and 26 nm for the silicon waveguide over a bounding box of 2.3 μm (X), 2 μm (Y) and 2.7 μm (Z) that contains 6 billion mesh cells. The temporal step for FDTD is set to 4.5×10^{-18} s. We use a perfectly matched layer (PML) made up of 8 individual layers as an artificial absorbing boundary. The metal dispersion is a 4th-order polynomial fit to the data of Johnson and Christy [25]. To reduce computation time, we use a distributed computing feature of the software package.

Figure 1.9a shows the transmission as a function of the cavity length (L). The figure shows multiple transmission peaks that appear at wavelengths that correspond to the longitudinal resonances of the cavity. As expected, the resonance positions strongly depend on the cavity length. The resonance peaks shift to a larger wavelength (red shift) as the cavity length increases. In the wavelength range from $\lambda = 1200$–1700 nm, the fundamental resonance disappears for cavity lengths smaller than 250 nm, which is due to phase-matching that cannot be obtained for cavity lengths smaller than $\lambda_{eff}/2$, where λ_{eff} is the effective wavelength of the fundamental guided mode inside the PSW. Moreover, below 1200 nm the silicon waveguide becomes multimoded

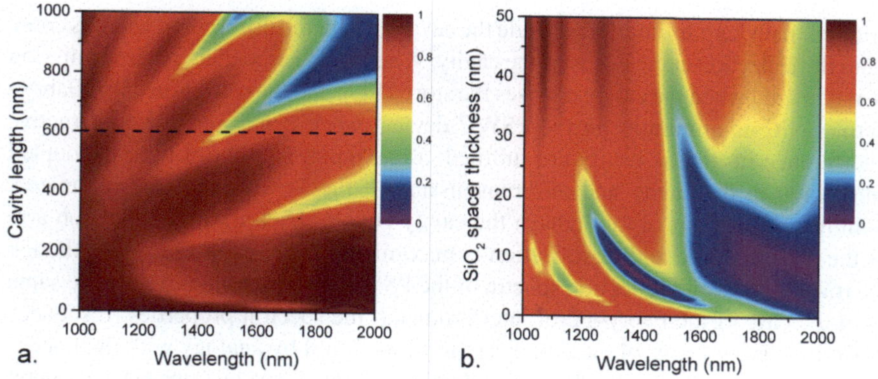

Fig. 1.9 **a** Dependence of the transmission spectra as function of the cavity length. **b** Dependence of the transmission spectra as function of the spacer thickness. Both figures exhibit typical Fabry–Perot resonance peaks. Reprinted with permission from [29] © The Optical Society

and the resonance starts to be affected by other modes. Figure 1.9a shows that higher order resonances appear for longer cavities. The dashed black line corresponds to the cavity length where the first harmonic of the resonance is located, at around $\lambda = 1.5\ \mu m$.

Figure 1.9b represents the transmission parametrically as a function of the silicon oxide spacer thickness and wavelength. The resonant wavelength position of the first harmonic occurs at around $\lambda = 1550$ nm and remains almost independent of changes in the spacer thickness. The thickness of the spacer that separates the PSWC and the waveguide core affects the coupling efficiency between the mode of the waveguide and the cavity mode. Due to the presence of the spacer all along the silicon waveguide, changing the spacer thickness also slightly affects the effective index of the guided mode present in the silicon waveguide. Due to fabrication restrictions, a slot width of $t_s = 50$ nm is chosen for our design. From these calculations, we opted for the following parameters: $t_m = 20$ nm, $w_s = 50$ nm, $L = 600$ nm, and $t_s = 40$ nm. Using the latter geometrical parameters, the first harmonic resonance is excited at a wavelength of around $\lambda = 1.5\ \mu m$ and has a high extinction ratio of about 3, which makes the resonance position easily measurable.

1.4.2.2 Fabry Pérot (FP) Model

The core of our sensor is a plasmonic slot waveguide cavity (PSWC) that behaves in a similar way to a Fabry–Pérot resonator. The effective refractive index of the modes supported depends on the geometry of the slot array and on the illumination wavelength. The analytical Fabry–Pérot model extracts the important properties of the PSWC sensor, such as the resonance position and the sharpness of the resonance.

From the calculation of the Fabry Pérot interferometer transmission, a resonance appears when the phase condition is satisfied according to (1.5). Physically, light with different path lengths interferes inside the cavity, where the different light paths come from the reflection at the edges of the cavity. When (1.5) is satisfied, the transmission is maximized when the cavity releases the stored light and the reflection of the Fabry–Pérot cavity is minimum. In the PSWC device, when conditions for constructive interference inside the cavity are fulfilled, a maximum amount of light is coupled and stored in the cavity and the transmission of the Si waveguide is minimized. Simultaneously, part of the light in the cavity couples back into the reflection arm of the Si waveguide and its reflection is maximized at the resonance. Even though the transmission and reflection spectra of the PSWC device are not exactly the same as in the case of the Fabry–Pérot interferometer, the spectral properties (resonance position, quality factor of the dip, etc.) can be analyzed by analogy with the Fabry–Pérot cavity. From (1.5), the effective index (n_{eff}) of the slot array is the key parameter in the calculation. In the present work, it depends on the geometrical dimensions of the cavity $n_{\text{eff}} = f(sw, L, tgold, tspacer, w)$. Figure 1.10 shows a comparison between a rigorous (FDTD) and an analytical study of the cavity length as a function of the Fabry–Pérot mode number. The results are in good agreement. The difference is mainly due to the phase difference between the analytical result and the FDTD result. In consequence some of the parameter studies should use rigorous methods.

As discussed previously, the shape of the resonance is important. The finesse F, which was introduced in (1.3), displays the sharpness of the peak. F is related to the reflection coefficient at the cavity edge and the absorption in the cavity. F is large for a Fabry–Pérot cavity with a highly reflective interface and a low absorption coefficient. The finesse F is correlated with the quality factor. The quality factor is one of the most important parameters in defining a resonance peak. In our device, the strong absorption of modes and small refractive index difference at the edges of the

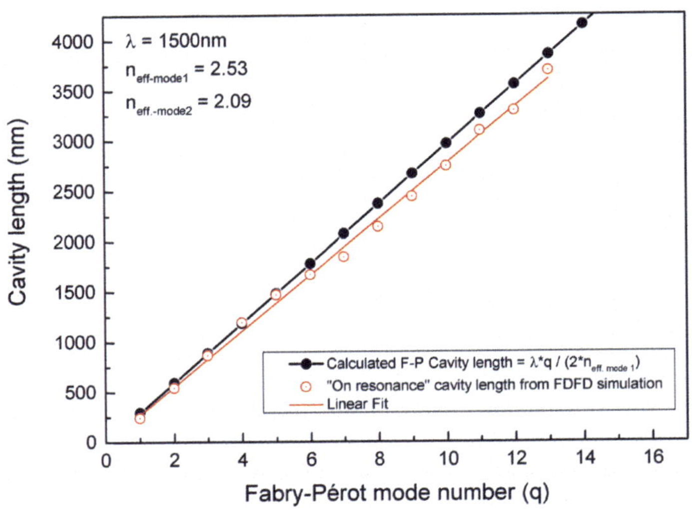

Fig. 1.10 Cavity length for different Fabry–Pérot mode number. Both, the analytical and the rigorous model show close result

cavity induce a low value of F for existing modes and their resonances. F increases slowly with decreasing wavelength, which is due to an increase in reflectivity at both ends of the Au cavity, as n_{eff} is increased. The simplified FP model extracts the important properties of the PSWC device: the definition of the resonance position and the sharpness of the resonance.

1.4.3 Light Management in the PSWC

1.4.3.1 Modal Analysis

It is desirable to confine strongly and enhance the E-field in the slot that is created by the two metallic pads. Since the PSWC is excited by a dielectric waveguide, we need to perform a modal analysis of the dielectric waveguide itself and of the coupled system, i.e. the dielectric waveguide coupled to the plasmonic slot waveguide in order to optimize it. We assume that the plasmonic cavity is excited by a single quasi-transverse-electric (qTE) mode travelling in a silicon wire waveguide at wavelengths in the range from 1200 to 1700 nm. The transverse electric (TE) case is the one where the polarization of the electric field vector is perpendicular to the y, z-plane, as shown in Fig. 1.11a. We perform a systematic analysis of the coupling in the weak coupling regime. From coupled-mode theory, a mode analysis of the uncoupled waveguide is made to describe the coupled waveguide, as suggested in reference

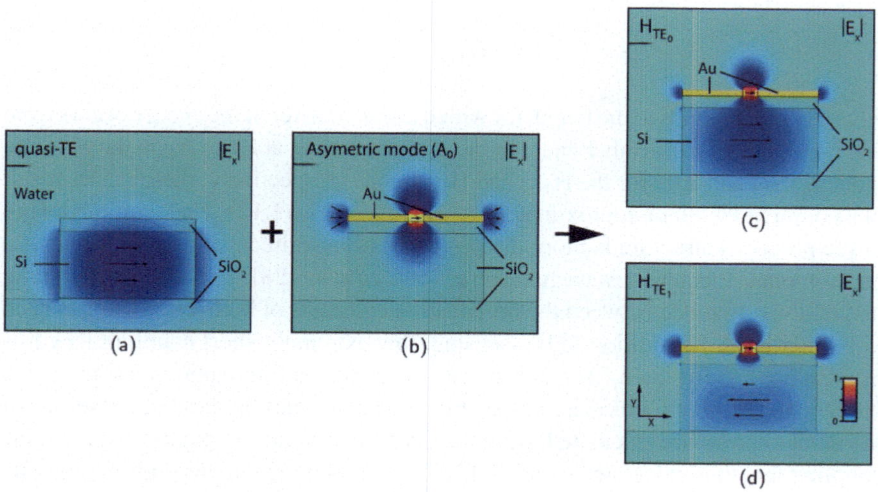

Fig. 1.11 2D E_x-field amplitude component at $\lambda = 1500$ nm. **a** Quasi-TE mode of the silicon waveguide with its SiO_2 spacer ($t_s = 40$ nm, $w = 450$ nm). **b** Uncoupled asymmetric PSW fundamental mode (A_0) with the waveguide core material replaced by silicon oxide ($t_m = 20$ nm). **c** and **d** H_{TE0} and H_{TE1} hybrid modes of the PSW coupled to the silicon waveguide. Reprinted with permission from [29] © The Optical Society

[26]. We calculated the uncoupled and coupled modes using the 2D mode solver from CST. Figure 1.11a shows the amplitude of the E_x-field component for the qTE mode of the silicon wire waveguide used to excite the PSWC. In Fig. 1.11b, we show the E_x amplitude for the uncoupled fundamental plasmonic slot waveguide mode, in which the silicon core has been replaced by silicon oxide. As a result of close mode matching between the qTE and the uncoupled asymmetric PSW fundamental A_0 mode, two hybrid modes, which are the superposition of the qTE and A_0 modes, are excited. The two calculated hybrid modes are depicted in Fig. 1.11c-d and are the result of the coupling between the PSW and the silicon waveguide (H_{TE0} and H_{TE1}). The two hybrid modes retain the high field localisation that is characteristic of the A_0 slot mode.

1.4.3.2 PSWC Coupling

Coupling of light from the dielectric waveguide to the PSW plays an important role in the performance of our sensor. The proposed cavity is in fact a PSW with a length of only 600 nm, which is shorter than the coupling length between the silicon waveguide and the PSW. In a similar way to the situation of two coupled dielectric waveguides, the two hybrid modes H_{TE0} and H_{TE1} of the structures have two different propagation constants βH_{TE0} and βH_{TE0}, which leads to interferences during their propagation. Consequently, for an infinitely long PSW, alternating energy transfer is observed—from the dielectric waveguide to the PSW and vice versa. The coupling length is equal to [27]:

$$L_c = \frac{\pi}{\beta H_{TE0} - \beta H_{TE1}} \tag{1.12}$$

where L_c corresponds to the length for which a total transfer of the energy occurs from one waveguide into the other one. For the given structure, at $\lambda = 1.6\ \mu m$ the effective index is 2.54 and 2.09 for the H_{TE0} and H_{TE1} modes respectively; these values result in a coupling length of approximately 1.78 μm (see Fig. 1.12). The coupling length of the proposed structure is more than an order of magnitude smaller than that of a conventional silicon wire waveguide directional coupler [28]. To reduce the coupling length, the difference between the propagation constants of both hybrid modes need to be increased. In addition to bio-sensing, this extremely short coupling length is of great interest for saving as much space as possible on future photonics integrated interconnects and circuits. The propagation constants and the field distributions of the modes shown are calculated using the CST Microwave 2D mode-solver and the coupling length is calculated using (1.12). The coupling length depends strongly on the geometrical parameters of the structure. When the spacer thickness is increased, the interaction of the evanescent tails of the two modes is reduced and therefore the coupling length is increased. On the other hand, shrinking the size of the silicon wire waveguide will shorten the coupling length because the evanescent tail of the silicon wire waveguide mode will interact more strongly with the PSW mode.

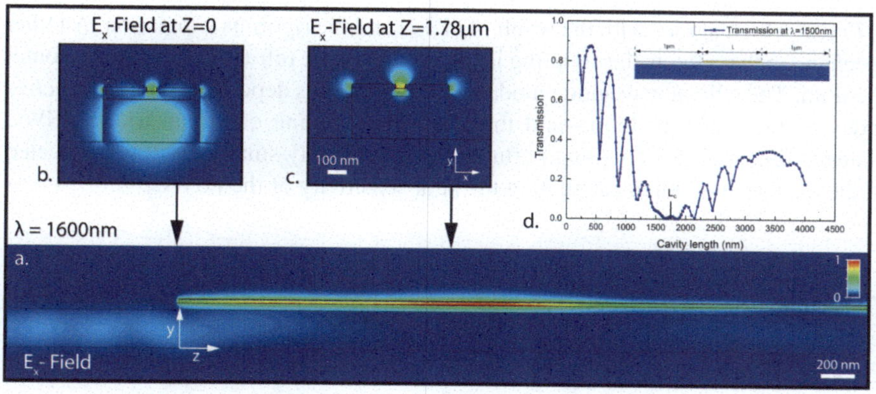

Fig. 1.12 **a** E_x-field amplitude showing the coupling between the dielectric waveguide and the metallic slot calculated using the 3D frequency-domain solver from CST Microwave. **b** and **c** E_x-field amplitude of the 2D mode profile at $z = 0$ and $z = 1.78$ μm. **d** Transmission of the excited cavity as a function of the cavity length at $\lambda = 1500$ nm. Reprinted with permission from [29] © The Optical Society

In order to confirm the coupling length calculations based on (1.12), a rigorous calculation based on the 3D frequency-domain solver version of CST Microwave has been carried out. Figure 1.12a shows the distribution of the E_x component of the electric field for a structure excited from the left side by the qTE mode of the silicon waveguide at a wavelength of 1.6 μm. After propagating for a distance of one micrometer, the mode starts to couple into an infinitely long PSW. The mode of the silicon waveguide is coupled into the metallic slot after a coupling length of about 1.78 μm. At the position of the coupling length (L_c), the total energy density is transferred to the slot where the electric field is confined and the energy is coupled back into the silicon waveguide at 2 L_c. The L_c value obtained shows good agreement with the mode coupling analysis described previously, which confirms that the device works in the weak coupling regime. The inset in Fig. 1.12b and c shows the modal field distribution at $z = 0$ and $z = L_c$, respectively. At $z = 0$, light is essentially confined in the silicon core—while, at L_c, light is confined in the sub-wavelength-sized slot. Figure 1.12d shows the transmission as a function of the cavity length at a wavelength of 1.5 μm. The inset of Fig. 1.12d shows a sketch of the structure, which is composed of an input and an output silicon waveguide of 1 μm length with the PSWC of length L in the middle. In contrast to a structure that is nominally infinitely long, the structure is now a PSW cavity and—therefore the metallic slot has a finite length that is then varied in order to observe its impact on transmission. Two types of oscillation appear in the transmission. The one with the smaller period corresponds to the cavity resonance produced by the reflection of the light at both ends of the slot. The one with a longer period originates from the interference between the two propagating modes, which have different propagation constants (βH_{TE0} and βH_{TE1}). The coupling length of the structure can therefore be retrieved from the transmission spectrum. At the coupling length, a transmission minimum occurs. In our case, $L_c =$

1.75 μm which agrees with the result of (1.12) and the rigorous simulation. Another mechanism that affects the coupling is the change of the refractive index of the outer medium. The silicon waveguide modal effective index is dependent on the refractive index of the outer medium—and therefore the coupling condition to the PSWC changes. Because this coupling perturbation is relatively small, it will be neglected in the next section when calculating the bulk sensitivity of the PSWC.

1.4.4 Field Enhancement and Sensitivity

1.4.4.1 Field Enhancement

The ability to detect small molecules in the vicinity of the PSWC is proportional to its field enhancement. Equation (1.13) expresses the average normalized E-field intensity enhancement due to the presence of the PSWC. It is obtained from the ratio of the integral of $|E|^2$ over a volume V including the PSWC to the same integral calculated over the same volume, but without the effect of the metallic cavity. The integration volume V is a box centered on the PSWC with dimensions: $[t_m + 40\ \mathrm{nm}](X)$, $[w + 40\ \mathrm{nm}](Y)$, and $[L + 40\ \mathrm{nm}](Z)$.

$$\mathrm{FE} = \frac{\iiint_V |E_{\mathrm{cav}}|^2 \mathrm{d}x\mathrm{d}y\mathrm{d}z}{\iiint_V |E_{\mathrm{wg}}|^2 \mathrm{d}x\mathrm{d}y\mathrm{d}z} \tag{1.13}$$

In Fig. 1.13a, we display the field enhancement of gold PSWC of 600 nm length (in red) as a function of the wavelength compared to its transmission spectrum (in

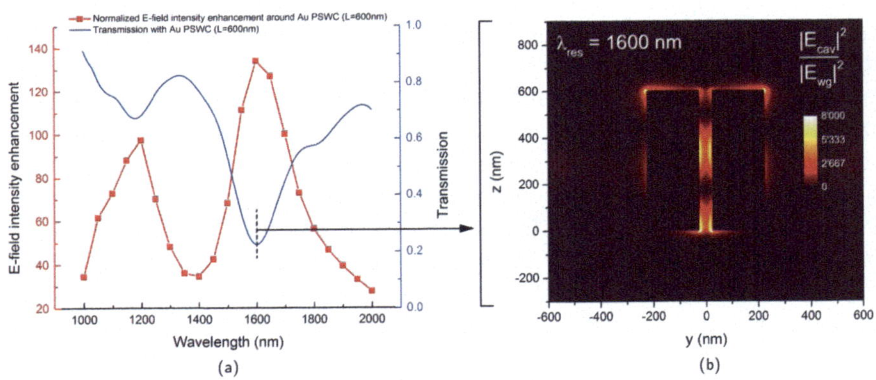

Fig. 1.13 **a** Transmission spectrum compared to the averaged E-field enhancement around a gold PSWC ($t_s = 40$ nm, $w = 450$ nm, $t_m = 20$ nm, $w_s = 50$ nm, $L = 600$ nm). **b** 2D normalized $|E|^2$ field distribution at half the metal thickness of the plasmonic cavity at resonance frequency. Reprinted with permission from [29] © The Optical Society

blue). The maximum field enhancement occurs at the longitudinal resonance of the H_{TE0} guided mode. In addition, we observe that at higher harmonics (i.e. for longer cavities at the same wavelength), the field enhancement increases. The increase is due to the length of the cavity, which is increasing and approaching the coupling length L_c, so that more light is coupled into the cavity. In Fig. 1.13b, the same electric field enhancement is shown in a plane cutting the PSWC at half of its thickness. It is observed that light is essentially confined between the two metallic interfaces, which is due to the excitation of a guided plasmonic mode inside the cavity that propagates from the bottom to the top in Fig. 1.13b. Nevertheless, higher field strengths are observable on the exterior of the cavity. They are mainly due to the presence of the H_{TE1} mode.

1.4.5 Bulk and Local Sensitivity

In a similar fashion, we have performed a sensitivity study of the structure with three different metals (gold, silver, and aluminum) to find the relationship between near-field enhancement and detection of localized changes of refractive index. It was shown previously that changing the spacer thickness does not vary the resonance wavelength but that it does significantly modify the coupling between the waveguide and the plasmonic cavity. Because we wish to keep the cavity as small as possible, we have studied the sensitivity of the PSWC as a function of the spacer thickness for the three different metals. The dispersion of the different metals is described using the data given by Johnson and Christy [25]. To calculate the bulk sensitivity, as explained earlier, in the beginning of Sect. 1.4.4, two FDTD simulations were performed with the refractive index of the outer medium changing from 1.32 (water at 1500 nm) to 1.35 (acetone at 1500 nm).

Figure 1.14a shows the bulk sensitivity, as a function of the spacer thickness, for a 300 nm long PSWC made from silver or gold. Figure 1.14b shows a similar result for a 600 nm long PSWC made from silver, gold or aluminum. The resonant wavelength of the PSWC in Fig. 1.13a, b are both around $\lambda_{res} = 1500$ nm. Therefore the fundamental resonance is excited for the 300 nm long PSWC, and the first harmonic is excited for the 600 nm long PSWC. Note that in Fig. 1.14a the aluminum PSWC sensitivity is missing because the extinction ratio of the resonance is too small. In Fig. 1.14a, b, we observe a general increase of the bulk sensitivity, for all the three metals, for increasing spacer thickness. For the silver PSWC in Fig. 1.14a, the bulk sensitivity reaches 615 nm/RIU for a spacer thickness of 120 nm and a cavity length of 300 nm. It shows that the fundamental resonance has higher bulk sensitivity than the first-order resonance. Moreover, when using the fundamental resonance, the detection volume becomes smaller, which is better for single molecule sensing at high concentration [24]. With increasing spacer thickness the fundamental resonance extinction ratio decreases, thus tracking the resonance becomes more difficult. The results of Fig. 1.14a, b show that the less coupled is the system, the stronger the resonance wavelength perturbation produced by a certain refractive index change of

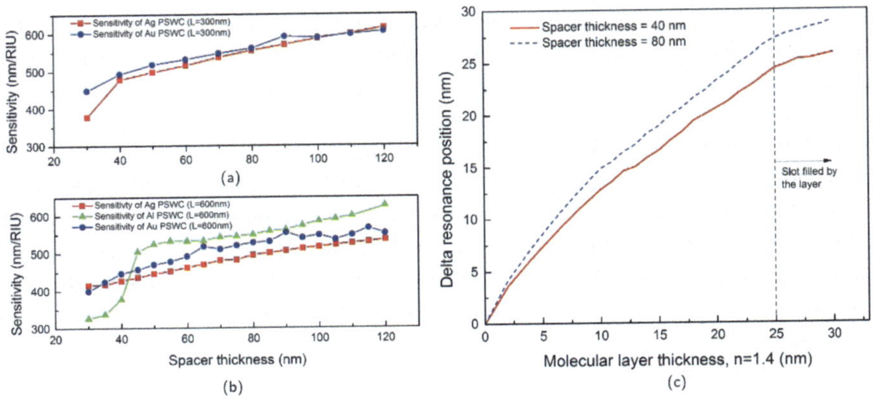

Fig. 1.14 **a** Bulk sensitivity for silver and gold PSWC of 300 nm length as function of the spacer thickness. **b** Bulk sensitivity for silver, gold and aluminum PSWC of 600 nm length as function of the spacer thickness. **c** Resonance wavelength shift of a 600 nm (*L*) long gold PSWC as a function of the thickness of a molecular layer of refractive index of 1.4 deposited on it. Reprinted with permission from [29] © The Optical Society

the bulk. However, for aluminum, even if the bulk sensitivity is higher, the quality factor and the extinction ratio are smaller because of the large amount of absorption loss; this strongly affects the detection limit of the sensor. Using silver for the cavity provides the highest quality factors and thus the highest photon life time in the cavity, as well as the highest local field enhancement.

The bulk sensitivity values obtained from our simulations are only 40% lower than the result shown in [25] where a periodic array of PSWCs is used on top of a slab waveguide—which also has a 30 times larger interaction volume. As the goal of this sensor is to detect the binding of small molecules to its surface, a high bulk sensitivity might create a lot of noise in the detection process, e.g. through temperature fluctuations. Therefore it is important to study how surface-sensitive this sensor is. In order to investigate surface or local sensitivity, we performed a resonance position tracking for different thicknesses of a molecular layer of refractive index equal to 1.4 around the gold PSWC. In Fig. 1.14c, the shift of the resonant wavelength is shown for a molecular layer thickness growing from 0 to 30 nm for a gold PSWC, with the same geometrical parameters as the one shown in Fig. 1.8. When the spacer thickness changes from 40 nm to 80 nm, an increase in the local sensitivity is observed. For thin layers the sensitivity is extremely high, since it is related to the field strength in the slot of the cavity. Field strengths and local sensitivities decrease, the further we are away from the metal surface. We believe that the high local sensitivity, combined with the unique excitation of this plasmonic sensor, can rival single-particle localized surface-plasmon resonances (LSPRs) and it can be multiplexed more easily.

The performance of the sensor can be benchmarked by using a figure of merit (FOM) expressed by the ratio of the sensitivity over the full width half maximum (FWHM) of the resonance. In the case of LSPR, the best FOM is around 3 and the sensitivity is around 200 nm/RIU. With a FOM of 4.1 for a spacer thickness of 40 nm, our sensor therefore outperforms LSPR sensors.

1.5 Conclusions

In the first part of this chapter, we have studied the basic principles of refractive index measurement. Widely used are interferometers such as Michelson, Fabry–Perot, and micro-ring resonators. The objective was to summarize the general formulas describing the performance of a sensor independently of the type of device. This is very helpful for the design and discussion of instruments. The resolving power of an instrument is equal to the optical path difference, OPD, divided by the wavelength, λ. In a resonator the resolving power can be increased by the number of round trips, also called the finesse, $\mathscr{F}$. It is important to know that the resolving power of a resonator is equal to the quality factor, Q. Consequently the resolving power, and therefore also Q, can be increased by increasing the cavity length. For miniaturized systems, i.e. short cavity lengths, a high finesse is needed, in order to achieve a high quality factor. The problem is that a high quality factor micro/nano system also needs a low coupling coefficient, which makes it difficult to couple light into the resonator. Another important factor is the sensitivity, which also depends on the cavity size. Unfortunately, the sensitivity cannot be increased by multiple round trips, i.e. by a high Q.

The key strength of a miniaturized sensor is neither the sensitivity nor the high value of Q, it is rather the possibility to measure a small volume locally, while accepting a rather weak sensitivity. Therefore the performance of such sensors is often given by [30]:

$$F_{\mathrm{P}} = \frac{3}{4\pi^2}\left(\frac{\lambda}{n}\right)^3 \frac{Q}{V} \tag{1.14}$$

where V is the mode volume. A high Q sensor is still of interest, because the detection limit is inversely proportional to the resonance Q-factor. It is usually limited by the noise equivalent power of the optical detector.

In the second part, we have presented a theoretical study of a plasmonic slot waveguide cavity (PSWC) sensor, which is a good candidate for detection of changes in refractive index in sub-attoliter volumes with a high surface sensitivity. The strength of the sensor compared with other plasmonic nanostructures [31, 32] is in the system architecture, enabling a fully integrated system. The sensor is excited by a silicon waveguide, which gives flexibility in multiplexing the cavity on the chip surface for higher throughput. The proposed structure behaves as a Fabry–Pérot resonator. The device has been studied using rigorous electromagnetic theory and compared with the analytical model. The resonant wavelength is sensitive to the surrounding refractive index and is also dependent on the geometrical parameters of the resonator.

We have investigated the dependence of the spectral transmission, and specifically the resonance wavelength, on several sensor geometrical parameters, such as the thickness of the spacer or the cavity length. Then we have performed a modal analysis based on coupled mode theory in order to describe the hybrid modes present in the cavity. Using this analysis we have discussed the transmission modulation of the plasmonic slot waveguide cavity. This extra modulation is due to the presence of two propagating hybrid modes inside the cavity that interfere with each other. We have also studied the bulk and local sensitivity of the sensor. Our simulations of the bulk sensitivity as a function of the spacer thickness showed that, for a weakly coupled PSWC, a higher quality factor and higher bulk sensitivity are expected. Finally, by simulation, we have shown that this sensor has high local sensitivity. A resonance wavelength shift of 3 nm is obtained for the deposition of a 1 nm thick monolayer over the PSWC. This last result is of great importance for the detection of very small molecules and even single-molecule sensing.

References

1. V. Saptari, *Fourier Transform Spectroscopy Instrumentation Engineering* (SPIE Press, 2004)
2. F.L. Pedrotti, L.S. Pedrotti, L.M. Pedrotti, *Introduction to optics* (Pearson Prentice Hall, Upper Saddle River, N.J., 2007)
3. S.G. Lipson, H. Lipson, *Optical Physics* (Cambridge University Press, 1969)
4. B.E.A. Saleh, M.C. Teich, *Fundamentals of Photonics*, 2nd edn. (Wiley-Interscience, Hoboken, N.J, 2007)
5. P. Connes, High Resolution and High Information Fourier Spectroscopy. Presented at the Aspen International Conference on Fourier Spectroscopy (1970), p. 121
6. G. Brooker, *Modern Classical Optics* (OUP Oxford, 2003)
7. D. Malacara, *Handbook of Optical Engineering* (CRC Press, 2001)
8. K.P. Koo, M. LeBlanc, T.E. Tsai, S.T. Vohra, Fiber-chirped grating Fabry-Perot sensor with multiple-wavelength-addressable free-spectral ranges. IEEE Photonics Technol. Lett. **10**(7), 1006–1008 (1998)
9. B.E. Little, S.T. Chu, H.A. Haus, Second-order filtering and sensing with partially coupled traveling waves in a single resonator. Opt. Lett. **23**(20), 1570–1572 (1998)
10. L. Zhou, X. Sun, X. Li, J. Chen, Miniature microring resonator sensor based on a hybrid plasmonic waveguide. Sensors **11**(7), 6856–6867 (2011)
11. D.G. Rabus, *Integrated Ring Resonators: The Compendium* (Springer, 2007)
12. H. Okamura, K. Iwatsuki, A finesse-enhanced Er-doped-fiber ring resonator. J. Lightwave Technol. **9**(11), 1554–1560 (1991)
13. B.E. Little, S.T. Chu, H.A. Haus, J. Foresi, J.-P. Laine, Microring resonator channel dropping filters. J. Lightwave Technol. **15**(6), 998–1005 (1997)
14. K. Tsia, *Understanding Biophotonics: Fundamentals, Advances, and Applications* (CRC Press, 2015)
15. C. Nylander, B. Liedberg, T. Lind, Gas detection by means of surface plasmon resonance. Sens. Actuators **3**, 79–88 (1982)
16. J. Homola, Surface plasmon resonance sensors for detection of chemical and biological species. Chem. Rev. **108**(2), 462–493 (2008)
17. J. Homola, Present and future of surface plasmon resonance biosensors. Anal. Bioanal. Chem. **377**(3), 528–539 (2003)
18. J. Homola, S.S. Yee, G. Gauglitz, Surface plasmon resonance sensors: review. Sens. Actuators B Chem. **54**(1–2), 3–15 (1999)

19. J. Homola, I. Koudela, S.S. Yee, Surface plasmon resonance sensors based on diffraction gratings and prism couplers: sensitivity comparison. Sens. Actuators B Chem. **54**(1–2), 16–24 (1999)
20. M. Piliarik, J. Homola, Surface plasmon resonance (SPR) sensors: approaching their limits? Opt. Express **17**(19), 16505–16517 (2009)
21. V. Kodoyianni, Label-free analysis of biomolecular interactions using SPR imaging. Biotechniques **50**(1), 32–40 (2011)
22. T.W. Ebbesen, H.J. Lezec, H.F. Ghaemi, T. Thio, P.A. Wolff, Extraordinary optical transmission through sub-wavelength hole arrays. Nature **391**(6668), 667–669 (1998)
23. C. Genet, T.W. Ebbesen, Light in tiny holes. Nature **445**(7123), 39–46 (2007)
24. Q. Tan, Subwavelength Metallic Structures for Sensing. EPFL, Lausanne (2012)
25. A. Cosentino, Q. Tan, M. Roussey, H.P. Herzig, Refractive index sensor based on slot waveguide cavity. J. Eur. Opt. Soc. Rapid Publ. **7** (2012)
26. Q. Tan, A. Cosentino, M. Roussey, H.P. Herzig, Theoretical and experimental study of a 30 nm metallic slot array. J. Opt. Soc. Am. B **28**(7), 1711 (2011)
27. A. Cosentino, Slot waveguide cavity: from fabrication to characterization and sensing applications. EPFL, Lausanne (2013)
28. G.D. Osowiecki, Excitation of resonant plasmonic cavities by integrated waveguides for sensing applications. EPFL, Lausanne (2016)
29. G. Osowiecki, E. Barakat, A. Naqavi, H.P. Herzig, Vertically coupled plasmonic slot waveguide cavity for localized biosensing applications. Opt. Express **22**(17), 20871–20880 (2014)
30. E. Purcell, Spontaneous emission probabilities at radio frequencies. Phys. Rev. **69**, 681 (1946)
31. ar800074d.pdf
32. A.A. Yanik et al., An optofluidic-nanoplasmonic biosensor for direct detection of live viruses from biological media. Nano Lett. **10**(12), 4962–4969 (2010)

Chapter 2
Long-Range Plasmonic Waveguide Sensors

Oleksiy Krupin and Pierre Berini

Abstract A novel biosensor utilizes long-range surface plasmon polariton (LRSPP) waveguides. LRSPPs are transverse magnetic (TM) surface plasmon waves that can propagate over considerable lengths along a metal slab or stripe bounded by dielectric regions. The LRSPP is a symmetric mode that is formed by combining single-interface surface plasmon polariton waves (SPPs) on a slab or stripe.

2.1 Introduction

A novel biosensor utilizes long-range surface plasmon polariton (LRSPP) waveguides. LRSPPs are transverse magnetic (TM) surface plasmon waves that can propagate over considerable lengths along a metal slab or stripe bounded by dielectric regions. The LRSPP is a symmetric mode that is formed by combining single-interface surface plasmon polariton waves (SPPs) on a slab or stripe. The mode is confined at the metal-dielectric interfaces in the plane transverse to the direction of propagation, which leads to the possibility of creating various structures—such as S-bends, Y-junctions, Mach–Zehnder Interferometers (MZIs) and couplers [1, 2]. The propagation length of LRSPPs is ~2000 μm, compared to that of a single-interface SPP of ~80 μm [3], thus the overall sensitivity can be larger due to a longer optical interaction length with the sensing medium. The penetration depth into the dielectric

O. Krupin
Department of Biological and Chemical Engineering, University of Ottawa, 161 Louis Pasteur, Ottawa, ON K1N 6N5, Canada

P. Berini (✉)
School of Electrical Engineering and Computer Science, University of Ottawa, 800 King Edward Avenue, Ottawa K1N 6N5, Canada
e-mail: pberini@uottawa.ca

Department of Physics, University of Ottawa, 150 Louis Pasteur, Ottawa K1N 6N5, Canada

Centre for Research in Photonics, University of Ottawa, 25 Templeton Street, Ottawa K1N 6N5, Canada

R. De La Rue et al. (eds.), *Biomedical Optical Sensors*,
Biological and Medical Physics, Biomedical Engineering,
https://doi.org/10.1007/978-3-030-48387-6_2

regions of the LRSPP wave is ~1 μm, much larger than that of a single-interface SPP (~200 nm). Increased penetration depth can be advantageous in the case where a matrix is attached to the waveguide, consequently increasing the loading capacity of small molecules and resulting in a larger sensitivity. The field penetration is also comparable with the dimensions of living cells (i.e. 1–2 μm for bacterial cells and 3–10 μm for animal cells), and potentially this penetration can be used for tracking changes within cells, or at the cytosol-membrane interface. Recently LRSPPs have been used in modified surface plasmon resonance (SPR) prism-based instruments, and found to improve the detection limit for detecting changes in the bulk refractive index [4] and for detecting bacteria [5].

The minimum insertion loss of the LRSPP propagating along a metal stripe occurs when the refractive indices (RI) of the dielectrics on both sides are equal. For sensing purposes, at least one surface of the waveguide has to be exposed to the sensing medium, where receptor chemistries would be applied and sensing occurs. Biologically-compatible fluids such as phosphate buffered saline (PBS) have an RI close to that of water ($n \sim 1.330$). Therefore, in order to maintain the RI symmetry around the waveguide, certain considerations for choosing a suitable dielectric medium below the stripe must be taken into account. Generally, dielectric polymers have a significantly higher RI—and the choice is limited to a few fluoropolymers of low index, such as Teflon (Dupont) or CYTOP (Asahi); indeed, both have been found suitable for sensing using prism-coupled SPR involving LRSPPs [4–9].

Dielectric waveguides have also been investigated for biosensing using photonic crystals [10], CYTOP [11], silicon-on-insulator [12] and silicon nitride [13–15]. Although dielectric waveguides have been successfully employed, gold is still generally preferable as a sensing surface, for a number of reasons. Gold is the most historically studied surface for sensing and remains the most used—primarily due to its application in commercial SPR sensors. A good understanding exists of the different chemistries that can be used to functionalize gold surfaces such as different alkanethiols for self-assembled monolayer (SAM) formation. Well-organized packing of alkanethiols can easily be achieved and the end groups can be further functionalized with sensing chemistry, or receptors, such as immunoglobulins (Igs). Furthermore, using alkanethiols of different chain length, SAMs can be designed such that functionalized receptors are separated from each other, thus increasing the avidity of the surface by removing steric hindrance [16]. Also, SAMs form on gold very rapidly (80–90% of a SAM is formed within minutes) and can be stable for a long period of time (weeks to months) [17]. Gold also does not react with many chemicals and is generally impervious to cleaning solvents used in the preparation of a sensing surface. SAM formation on silica (SiO_2) substrates, however, is more troublesome due to the moisture sensitivity of alkylsalines and the organization is not as well-packed and controlled as for gold. Frequently, the SiO_2 based sensing waveguides are functionalized by adsorption of the receptor directly onto the waveguide, which diminishes the specificity due to an increase of non-specific binding and reduction of overall surface avidity. Application of protein G for immunoglobulin (Ig) functionalization on Au surfaces is also rapidly emerging. Protein G has an affinity for the Fc (fragment crystallizable region) of Ig—and therefore it favours a vertical orientation of Igs on the surface, thereby increasing surface avidity due to Fab accessibility [18, 19]. Another advantage of using gold as a substrate is its conductive property, which can be used for topo-selective functionalization of complex structures, such as MZIs—where the two MZI arms should have different chemistries, again, for biosensing applications [20, 21]. Thermo-optic modulation [22, 23] on gold MZIs can significantly improve signal-to-noise ratios through bandwidth-limited signal processing.

In this chapter we discuss various aspects of novel LRSPP waveguide biosensors: we describe the structure, its fabrication and its ability to detect changes in the RI of solutions—as well as to detect cells and proteins. In addition, several applications for clinical detection problems such as Dengue fever and B-cell leukemia detection are discussed. MZIs are also demonstrated for bulk sensing—and exhibit high sensitivity.

2.2 Sensing Platforms

2.2.1 Sensor Structure and Fluidic Integration

The sensors are fabricated using photolithography following the process described in Sect. 2.2.2. After fabrication, the entire wafer (diameter 4 inch), which hosts ~300 dies, is covered with dicing photoresist and diced into individual dies (3.8 mm wide by 6.4 mm long). Each device contains 16 straight waveguides (SWGs), each of which consists of 5 μm wide, ~25–35 nm thick Au stripes, embedded in CYTOP (Fig. 2.1). The top cladding of CYTOP is etched down to the Au waveguide level, creating a cavity (fluidic channel), where the waveguide surface is exposed to the fluid. The active sensing length of the waveguide is ~1.65 mm. The non-etched (clad) parts at the ends of the sensing section serve as supports for integration with fluidics. A custom-made Plexiglas jig with a fluorocarbon O-ring attached to the bottom was designed so that the O-ring rests on the clad part—providing a seal for the fluid in the channel [24]. Tubing is attached to two holes for fluid injection by a syringe pump. The device is integrated into the fluidic cell by placing it onto the custom-made metal base and affixing the Plexiglas jig. The LRSPP wave is excited by butt-coupling a polarization-maintaining optical fibre to the waveguide at the input facet.

A device can be functionalized with sensing chemistry either prior to insertion in the flow cell or in situ—depending on the requirements. For example, the O-ring is attached to the Plexiglas jig by silicone glue. Since organic solvents such as isopropanol (IPA) might slightly dissolve silicone glue, it would not be recommended to form a SAM on waveguides in situ—so incubating the device in an alkanethiol solution of IPA prior to insertion into the flow cell would be preferable. Once the surface is functionalized for a specific sensing application, the device is installed in the fluidic set-up while dry, a solution of interest is introduced, and the whole assembly is integrated with the optical set-up as described in Sect. 2.2.3.

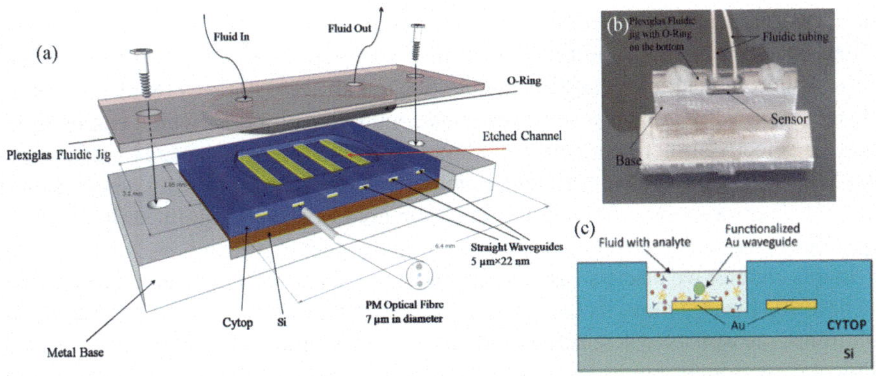

Fig. 2.1 Sensing device with integrated fluidics: **a** A schematic diagram of the sensor resting on the metal base, with Plexiglas jig prior to assembly; the volume of the fluidic cell is 20 μL. Reproduced with permission from [24]; **b** image of the sensor after integration with fluidics. Reproduced with permission from [24]; **c** a cross-section of the device showing a functionalized Au waveguide and a waveguide embedded in CYTOP. Reprinted with permission from [25]. Copyright (2014) American Chemical Society

2.2.2 *Sensor Fabrication and Properties*

The process for device fabrication was reported in [26, 27] and is summarized in Fig. 2.2a. After a 4 inch diameter Si wafer is stripped of native SiO_2, the first layer of CYTOP is formed by spinning M-Grade CYTOP onto the wafer. This grade of CYTOP incorporates a salinated end-group to facilitate adhesion by covalently bonding to the Si (Fig. 2.2b). Furthermore, multilayers are formed by multiple spinning/curing steps of S-grade CYTOP (optical grade), thereby creating a bottom cladding ~8 μm in thickness. Since a CYTOP surface is highly hydrophobic, it

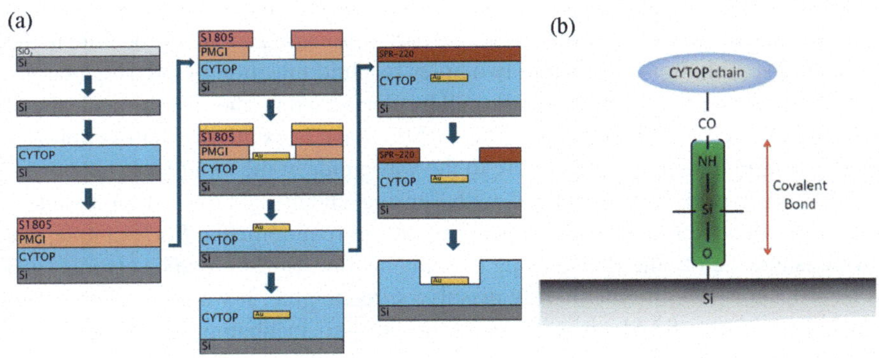

Fig. 2.2 Device fabrication: **a** simplified fabrication flow diagram for LRSPP biosensor fabrication; **b** covalent attachment of salinated end-groups of CYTOP to the Si substrate. Reproduced with permission from [27]

is primed in O_2 plasma, then coated with HMDS, a lift-off resist (PMGI), and a photoresist (S1805) for photolithographic development. The wafer is then exposed to UV light through a mask that defines the desired waveguide features—and Au is evaporated up to the desired thickness of ~35 nm. S1805 is a positive tone photoresist—and exposed areas are therefore susceptible to the developer, thereby leaving only the desired gold structures post lift-off. In order to create fluidic channels, the second cladding of CYTOP is deposited by multiple spin-coating/curing steps to a total thickness of ~7–8 μm and covered with positive tone photoresist SPR-220. A second mask, defining fluidic channels, is applied and the resist developed, then O_2 plasma etching is used in the exposed areas of SPR-220 to selectively etch CYTOP, thereby defining the fluidic channels. The O_2 plasma etch is performed in several steps while confirming the depth of etch after each step, using a profilometer. The process is stopped when a slight pedestal under the Au features is observed (<500 nm). The leftover SPR-220 is removed using acetone/IPA and de-ionized water, the wafer is dried and SPR-220 is spin-coated again to protect from damage during dicing.

While the thickness of the CYTOP claddings was verified during fabrication, the actual thickness and quality of the gold used in the waveguides in the sensing channels were investigated using an atomic force microscope (AFM), after dicing (Fig. 2.3). The device was randomly picked from the wafer and, using an ultrasonic bath, was damaged until part of the Au stripe is removed (in order to create an Au step). The waveguide was profiled by AFM in two places and the thickness was found to be between 34 and 35 nm (Fig. 2.3b–d), which is a good thickness for an

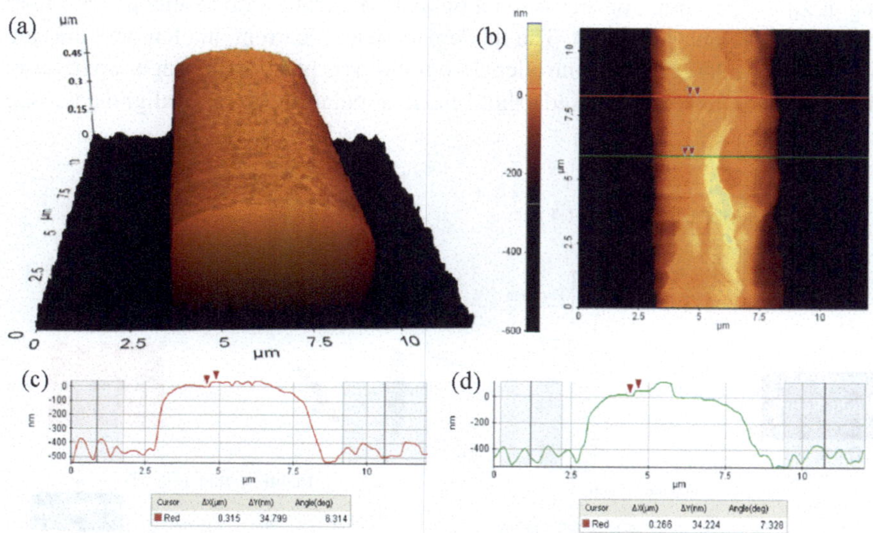

Fig. 2.3 AFM profiles of a waveguide: **a** 3D representation of an over-etched waveguide creating an Au stripe on a CYTOP pedestal; **b** gold thickness measurement taken at two locations (shown in red and green); **c** AFM profile (red curve) showing a Au thickness of 34.799 nm, and **d** (green curve) showing a Au thickness of 34.224 nm. Reproduced with permission from [27]

LRSPP biosensing waveguide at an operating wavelength of 1300 nm [3]. Figure 2.3a shows a 3D representation of an over-etched waveguide with a ~500 nm pedestal. A mild curvature of the top surface is observed due to slight deformation during upper cladding formation [26].

Several devices on a wafer have a structure consisting of interconnected waveguides that have individual fluidic channels of different lengths. Prior to using the wafer for biosensing, three of these waveguides of varying sensing length are used to investigate the optical properties of the devices on the particular wafer. Properties such as waveguide attenuation and coupling losses at the input facet and fluidic channel input/output are further used to extract data from the biosensing experiments. Estimation of the protein mass load on the waveguide (Sect. 2.3.3) and the theoretical fitting process for MZI bulk measurements (Sect. 2.3.5) exploited data from such optical measurements.

Experimental coupling losses at the input facet are relatively low (0.5–2.5 dB) and are usually in agreement with the theoretical value (0.89 dB/facet) [28]. The major source of the modal power attenuation resides in the fluidic channel: ~10 dB/mm, as determined experimentally—whereas the theoretical value is 7.2 dB/mm [29].

2.2.3 *Optical and Fluidic Integration*

The optical set-up (Fig. 2.4) includes the light source, which is a diode laser emitting at $\lambda_0 = 1310$ nm, connected to an optical polarization-maintaining (PM) fibre with a core diameter of 7 μm. The diode parameters (current and temperature) are controlled by a laser diode controller. Two multi-axis positioning stages are used in the setup. A six-axis stage is used to hold and manipulate the fibre for alignment to the

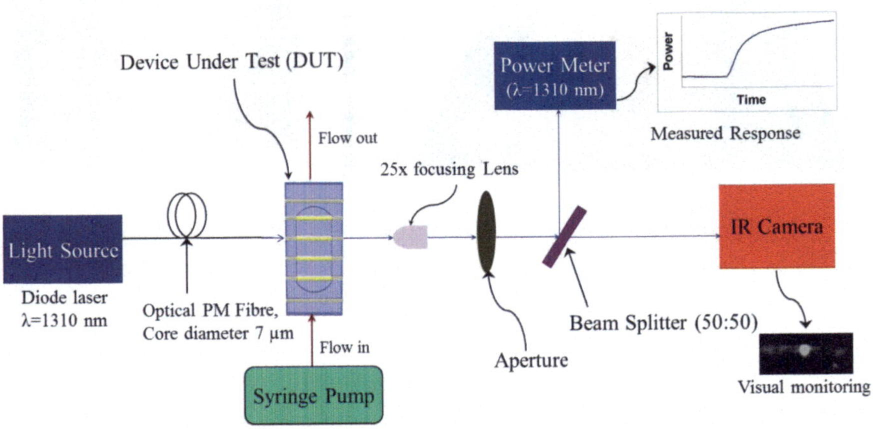

Fig. 2.4 Schematic representation of the optical setup for biosensing using LRSPP waveguides. Reproduced with permission from [24]

waveguide. After the device is incorporated into the fluidic cell, the assembly is fixed onto a 3-axis stage. By manipulating the two stages, the fibre is butt-coupled to the waveguide at the input facet and aligned to produce maximum output power, together with low background radiation. Improper alignment results in drift and power fluctuations of the output signal over time, which can mask actual binding events. In order to avoid Fabry–Perot interference at the input, glass index-matching oil is introduced between the facet and the fibre. A 25× collimating lens, firmly affixed to the optical table, is used to approximately collimate the beam exiting the device and to maintain collimation throughout the whole optical path. A syringe pump provides fluids to the system. Most of the experiments are performed by fluid suction, since experimentally this has been noted to produce smaller changes in pressure during stops, when changing fluids. An aperture is used to remove background radiation that in some cases can be strong enough to cause either signal instability or masking of the sensing event. In order to read the output power and visually monitor the mode quality simultaneously, a beam splitter is installed in the optical path, which splits the signal into two parts (~50:50). One part enters a power meter connected to the computer and the data is recorded using Labview software. The second portion of the output signal enters an IR camera connected to the monitor. Visual observation is highly important in enabling fibre-waveguide alignment. For example, if the alignment is not ideal, the power of the mode will be low, but the power reading might still be high due to greatly increased background radiation. As a result, the reading will be misleading and the sensitivity will drastically decrease.

2.2.4 *Experimental Procedures*

Prior to assembly with fluidics, each device undergoes extensive cleaning. Firstly, a 5 min ultrasonic bath in octane is applied to clean the device facets from possible debris accumulated during wafer dicing. Improper cleaning of the facets has been experimentally associated with instability of the signal and significant drift. Ultrasonication has a damaging effect on waveguides, resulting in a breakup of the integrity of the waveguide, or even a complete removal from the sensing channel. Therefore octane was specifically chosen for ultrasonic cleaning, since it does not dissolve the dicing photoresist that covers the surface of the sensor, and therefore keeps the waveguides protected during ultrasonication. The next cleaning step consists of two immersions in acetone, of 5 and 20 min each, to remove the photoresist—followed by an extensive wash in IPA and distilled/deionized water (DDI H_2O) and drying with nitrogen gas. The final step in sensor preparation is the removal of possible organic contamination from the waveguides via UV/Ozone cleaning for 30 min. UV exposure breaks organic covalent bonds, while at the same time converting available O_2 into ozone (O_3). The UV lamp is turned on for 5 min, and the rest of the organic compounds are broken down by ozone over the remaining 25 min. Further functionalization of the device varies upon the requirements of the experiment to be performed. Detailed functionalization procedures will be discussed further—in each section concerning the individual experiments.

Once the device is properly cleaned and functionalized, it is assembled using the fluidics jig—and sensing buffer solution is injected into the fluidic cell, fully covering the etched part of the sensor. Prior to device integration into the optical set-up, the stability and polarization of the incident light are verified by focusing the fiber on to the lens—and then recording data over a 10 min period. After the device is installed into the optical setup, the fibre is aligned with the waveguide. A baseline signal is established by running the sensing buffer solution through the system at 20 μl/min over ~10 min. This data is further used to determine the baseline noise, which is then used to estimate the signal-to-noise ratio (SNR) of the sensing response. At this point, the buffer containing the analyte is injected into the system and the response is observed and recorded. Each experiment has its own flow-rate and duration—as will be discussed in the following sections.

In order for the LRSPP to propagate along the waveguide, the RI of the sensing solution has to be matched to that of CYTOP (n = 1.3348). This is achieved by doping the biological buffer solution (phosphate buffered saline, or PBS, n = 1.290) with 7–12% glycerol w/w (n = 1.46). In our particular system of experiments, we have used two mixtures of PBS/Gly solutions: 7.25% w/w (n = 1.330) and 16.5% w/w (n = 1.338). Slight deviations from the exact match to CYTOP are discussed in Sects. 2.3.1 and 2.3.4.

2.3 Sensing Demonstrations

2.3.1 Bulk Sensing Using Straight Plasmonic Waveguides

Before using the devices for actual biosensing, it is necessary to find the limit of detection (LOD) for bulk refractive index changes and observe the general behaviour of our sensor under fluidic conditions. The Au waveguides were functionalized with 16-mercaptohexadecanoic acid (16-MHA) to form a carboxyl-terminated self-assembled monolayer (SAM) via incubating the sensing device in 2 mM of 16-MHA in isopropanol (IPA) for more than two hours. In order to maintain optical symmetry around the waveguide, the refractive index (RI) of the top cladding (in our case a sensing solution) has to be close to that of the bottom cladding (i.e. CYTOP, n = 1.3346). Therefore, distilled/deionized water (DDI H_2O, $n \sim 1.328$ at $\lambda_0 = 1310$ nm) was doped with glycerol (Gly) ($n \sim 1.470$) to raise the RI of the H_2O/Gly tested solutions. Six H_2O/Gly solutions of varying refractive indices (with increments of ~2×10^{-3} RIU) were sequentially injected over the 16-MHA functionalized surface twice: from low to high RI and then back to low RI. The intention was to: (a) observe the long-term stability of the system under fluidic conditions, (b) test how it behaves when the solutions are being exchanged and (c) confirm the repeatability. The refractive index of the prepared solutions was determined independently by a

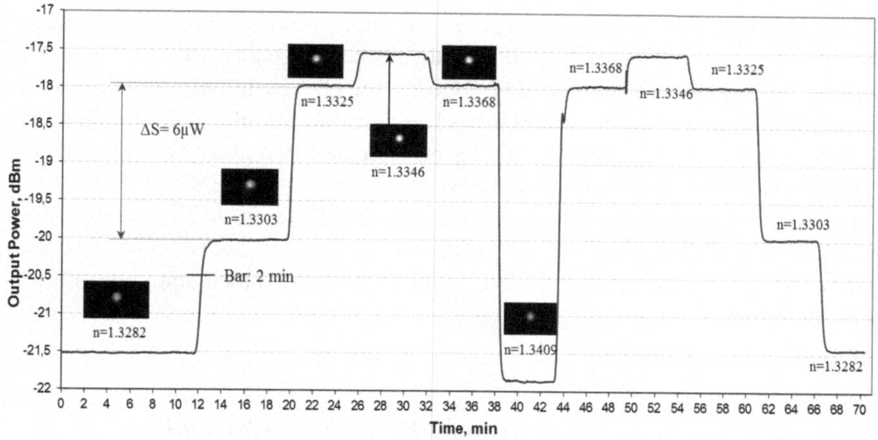

Fig. 2.5 Response of a straight Au plasmonic waveguide (22 nm thick) to six solutions of different refractive index following an increment of ~2 × 10^{-3} RIU. The input power was 5 dBm, $\lambda_0 =$ 1310 nm and the flowrate was set to 20 μl/min. Reproduced with permission from [24]

Metricon prism-based refractometer. The whole experiment was performed under flow at a 20 μl/min flow-rate with an optical power input of 5 dBm—and is presented in Fig. 2.5.

Since the RI of CYTOP is $n = 1.3348$, the highest output response is expected for the solution with $n = 1.3346$ (and confirms the symmetry of the LRSPP waveguides). Although not very apparent from the images of the mode, the background radiation is the lowest amount in this case as well. Background radiation can be caused by asymmetry between the top and bottom dielectrics, which results in light radiation. Although the increments in the RI of the solutions are relatively constant, the change of the response is not linear. The highest power output change (ΔS) was observed for a step from $n = 1.3303$ to $n = 1.3325$ ($\Delta S = 6$ μW) resulting in the highest bulk sensitivity. For this reason and for further biosensing experiments, the sensing solution was chosen to have a refractive index of $n = 1.3303$. The standard deviation σ for the baseline noise was $\sigma = 6.4$ nW over a 9 min period of buffer running at $n = 1.3282$, yielding a signal-to-noise ratio (SNR) of ~942 and a limit-of-detection (LOD) of 2.3 × 10^{-6} RIU near $n = 1.3303$ [24].

Another important observation from the experiment is the fluid exchange time, which was found to be ~2 min. This knowledge is important for biosensing experiments when, for technical reasons, the solution containing an analyte cannot be index-matched to the baseline solution. There are two such technical causes: (a) a high concentration of analyte added to the sensing buffer, which will change the RI of the solution (e.g. >100 μg/ml of protein, See Sect. 2.3.3)—and (b) purchased analyte samples often contain buffers with high concentrations of various stabilizers that can also result in an RI that is very different from the sensing solution (e.g. antibodies), and (c) the refractive index of complex fluids such as plasma or serum also differ from $n = 1.3303$. Altering the RI of such solutions to reach $n = 1.3303$ is practically impossible because waveguides are much more sensitive than commercial refractometers, so the bulk step will always be noticeable. Ideally, for protein sensing, lyophilized protein should be added to the phosphate-buffered saline glycerol (PBS/Gly) buffer with concentrations less than 100 μg/ml. Otherwise, knowing the fluid exchange time to be ~2 min, it is possible to differentiate between the bulk step and association and disassociation events.

In terms of stability, a very slight change (~0.05 dBm) was observed at the end of the whole experiment, when the initial solution was re-injected (70 min). Repeatability is also good, since the re-injected solutions have the same output power levels as the initial ones.

2.3.2 *Cell Sensing Using Straight Plasmonic Waveguides*

Selective cell sensing is important for various applications, such as: detection of bacterial contamination in food [30] and in blood products [31], and the detection of rare circulating tumour cells [32]. In order to demonstrate selective cell capture, a strategy to differentiate human red blood cells (RBCs), based on their ABO groups, was developed.

2.3.2.1 Selective Capture of Human RBCs

The gold surface of the SWGs was functionalized with antibodies against blood group A through use of carbodiimide chemistry and 16-MHA SAM [24, 33] . Firstly, after proper cleaning, the device was immersed in a 2 mM solution of 16-MHA for more than 2 h, following cleaning with IPA and DDI H_2O—and placed in a DDI H_2O solution of 0.1 M EDC/NHS (1:1) for 15 min to activate the carboxyl groups. After rinsing with DDI H_2O, the device was placed in a solution of Anti-A antibodies After rinsing with DDI H_2O, the device was placed in a solution of Anti-A antibodies (supplied by Mount Sinai Hospital collaborators, Toronto, who obtained these from the Immucor company) for ~2 h to complete the surface functionalization. Anti-A antibodies from Immucor are chimera antibodies that were specifically developed and are being used in clinics for blood grouping. The final step consisted of a wash with the PBS/Gly solution to remove excess antibodies from the surface; the device was dried with N_2 gas and installed into the fluidic and optical set-up.

Figure 2.6 demonstrates the selective sensing of human red blood-cells (RBCs) containing A-antigen (groups A and AB), where O-group RBCs and B-group RBCs were used as negative controls. The different types of RBCs (at a flow-rate of 5 $\times$ 10^7 cells/ml) were injected into the system over the same Anti-A functionalized waveguide. The flow was stopped for a few minutes to allow the cells to settle—and then a PBS/Gly buffer wash followed. In all cases, after cell settling, the signal dropped to practically zero (from ~ −27 dBm to ~ −48 dBm) due to cells covering the waveguide. In the cases of O-group RBCs and B-group RBCs, the signal recovered

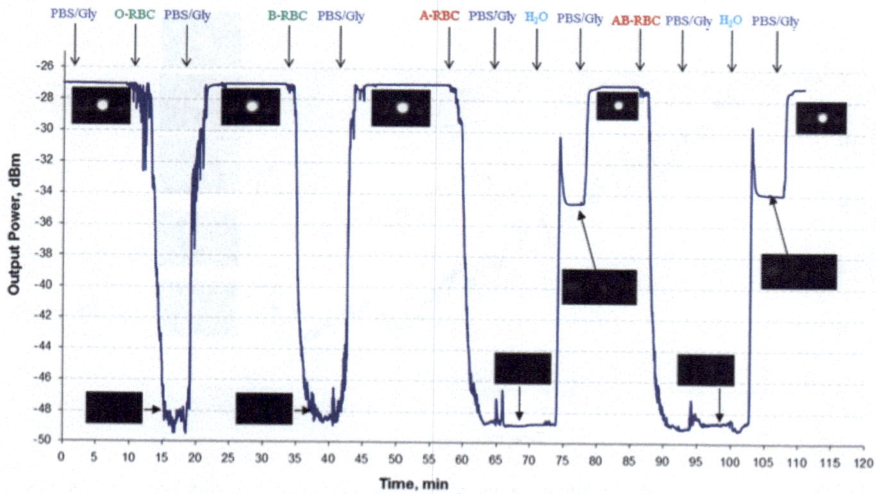

Fig. 2.6 Response of a straight Au plasmonic waveguide functionalized with Anti-A antibodies to PBS/Gly solutions carrying four different group types of human red blood cells (RBCs): O, B, A and AB. The waveguide thickness was ~35 nm, the input power was 10 dBm, the cell concentration was ~5×10^7 cells/ml, $\lambda_0 = 1310$ nm and the flowrate was set to 65 μl/min. Reproduced with permission from [34], *licence number 3713741431933*

with buffer injection to about −27 dBm, whereas for A-group RBCs and AB-group RBCs the signal recovery failed. In order to remove the cells, an additional injection of DDI water was required, which causes cell lysis. Based on this data, the following can be concluded: (a) antibody functionalization of the surface was successful, and (b) a monolayer of cells entirely blocks the optical pathway (after the PBS/Gly wash, all the cells were washed away—leaving only cells attached to the surface).

2.3.2.2 LOD for RBC Detection

In order to determine the limit of detection for cells, five solutions of varying A-group RBC concentration ranging from 1.14×10^5 to 1.83×10^5 cells/ml were tested on SWGs functionalized with Anti-A group (Fig. 2.7). The same waveguide was used for all experiments and the solutions were introduced in a random order. The surface was regenerated each time using DDI H_2O, as described in the previous experiment (Sect. 2.3.2.1). Anti-A group surface regeneration using DDI H_2O has been established to be repeatable for ten cycles without having a significant effect on cell binding [34].

As is evident from Fig. 2.7, the LOD for RBC capture is $<3.0 \times 10^5$ cells/ml, and is lower than that found for conventional SPR using a similar detection strategy (3.3×10^8 cells/ml) [35]. The observed spikes are due to cells passing across the waveguide with a velocity that is too high for them to become specifically attached. Another remarkable observation is a step-like response in all curves—the most prominent

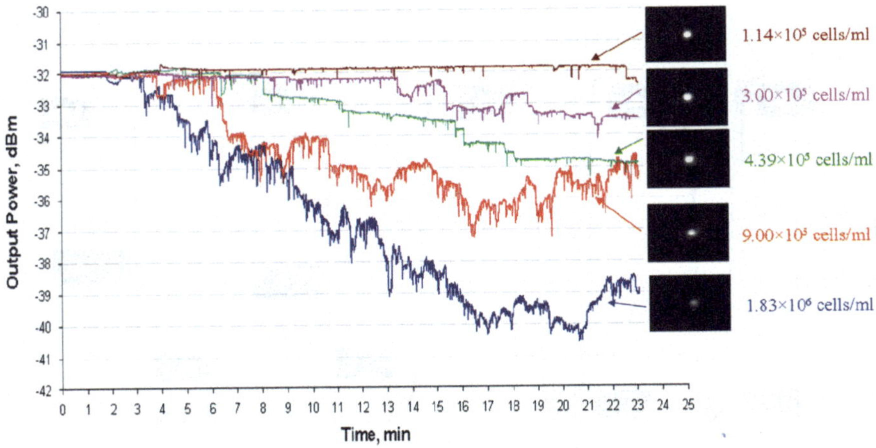

Fig. 2.7 Response of a straight Au plasmonic waveguide functionalized with Anti-A antibodies to PBS/Gly solutions carrying A-group RBCs of different concentrations to determine the LOD. The waveguide thickness was ~35 nm, the input power was 10 dBm, $\lambda_0 = 1310$ nm, and the flowrate was set to 20 μl/min. Reproduced with permission from [34], *licence number 3713741431933*

being seen on the curve corresponding to 4.39 × 10^5 cells/ml (green). This behavior is most likely due to individual cells binding to the waveguide. Also, a sudden increase in power can be attributed to weakly bound cells becoming detached from the waveguide. This step-like response suggests the possibility of single-cell detection. The fluidic channel designed for this sensor is large and the introduced cells flow across the waveguide, which diminishes the sensitivity of the device. New generation sensors will consist of individual microfluidic channels (~15 μm wide) for a single waveguide, where cells will flow along the waveguide, implying that all cells introduced into the system will have a much greater chance of encountering the waveguide surface. Hence, dedicated channeling should improve the LOD for cell sensing.

2.3.2.3 Single Cell Detection

The step-like behaviour of cell binding in the previous experiment (Sect. 2.3.2.2) suggests the possibility of single cell detection. In order to investigate this, A-group RBCs (at 8 × 10^5 cells/ml) were injected into the system with Anti-A group functionalized SWGs; the cells were allowed to settle and the fluidic chamber was washed with PBS/Gly (Fig. 2.8a). Images of the waveguide were taken before and after the experiment (Fig. 2.8b). Seven cells were bound to the waveguide, producing a signal change of 3.05 μW (0.43 μW/cell). The baseline noise was calculated by taking the standard deviation of the response over the first 6 min (before cell injection) and was found to be $\sigma = 4.5$ nW, so the signal-to-noise ratio was estimated to be SNR ~ 95 for a single cell. Dedicated channeling for SWGs would not increase the SNR, but rather would improve the chances of selective cell capture from a pool of

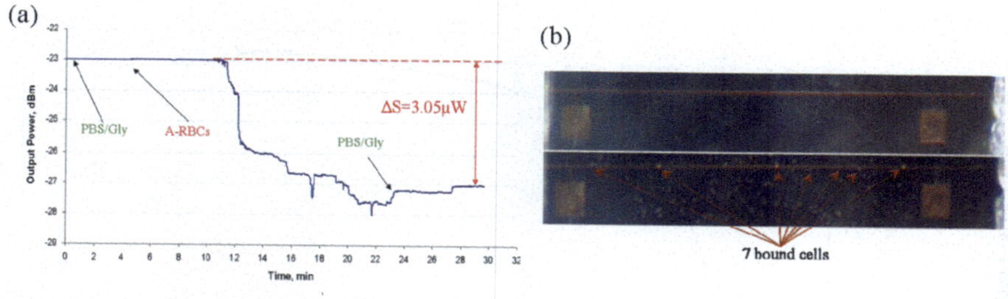

Fig. 2.8 Single cell detection using a plasmonic Au SWG: **a** Response of a Au waveguide functionalized with Anti-A to a PBS/Gly solution carrying A RBCs; **b** images of the waveguide taken before and after cell capture. The cell concentration was 8×10^5 cells/ml, the waveguide thickness ~35 nm, the input power was 10 dBm, and $\lambda_0 = 1310$ nm. Reproduced with permission from [34], *licence number 3713741431933*

cells. Single-cell detection has a particular significance for detecting rare circulating tumour cells (CTCs), which are cancer cells that leach into the blood stream in cases of aggressive cancers [32].

2.3.3 *Protein Sensing Using Straight Waveguides (SWGs)*

Protein sensing is probably the most important type of sensing application for biosensors. Protein-protein interactions are among the most abundant in biological systems and are being used for disease detection as well as drug discovery.

In order to demonstrate the capability of SWGs to sense proteins, two devices were functionalized with chemistries having different protein affinities and were tested for bovine serum albumin (BSA) adsorption. A carboxyl-terminated surface has a high affinity for proteins, due its negative charge [36], and a polyethyl glycol terminated (PEG-terminated) surface has a low protein affinity. Moreover, different chemical modifications of PEGs of varying chain lengths are primary targets for investigation in preventing protein adsorption [37]. The first device was functionalized with 16-MHA (2 mM 16-MHA in IPA for 2 h) and the second device was functionalized with T-PEG (thiolated 11-carbon with 3PEG units chain) by incubation in 2 mM T-PEG in IPA for 2 hand tested for BSA adsorption.

The addition of protein to the sensing buffer increases the overall bulk RI of the solution since the RI of proteins is ~1.5. The effect is concentration-dependant and therefore, taking into account $\partial n/\partial c = 0.185$ mm^3/mg for proteins [38], and the LOD for bulk sensing (2.3×10^{-6} RIU) found in Sect. 2.3.1, a concentration of 100 μg/ml of BSA should produce only a slight increase in bulk index.

A 100 μg/ml solution of BSA in PBS/Gly ($n = 1.3303$) was injected into the system for 25 min, with subsequent washing of the fluidic channel, and with PBS/Gly for another 8–10 min after the maximum response was reached. Figure 2.9 shows the response for BSA on both surfaces. The behaviour of both curves is very typical

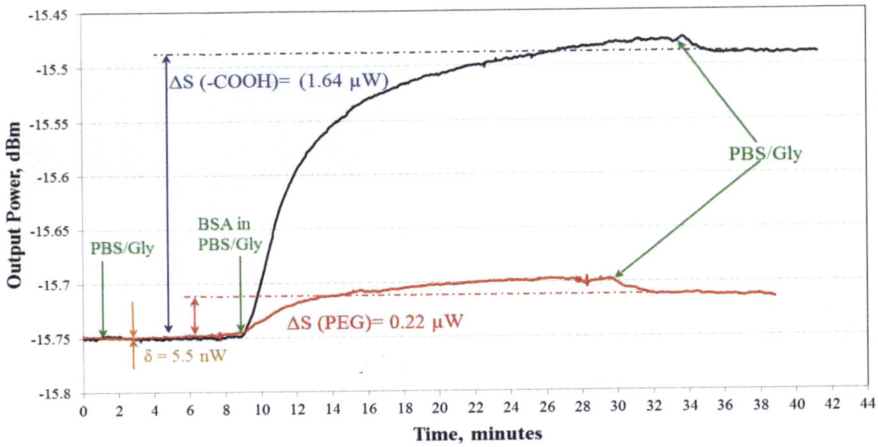

Fig. 2.9 Response of a straight Au plasmonic waveguide for BSA adsorption on two surfaces: carboxyl (–COOH) and PEG-terminated. The waveguide thickness was ~22 nm, the input power was 10 dBm, $\lambda_0 = 1310$ nm and the flow-rate was set to 20 μl/min. Reproduced with permission from [24]

of the binding process—and suggests adsorption of BSA on the waveguide. A slight decrease in power after the PBS/Gly wash can be explained by a bulk effect—and by BSA dissociation from the surface. The responses for carboxylic acid COOH- and PEG-terminated surfaces were determined to be 1.64 μW and 0.22 μW respectively, with corresponding baseline noise levels δ being $\delta = 5.5$ nW and $\delta = 6.5$ nW. A stronger signal change is observed for COOH-terminated surface with signal-to-noise ratio $\Delta S/\delta = 297$ compared to the PEG-terminated surface ($\Delta S/\delta = 45$), which is in agreement with previous studies performed on similar surfaces for BSA adsorption [36]. Although it was expected that there would be no response for the PEG-surface, significant BSA adsorption still occurred. The T-PEG used in the experiment was comprised of 11-carbon chains and 3 PEG units, which results in a SAM that is very rigid and not optimal as a protein blocking surface. Generally, research suggests the use of much longer (3–5 kDa) oligoethylene glycols to form a SAM, in order to prevent protein adsorption [37].

The output power increases for BSA adsorption on both surfaces. Since the sensing buffer solution has $n = 1.3303$ and is lower than the index of CYTOP ($n = 1.3348$), it creates a slight asymmetry around the waveguide. When the sensing fluid (buffer + BSA) is introduced, BSA, having a higher refractive index of $n = 1.5$ adsorbs as an adlayer onto the stripe—thereby pulling the waveguide into symmetry and thereby lowering the insertion loss.

In order to verify this interpretation, mode computations were performed using the following parameters: (1) waveguide dimensions (5 μm wide and 22 nm thick, as measured by AFM), (2) sensing length of the waveguide (1.65 mm), (3) location of the stripe on a 400 nm CYTOP pedestal due to slight over-etching [26], (4) thickness of the lower and upper CYTOP claddings (8 μm and 6.7 μm respectively),

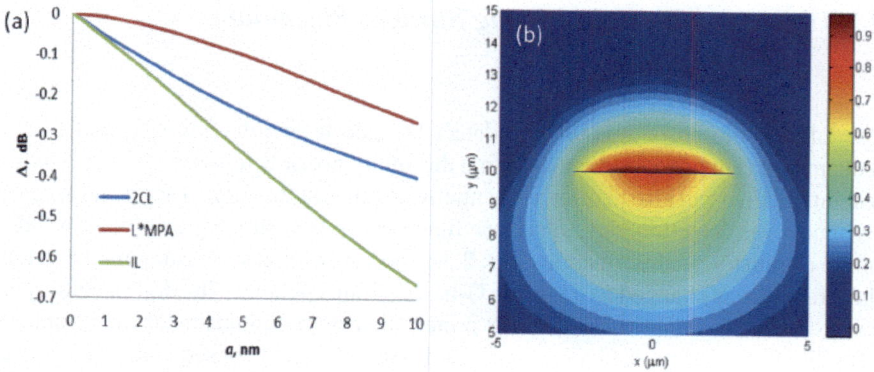

Fig. 2.10 Theoretical response of an SWG to ad-layer formation: **a** modelled insertion loss of the waveguide for the formation of an ad-layer with $n = 1.5$; **b** Distribution of the E_y field component of the LRSPP. Reproduced with permission from [24]

and 5) refractive indices of CYTOP ($n = 1.3348$) and sensing fluid ($n = 1.3303$). A more detailed description can be found in [24]. The insertion losses (IL) were plotted against the thickness of the ad-layer formed on top of the waveguide in Fig. 2.10a. Previous studies on the formation of a close-packed BSA monolayer on an Au surface suggested a thickness of $a = 4$ nm [39, 40]. From Fig. 2.10a, an ad-layer of 4 nm corresponds to a decrease in insertion loss of 0.27 dB, and is consistent with the experimental value for BSA adsorption on a COOH-terminated SAM of 0.26 dB (Fig. 2.9). The response for the PEG-terminated surface was 0.036 dB, which suggests sub-monolayer adsorption of BSA with an equivalent thickness of ~0.5 nm, and confirms the blocking abilities of a PEG SAM.

The distribution of the field mode (E_y) was modeled for an ad-layer thickness of $a = 4$ nm and was found to penetrate 2 μm into the sensing medium (Fig. 2.10b).

The surface mass density Γ (g/m^2) of the adlayer can be computed using the following expression:

$$\Gamma = \frac{a(n_a - n_c)}{\partial n/\partial c} \tag{2.1}$$

where a is the thickness of the ad-layer (4 nm), n_a is the RI of the ad-layer (1.5), n_c is the RI of the sensing solution (1.3303) and $\partial n/\partial c = 0.185$ mm^3/mg. Using these parameters, (2.1) yields $\Gamma = 3669$ pg/mm^2. The detection limit for our system can be estimated using a signal-to-noise ratio for BSA adsorption on COOH-terminated surface of $\Delta S/\delta = 297$, since $\Delta\Gamma$ is ~12 pg/mm^2.

2.3.4 Disease Detection Using Straight Plasmonic Waveguides

Our LRSPP biosensor was used to detect Dengue fever infection [25] and B-cell leukemia [41] in patients, demonstrating the ability not only to sense specific biological interactions but also to offer a potentially useful tool for clinical applications. For Dengue fever detection, the Au surface functionalization strategy involved carbodiimide chemistry, as described in Sect. 2.3.1; however for B-cell leukemia detection the strategy was changed to a Protein G-based affinity scheme. In addition, a deeper analysis of waveguide sensitivity for protein detection demonstrated the optimum PBS/Gly buffer refractive index to be $n = 1.338$ [42]—and consequently this buffer was used in leukemia detection experiments.

2.3.4.1 Dengue Virus Detection

Dengue fever is a common mosquito-borne viral disease in tropical countries, with 390 million people being infected every year worldwide [43]. Common dengue fever (DF) can develop into severe dengue hemorrhagic fever (DHF) and cause 44% mortality. Clinical tests are usually performed with an enzyme-linked immunosorbent assay (MAC-ELISA) to test for the presence of IgM against envelope protein E of the virus, and the results are commercially available.

Two detection strategies, also common for clinical tests using ELISA, were applied: (a) the surface was functionalized with neutralized dengue virus serotype 2 (DENV-2)—and patient plasma containing anti-dengue IgM was introduced into the system for IgM detection—and (b) patient plasma was functionalized on the surface and then DENV-2 was used as an analyte. Surface preparation and functionalization was performed as described previously, but actual antigen functionalization (with DENV-2 or blood plasma) was performed in situ after COOH-terminated SAM formation with 16-MHA. After surface functionalization, a 100 μg/ml BSA solution was used to block non-specific binding sites, but only for experiments where DENV-2 was functionalized on the surface. Three anti-dengue IgM positive patient samples were tested on both surfaces, with three repeats for each. After each run, the surface was regenerated with sodium dodecyl sulfate (SDS) down to the antigen level, allowing for another test of the same sample fluid (blood plasma or DENV-2). Previously it had been shown that SDS, which is an organic detergent, can be used to regenerate the surface without it losing its immuno-reactivity [44]. As a negative control, normal patient plasma was used. After each (bio)chemical reaction, the surface was washed with PBS/Gly buffer and the differences in response, before and after the reaction, were observed. The response for the first iteration for both detection approaches is presented in Fig. 2.11.

The significant power changes observed for injections of NHS/EDC, SDS, DENV-2 and plasma fluids are due to the bulk step caused by the solutions having unmatched indices. In the DENV-2 functionalization approach (Fig. 2.11a), a considerable

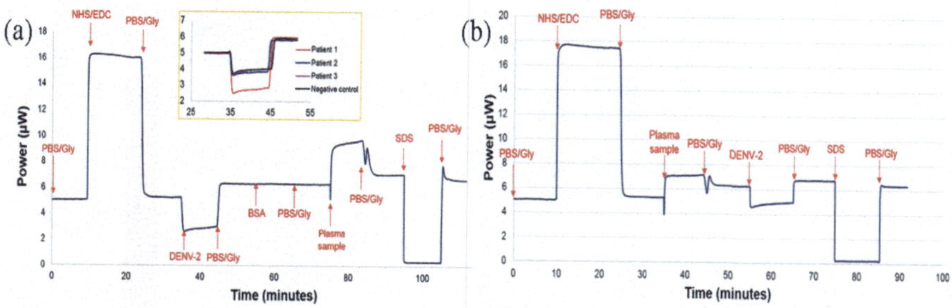

Fig. 2.11 Dengue fever detection with SWGs using two different approaches: **a** DENV-2 surface functionalization followed by detection in patient plasma containing anti-dengue IgM and, **b** patient plasma functionalization followed by detection of DENV-2. Reprinted with permission from [25]. Copyright (2014) American Chemical Society

response is observed for DENV-2 functionalization (35–45 min) and the detection of anti-dengue IgM from the patient sample (75–90 min). BSA blocking did not produce an apparent change in power level, suggesting that the surface was fully covered with viral particles. For the plasma-functionalization strategy (Fig. 2.11b), significant responses are noted for surface functionalization with patient plasma (35–50 min) and for detecting DENV-2 (55–65 min). For both cases, a decrease in the power level was observed for the SDS regeneration steps, indicating removal of biological material from the waveguide.

Similarly to the BSA adsorption experiment (Sect. 2.3.3), the mass load of analyte on the waveguide surface was calculated from all of the responses. In order to compare with data from MAC-ELISA obtained for the same samples, the mass load was further normalized to determine P/N ratios, which are the optical density (OD) ratios of positive-to-negative samples—and are directly proportional to the concentration of analyte in solution. A comparison of all the experiments with the MAC-ELISA data is presented in Fig. 2.12. For diagnostics, a P/N ratio greater than 2 is considered positive. Generally, the plasma surface functionalization approach showed a stronger response than the DENV-2 approach. The non-specific binding was also smaller—possibly due to the DENV-2 solution being pure. The results from MAC-ELISA show a descending pattern from patient 1 to patient 3, and the same pattern is observed for our sensor. Although, for patient 3 in iterations 1 and 2, the P/N ratio is less than that determined from MAC-ELISA, the number is still greater than 2—suggesting that the experiment has clinical relevance. In the third iteration, the P/N ratio is significantly better than for MAC-ELISA. Compared to MAC-ELISA, the sensor provides a faster label-free detection, and requires a very small amount of patient sample (~10 μL), even for three repeated experiments. Considering the largest signal change and the baseline noise, the signal-to-noise ratio was calculated to be ~135. The detection limit in terms of mass load was $\Delta\Gamma$ ~22 pg/mm^2, worse than that found in the BSA adsorption experiment ($\Delta\Gamma$ ~12 pg/mm^2), suggesting the possibility of improving the sensitivity for dengue detection.

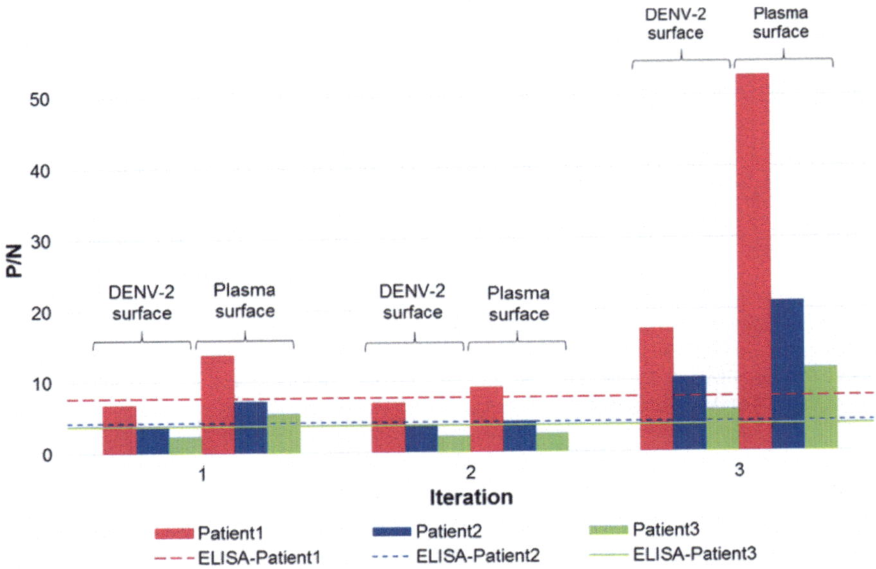

Fig. 2.12 P/N ratios for dengue detection in three positive patient plasma samples versus iteration. MAC-ELISA test results of the same patient samples are presented for comparison. Reprinted with permission from [25]. Copyright (2014) American Chemical Society

2.3.4.2 B-Cell Leukemia Detection

Damage in a replication gene in a B-cell results in the overproduction of B-cell clones, which leads to a variety of B-cell tumors—such as Waldenström's disease, plasma cell neoplasms, chronic lymphocytic leukemia (CLL) and lymphomas. B-cells produce and secrete antibodies (immunoglobulin G, IgG) into the blood stream. Two variations of light chains in IgGs are known in mammals: IgG-kappa (IgGκ) and IgG-lambda (IgGλ)—and each B-cell produces either kappa or lambda immunoglobulin. Normal serum contains polyclonal IgGs with a ratio of IgGκ:IgGλ that ranges from 1.4:1 to 2.0:1 [45]. In patients with B-cell tumors, either kappa or lambda IgG can dominate—due to massive overproduction of monoclonal IgGs by B-cell clones. There are a number of techniques that are used in hospitals for early B-cell leukemia detection, such as blood cell morphology, bone marrow biopsy or flow cytometry. However, recent studies have demonstrated the possibility of B-cell leukemia diagnostics based on the IgGκ:IgGλ (κ:λ) ratio in serum [45–47]. These new findings encourage a detection strategy centered around the determination of IgGκ:IgGλ ratio abnormalities in human serum.

Three patient sera samples of known IgGκ:IgGλ ratios (determined by densitometry) were tested using straight LRSPP waveguides: high IgG-kappa content serum (HKS, κ:λ ~ 12.7:1), high IgG-lambda content serum (HLS, λ:κ ~ 6.9:1) and normal serum (NS, κ:λ ~ 1.7:1). The surface functionalization strategy was altered from the usual carbodiimide chemistry to Protein G (PG) for the selective capture of immunoglobulins for further analysis. The type of human immunoglobulins in patient sera was detected by goat anti-human IgG-kappa light chain (AK) and goat anti-human IgG-lambda light chain (AL) antibodies. The bare Au surface was functionalized with Protein G by direct adsorption from a 50 μg/ml solution of Protein G in PBS/Gly buffer. After this step, two detection approaches were taken: (a) the reverse approach, where the Protein G surface was functionalized with patient serum, then tested against the goat antibodies, and (b) the direct approach, where the Protein G surface was functionalized with goat antibodies—and then patient serum was introduced. The schematics of the surface functionalization and detection protocols applied for both approaches are given in Fig. 2.13. Preliminary results demonstrated the reverse approach to be more sensitive and robust—and therefore it was decided to perform triplets of the experiments using this technique, in order to assess repeatability and gather more data points for accuracy.

A single waveguide was used for testing both kappa and lambda IgGs from the same patient serum to avoid the effects of variations that can arise from slight differences in devices, because of fabrication imperfections. However, different devices were used to provide replicate data (in the case of the reverse approach). Between the experiments, the sensor surface was regenerated down to the Au level using sodium dodecyl sulfate (SDS) and UV/Ozone cleaning. An example of three repeats of the full real-time experimental responses for the reverse approach is presented in Fig. 2.14a.

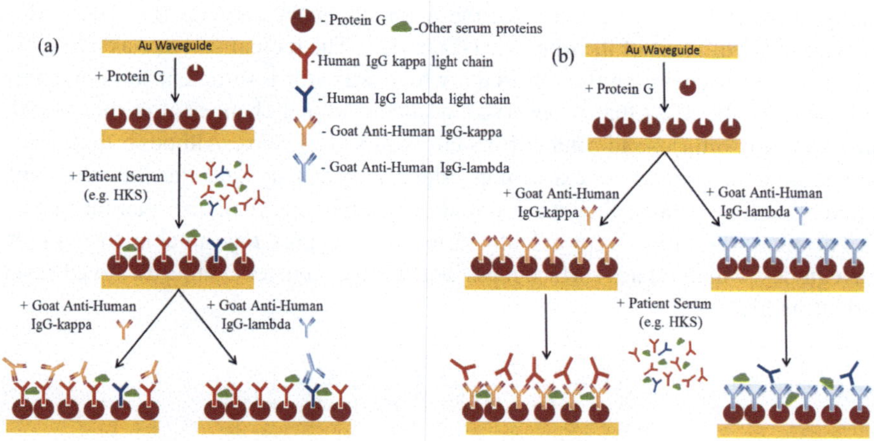

Fig. 2.13 Schematic illustration of two approaches for patient sample testing. **a** Reverse approach: The Au waveguide was functionalized with Protein G, followed by the immobilization of patient serum (an example with HKS is shown), and, finally, sensing with AK. The same waveguide was cleaned down to the Au level, and the whole experiment was repeated but the patient-functionalized surface was then tested against AL. **b** Direct approach: The Au waveguide was functionalized with Protein G, followed by the immobilization of AK, and, finally, sensing of patient serum (e.g. HKS). The same waveguide was cleaned down to the Au level—and the experiment was repeated, but with AL on the Protein G surface. Adapted by permission of The Royal Society of Chemistry from [41]

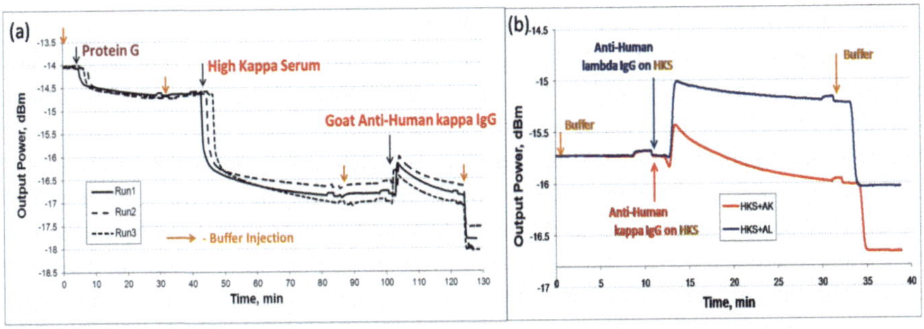

Fig. 2.14 Experimental responses: **a** full experimental runs for testing HKS using the reverse approach showing three repeats; **b** Response of HKS-functionalized device for AK and AL. Reproduced by permission of The Royal Society of Chemistry from [41]

The three repeats shown in Fig. 2.14a demonstrate a very satisfactory repeatability of the experiments, where only slight deviations at the end of the response can be observed. The smaller response for Protein G compared to that of immunoglobulins is also in agreement with their weight differences: ~65 kDa for Protein G and ~150 kDa for IgG.

The actual real-time responses of an HKS-functionalized sensor to goat anti-human-kappa and anti-human-lambda are shown in Fig. 2.14b (reverse approach). The two responses are representative of all other tests, demonstrating a significantly stronger binding response for the complementary analytes (i.e. HKS-AK), as compared with that of the opposite (HKS-AL). The patient serum was diluted to 1:150 with the sensing buffer PBS/Gly, which was still insufficient to completely remove the bulk effect that is observed as sudden signal changes after the AK/AL injection (~13 min)—and after a PBS/Gly wash (~35 min). Although these bulk effects are of no concern for extracting surface mass density data, they interfere with kinetics studies, because they mask initial association and final dissociation steps.

The final $\kappa{:}\lambda$ ratios were calculated using an expression suitable for straight waveguides, which relates surface mass density to the power change due to ad-layer formation [42]:

$$\Delta\Gamma = \frac{1}{k_2}\frac{(n_a - n_c)}{\partial n/\partial c}\left(\frac{P_{out}(a_1)}{P_{out}(a_0)} - 1\right) \tag{2.2}$$

where $\Delta\Gamma$ is a change in surface mass density; k_2 is a constant that is specific for an individual sensor—and is a function of variations in fabrication; $P_{out}(a_0)$ and $P_{out}(a_1)$ are the output powers (in watts) before and after ad-layer formation respectively; n_a, n_c, and $\partial n/\partial c$ are the same parameters as in (2.1). The $\kappa{:}\lambda$ ratios were then computed on the basis of the mass of analyte bound on to the waveguide surface.

Experimental results for both approaches are compared to the densitometric data in Fig. 2.15, where the yellow stripes indicate the normal region. (Note: for HLS, the data is compared with the $\lambda{:}\kappa$ ratio). Both approaches have demonstrated the

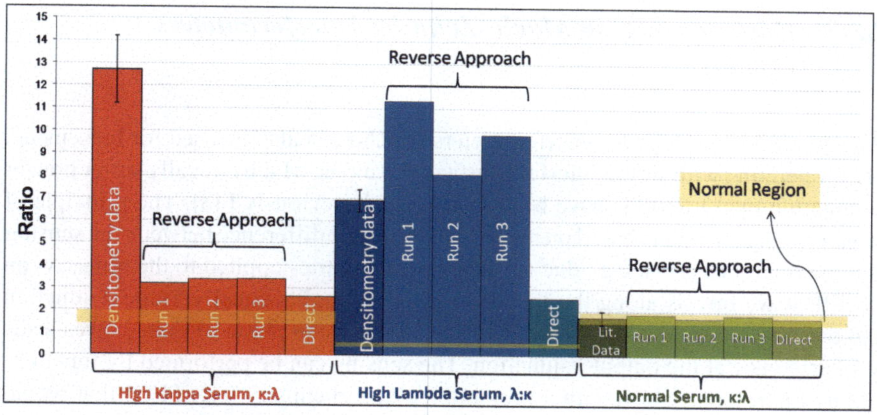

Fig. 2.15 Kappa-lambda ratios for HKS, HLS and NS obtained using LRSPP sensors. HKS and HLS data are compared to the densitometric data provided by Mount Sinai Hospital laboratory, and NS data is compared to the literature data. Adapted by permission of The Royal Society of Chemistry from [41]

ability of the LRSPP waveguide to detect abnormalities in kappa:lambda ratios in patient serum. In addition, the results for normal serum were within the range of the literature data. The reverse approach provided higher κ:λ and λ:κ ratios compared to the direct approach. This is explained by a considerable reduction of non-specific binding because the Protein G surface acts as an "immunological filter" and captures exclusively immunoglobulins out of the pool of serum proteins, followed by detection using *pure* goat anti-human IgGs. In general, the main application of Protein G is for immunoglobulin purification in a separation column where Protein G functionalized beads selectively capture IgGs. However, in biosensing, this is the first known example where a protein G surface was used to extract IgGs from a complex fluid for further surface analysis, in situ.

When compared with the actual densitometry data, the results obtained by our LRSPP biosensor differed from the ones provided by the laboratory. Generally, for the reverse approach, the kappa detection is under-estimated in comparison with the lambda detection. As a result, a lower response for HKS and higher response for HLS are observed. There are some considerations on this issue: (a) there is a significant difference in the dissociation kinetics between kappa:anti-kappa and lambda:anti-lambda interactions that cannot be estimated due to the bulk effect—and (b) the densitometric analysis of immunoglobulins may not be as precise (see [41] for further discussion). Nevertheless, at present, the sensor can provide a solution to B-cell leukemia screening.

2.3.5 *Bulk Sensing on Mach–Zehnder Interferometers (MZIs)*

Waveguide Mach–Zehnder interferometers (MZIs) can also be used for biosensing. MZIs operating with single-interface surface plasmons, which typically suffer greater attenuation than LRSPPs, have been investigated previously [48]. The principle of sensing using an MZI is based on measuring the phase difference between the sensing and the reference arms, e.g., due to different chemistries applied to the arms. As an LRSPP wave travels along the input waveguide, it splits equally at the Y-junction and recombines at the output junction. The difference in phase of these waves leads to interference at the output Y-junction. The sensing can be performed by functionalizing the reference arm with a protein-blocking chemistry (e.g. PEG) that would prevent protein adsorption, while the sensing arm is functionalized with receptors to capture the analyte. Slight changes in the phase along the sensing arm, due to analyte binding, relative to the reference arm produce constructive and destructive interference, resulting in the MZI amplitude response being sinusoidal. Our interest in using plasmonic MZIs for biosensing is primarily due to their lower theoretical detection limit: 0.25 pg/mm^2 versus 1 pg/mm^2 for commercial SPR systems [3].

A sensor containing the MZI structures was manufactured using the same lithographic procedures as described in Sect. 2.2.2; however a different mask was applied to alter the fluidic patterns. The sensor chip consisted of alternating MZIs and SWGs. Two of the MZIs were manufactured for actual sensing, where one arm (i.e. the reference arm) was fully clad (not etched) and the other arm fully exposed to the sensing medium via the microfluidic channel (Fig. 2.16a). The waveguides were designed for operation at 1310 nm (5 μm wide and 34 nm thick Au stripes) [29].

Initially, we carried out measurements to study the response of an MZI to solutions with different refractive indices, near the "ideal" RI point (i.e., near the RI of CYTOP, $n \sim 1.3348$), then away from the RI of CYTOP ($n \sim 1.3300$). In both cases, the range of RI of the solutions was chosen to cover a full period of the MZI transfer characteristic (Fig. 2.16b, c).

For the "near-CYTOP" experiments, sixteen solutions ranging in refractive index from 1.3337 to 1.33525 with an increment of 1.1×10^{-4}, were sequentially injected into the microfluidic channel. The experimental results and theoretical fit are presented in Fig. 2.16b and, as expected, the response exhibits a sinusoidal behaviour. The detection limit for this case was estimated to be $\sim 9 \times 10^{-7}$ RIU—calculated from two points near the maximum slope and taking the signal-to-noise ratio to be SNR = 2 (for $\sigma = 15.5$ nm). A more detailed description of the theoretical fit can be found in [29].

The relationship between the input and output power for an MZI is given by (2.3):

$$P_{\text{out}} = (1/2) P_{\text{in}} T_{\text{C}} e^{-2\alpha_{\text{C}}(L_0+L_{\text{F}})} W(1 + V \cos(\Phi_{\text{D}})) \tag{2.3}$$

Two parameters that have a significant effect on the sensitivity of the MZI are W and V, where V defines the visibility of the response and W is a power penalty;

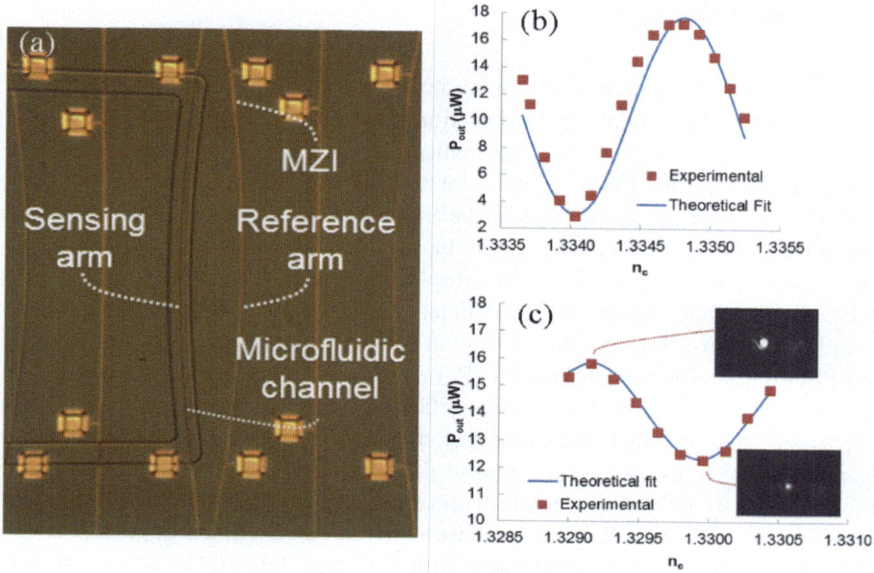

Fig. 2.16 Bulk sensing using Mach–Zehnder Interferometer (MZI): **a** a microscope image of the portion of the device bearing an MZI with an etched microfluidic channel defining the sensing arm; **b** MZI response at $\lambda_0 = 1310$ nm for "near CYTOP RI" solutions with different indices and increment of ~1.1×10^{-4} and **c** MZI response at $\lambda_0 = 1310$ nm for "away from CYTOP RI" solutions with an increment of ~1.6×10^{-4} Adapted from [29], [*licence number 3713750643339*]

both are related to the internal characteristics of the MZI structure, such as the difference in the attenuation of the two arms. Ideally, when the optical performance of the sensing and reference arms is identical, $V = 1$ and $W = 1$, thereby generating maximum sensitivity. Structural imperfections within the microfluidic channel result in additional attenuation at the CYTOP/liquid interface that is absent in the cladded arm, thereby lowering V and W. From the experimental data, $V = 0.7$ and $W = 0.6$, thereby limiting the sensitivity of this structure. If the MZI were perfectly balanced (i.e. $V = W = 1$) and considering our lowest baseline noise from previous experiments (σ ~ 5 nW), the limit of detection can potentially be improved to ~1.7×10^{-7} RIU.

Figure 2.16c shows a plot of the MZI response for the solutions with a lower RI than that of CYTOP, ranging from 1.32900 to 1.33044, with an increment of ~1.6×10^{-4} RIU. Lower solution indices increased the difference in the attenuation of the two arms, resulting in $W = 0.5$ and $V = 0.12$. Nevertheless, the MZI performance at lower refractive indices is satisfactory and remains useful.

2.4 Conclusions

A novel LRSPP waveguide biosensor has been described in this chapter and demonstrates significant potential for the detection of various biological and biochemical entities. Two different plasmonic waveguide sensors have been presented: straight waveguide (SWG) and Mach–Zehnder Interferometer (MZI) sensors.

Extensive experimental research on the biosensing capabilities of SWGs has determined their ability to detect changes in the refractive index of bulk solutions with a detection limit of 2.3×10^{-6} RIU, to selectively capture human red blood cells (RBCs) with a SNR ~ 95 for single-cell detection—and to detect proteins in clear and complex solutions. Non-specific adsorption of BSA was demonstrated for carboxyl- and PEG-terminated surfaces with SNR values of ~297 and ~45 respectively. The detection limit in terms of the protein surface mass density was found to be $\Delta\Gamma \sim 12$ pg/mm^2. As a practical application, patient plasma samples containing antibodies against dengue virus were tested against the neutralized dengue virus serotype 2. The results were either comparable with or better than those obtained from standard MAC-ELISA tests. Another demonstration relevant to clinical applications was performed by testing two patient sera with abnormal IgGκ:IgGλ ratios (patients with B-cell leukemia) and compared with serum having a normal IgGκ:IgGλ ratio (healthy patient). Two detection approaches were evaluated: the reverse approach where a Protein G surface is functionalized with serum first and the direct approach where recognition antibodies are firstly attached to Protein G. The reverse approach has been found to be more robust and more sensitive—mainly because of the significant reduction in non-specific binding. The overall results have demonstrated the ability of the sensor to perform as a screening device for B-cell leukemia patients.

An MZI structure with an etched sensing arm and a fully clad reference arm has been investigated for the detection of changes in the bulk refractive index of solutions. The detection limit was found to be 9×10^{-7} RIU, with potential improvements leading to a LOD of 1.7×10^{-7} RIU.

Overall, the sensor platform provides a low-cost and compact solution for detecting changes in the bulk refractive index and specific detection of biological entities over a wide range of masses (from cells to proteins). The ability to detect a desired protein in complex solutions (i.e. patient plasma) indicates the potential importance of this sensor for clinical diagnostic applications.

References

1. R. Charbonneau, C. Scales, I. Breukelaar, S. Fafard, N. Lahoud, G. Mattiussi, P. Berini, Passive integrated optics elements based on long-range surface plasmon polaritons. J. Lightwave Technol. **24**(1), 477–494 (2006)
2. A. Boltasseva, T. Nikolajsen, K. Leosson, K. Kjaer, M.S. Larsen, S.I. Bozhevolnyi, Integrated optical components utilizing long-range surface plasmon polaritons. J. Lightwave Technol. **23**(1), 413–422 (2005)
3. P. Berini, Long-range surface plasmon polaritons. Adv. Opt. Photonics **1**, 484–588 (2009)

4. R. Slavík, J. Homola, Ultrahigh resolution long range surface plasmon-based sensor. Sens. Actuators B Chem. **123**, 10–12 (2007)
5. M. Vala, S. Etheridge, J.A. Roach, J. Homola, Long-range surface plasmons for sensitive detection of bacterial analytes. Sens. Actuators B Chem. **139**(1), 59–63 (2009)
6. A.W. Wark, H.J. Lee, R.M. Corn, Long-range surface plasmon resonance imaging for bioaffinity sensors. Anal. Chem. **77**(13), 3904–3907 (2005)
7. Y.H. Joo, S. Song, R. Magnusson, Demonstration of long-range surface plasmon-polariton waveguide sensors with asymmetric double-electrode structures. Appl. Phys. Lett. **97**(20), 201105 (2010)
8. J. Dostálek, A. Kasry, W. Knoll, Long range surface plasmons for observation of biomolecular binding events at metallic surfaces. Plasmonics **2**(3), 97–106 (2007)
9. J. Guo, P.D. Keathley, J.T. Hastings, Dual-mode surface-plasmon-resonance sensors using angular interrogation. Opt. Lett. **33**(5), 512–514 (2008)
10. N. Skivesen, A. Têtu, M. Kristensen, Photonic-crystal waveguide biosensor. Opt. Express **15**(6), 3169–3176 (2007)
11. B. Agnarsson, J. Halldorsson, N. Arnfinnsdottir, S. Ingthorsson, T. Gudjonsson, K. Leosson, Fabrication of planar polymer waveguides for evanescent-wave sensing in aqueous environments. Microelectron. Eng. **87**(1), 56–61 (2010)
12. D.X. Xu, A. Densmore, A. Delâge, P. Waldron, R. McKinnon, S. Janz, J. Lapointe, G. Lopinski, T. Mischki, E. Post, P. Cheben, J.H. Schmid, Folded cavity SOI microring sensors for high sensitivity and real time measurement of biomolecular binding. Opt. Express **16**(19), 15137–15148 (2008)
13. N. Kinrot, Analysis of bulk material sensing using a periodically segmented waveguide Mach-Zehnder interferometer for biosensing. J. Lightwave Technol. **22**(10), 2296–2301 (2004)
14. B.Y. Shew, Y.C. Cheng, Y.H. Tsai, Monolithic SU-8 micro-interferometer for biochemical detections. Sens. Actuators A Phys. **141**(2), 299–306 (2008)
15. R.G. Heideman, P.V. Lambeck, Remote opto-chemical sensing with extreme sensitivity: design, fabrication and performance of a pigtailed integrated optical phase-modulated Mach-Zehnder interferometer system. Sens. Actuators B Chem. **61**(1–3), 100–127 (1999)
16. W.-C. Tsai, I.-C. Li, SPR-based immunosensor for determining staphylococcal enterotoxin A. Sens. Actuators B Chem. **136**, 8–12 (2009)
17. J.C. Love, L.A. Estroff, J.K. Kriebel, R.G. Nuzzo, G.M. Whitesides, Self-assembled monolayers of thiolates on metals as a form of nanotechnology. Chem. Rev. **105**(4), 1103–1170 (2005)
18. K. Saha, F. Bender, E. Gizeli, Comparative study of IgG binding to proteins G and A: nonequilibrium kinetic and binding constant determination with the acoustic waveguide device. Anal. Chem. **75**, 835–842 (2003)
19. B.N. Johnson, R. Mutharasan, pH Effect on protein G orientation on gold surfaces and characterization of adsorption thermodynamics. Langmuir **28**, 6928–6934 (2012)
20. M. Tencer, H.-Y. Nie, P. Berini, Electrochemical differentiation and TOF-SIMS characterization of thiol-coated gold features for (bio)chemical sensor applications. J. Electrochem. Soc. **156**(12), J386–J392 (2009)
21. M. Tencer, A. Olivieri, B. Tezel, H.-Y. Nie, P. Berini, Chip-scale electrochemical differentiation of SAM coated gold features using a probe array. J. Electrochem. Soc. **159**(3), J77–J82 (2012)
22. G. Gagnon, N. Lahoud, G.A. Mattiussi, P. Berini, Thermally activated variable attenuation of long-range surface plasmon-polariton waves. J. Lightwave Technol. **24**(11), 4391–4402 (2006)
23. T. Nikolajsen, K. Leosson, S.I. Bozhevolnyi, Surface plasmon polariton based modulators and switches operating at telecom wavelengths. Appl. Phys. Lett. **85**(24), 5833–5835 (2004)
24. O. Krupin, H. Asiri, C. Wang, R.N. Tait, P. Berini, Biosensing using straight long-range surface plasmon waveguides. Opt. Express **21**, 698–709 (2013)
25. W.R. Wong, O. Krupin, S.D. Sekaran, F.R.M. Adikan, P. Berini, Serological diagnosis of dengue infection in blood plasma using long-range surface plasmon waveguides. Anal. Chem. **86**, 1735–1743 (2014)

26. C. Chiu, E. Lisicka-Skrzek, R.N. Tait, P. Berini, Fabrication of surface plasmon waveguides and devices in CYTOP with integrated microfluidic channels. J. Vac. Sci. Technol. B **28**(4), 729–735 (2010)
27. H. Asiri, Fabrication of surface plasmon biosensors in CYTOP. Master's Thesis, Department of Chemical and Biological Engineering, University of Ottawa, Ottawa (2012)
28. R. Daviau, A. Khan, E. Lisicka-Skrzek, R.N. Tait, P. Berini, Fabrication of surface plasmon waveguides and integrated components on CYTOP. Microelectron. Eng. **87**, 1914–1921 (2010)
29. A. Khan, O. Krupin, E. Lisicka-Skrzek, P. Berini, Mach-Zehnder refractometric sensor using long-range surface plasmon waveguides. Appl. Phys. Lett. **103**, 111108 (1–4) (2013)
30. K.S. Gracias, J.L. McKillip, A review of conventional detection and enumeration methods for pathogenic bacteria in food. Can. J. Microbiol. **50**, 883–890 (2004)
31. C.D. Hillyer, C.D. Josephson, M.A. Blajchman, J.G. Vostal, J.S. Epstein, J.L. Goodman, Bacterial contamination of blood components: risks, strategies, and regulation. Hematology 575–589 (2003)
32. D.F. Hayes, M. Cristofanilli, G.T. Budd, M.J. Ellis, A. Stopeck, M.C. Miller, J. Matera, W.J. Allard, G.V. Doyle, L.W.W.M. Terstappen, Circulating tumor cells at each follow-up time point during therapy of metastatic breast cancer patients predict progression-free and overall survival. Clin. Cancer Res. **12**(14), 4218–4224 (2006)
33. T.G. Hermanson, *Bioconjugate Techniques*, 2nd edn. (Academic Press, 2008) (Chap. II(3))
34. O. Krupin, C. Wang, P. Berini, Selective capture of human red blood cells based on blood group using long-range surface plasmon waveguides. Biosens. Bioelectron. **53**, 117–122 (2014)
35. J.G. Quinn, R. O'Kennedy, M. Smyth, J. Moulds, T. Frame, Detection of blood group antigens utilising immobilised antibodies and surface plasmon resonance. J. Immunol. Methods **206**, 87–96 (1997)
36. V. Silin, H. Weetall, D.J. Vanderah, SPR studies of the nonspecific adsorption kinetics of human IgG and BSA on gold surfaces modified by self-assembled monolayers (SAMs). J. Colloid Interface Sci. **185**(1), 94–103 (1997)
37. M.J. Felipe, P. Dutta, R. Pernites, R. Ponnapati, R.C. Advincula, Electropolymerized bioresistant coatings of OEGylated dendron carbazoles: design parameters and protein resistance SPR studies. Polymer **53**(2), 427–437 (2012)
38. J.A. De Feijter, J. Benjamins, F.A. Veer, Ellipsometry as a tool to study the adsorption behavior of synthetic and biopolymers at the air-water interface. Biopolymers **17**(7), 1759–1772 (1978)
39. R. Charbonneau, M. Tencer, N. Lahoud, P. Berini, Demonstration of surface sensing using long-range surface plasmon waveguides on silica. Sens. Actuators B Chem. **134**(2), 455–461 (2008)
40. M. Tencer, R. Charbonneau, N. Lahoud, P. Berini, AFM study of BSA adlayers on Au stripes. Appl. Surf. Sci. **253**(23), 9209–9214 (2007)
41. O. Krupin, C. Wang, P. Berini, Detection of leukemia markers using long-range surface plasmon waveguides functionalized with protein G. Lab Chip **15**, 4156–4165 (2015)
42. W.R. Wong, O. Krupin, F.R.M. Adikan, P. Berini, Optimization of long-range surface plasmon waveguides for attenuation-based biosensing. J. Lightwave Technol. **33**(15), 3234–3242 (2015)
43. S. Bhatt, P.W. Gething, O.J. Brady, J.P. Messina, A.W. Farlow, C.L. Moyes, J.M. Drake, J.S. Brownstein, A.G. Hoen, O. Sankoh, The global distribution and burden of dengue. Nature **496**, 504–507 (2013)
44. R.F. Dutra, R.K. Mendes, V. Lins da Silva, L.T. Kubota, Surface plasmon resonance immunosensor for human cardiac troponin T based on self-assembled monolayer. J. Pharm. Biomed. Anal. **43**, 1744–1750 (2007)
45. J.A. Katzmann, R.J. Clark, R.S. Abraham, S. Bryant, J.F. Lymp, A.R. Bradwell, R.A. Kyle, Serum reference intervals and diagnostic ranges for free κ and free λ immunoglobulin light chains: relative sensitivity for detection of monoclonal light chains. Clin. Chem. **48**(9), 1437–1444 (2002)
46. A. Bradwell, S. Harding, N. Fourrier, C. Mathiot, M. Attal, P. Moreau, J.L. Harousseau, H. Avet-Loiseau, Prognostic utility of intact immunoglobulin $Ig'\kappa/Ig'\lambda$ ratios in multiple myeloma patients. Leukemia **27**(1), 202–207 (2013)

47. J.P. Campbell, M. Cobbold, Y. Wang, M. Goodall, S.L. Bonney, A. Chamba, J. Birtwistle, T. Plant, Z. Afzal, R. Jefferis, M.T. Drayson, Development of a highly-sensitive multiplex assay using monoclonal antibodies for the simultaneous measurement of kappa and lambda immunoglobulin free light chains in serum and urine. J. Immunol. Methods **391**(1–2), 1–13 (2013)
48. Y. Gao, Q. Gan, Z. Xin, X. Cheng, F.J. Bartoli, Plasmonic Mach-Zehnder interferometer for ultrasensitive on-chip biosensing. ACS Nano **5**(12), 9836–9844 (2011)

Chapter 3
Multimode Spectroscopy in Optical Biosensors

Farshid Bahrami, J. Stewart Aitchison, and Mo Mojahedi

Abstract Surface plasmon resonance (SPR) sensors are one of the most widely used optical sensors for studying processes at the nanoscale. The strong confinement of the surface plasmon (SP) wave makes it extremely sensitive to perturbations in material properties (e.g. changes in refractive index) that occur along the electromagnetic field penetration depth. The penetration depth of the SP wave is on the order of a few hundred nanometers.

3.1 Introduction

Surface plasmon resonance (SPR) sensors are one of the most widely used optical sensors for studying processes at the nanoscale. The strong confinement of the surface plasmon (SP) wave makes it extremely sensitive to perturbations in material properties (e.g. changes in refractive index) that occur along the electromagnetic field penetration depth. The penetration depth of the SP wave is on the order of a few hundred nanometers.

One of the most important applications of the SPR sensor is in biosensing—which requires the incorporation of bio-recognition elements, immobilized on the sensing surface, to interact with a target analyte. The dimension of the bio-molecular target analyte (e.g. proteins) is typically in the range of a few nanometers, which is almost two orders of magnitude smaller than the penetration depth of the conventional SP wave. As a consequence, the SP wave is not only susceptible to the surface binding, within a few nanometers of the sensor surface, but is also sensitive to any other variations in material properties within the field profile [1]. This results in cross-sensitivity of the SP wave to the binding of target analyte and the background variations of the index that originate from fluctuations in the composition and temperature of the analyte.

F. Bahrami · J. Stewart Aitchison · M. Mojahedi (✉)
Department of Electrical and Computer Engineering, University of Toronto, Toronto, ON M5S 3G4, Canada
e-mail: mojahedi@ece.utoronto.ca

R. De La Rue et al. (eds.), *Biomedical Optical Sensors*,
Biological and Medical Physics, Biomedical Engineering,
https://doi.org/10.1007/978-3-030-48387-6_3

The most common method for distinguishing between the refractive index changes caused by specific interaction of bio-recognition elements with the target analyte (also called the adlayer) and those due to variations of the background refractive index is to adopt a *reference channel* beside the *sensing channel* [2]. In this approach, the reference channel is covered with receptors that have no affinity with the target analyte. Ideally, by subtracting the signal of the reference channel from that of the sensing channel, the background index variations can be removed. Although this method is the most common technique for compensation of the background index variations, it is based on an assumption that limits the efficiency of the method. This assumption is that both the reference and sensing channels are exactly identical, with the exception of the adlayer, in order to extract only the material properties of the adlayer. Therefore, the reference channel should be covered with receptors that have a similar refractive index to that of the receptors covering the sensing arm, but with no affinity to the target analyte. A requirement that is not easily met in practice. Other practical issues that make this task even more challenging are: fabrication imperfections, the temperature difference between the reference and sensing channels, the different concentrations of the solutions flowing over the channels, and the different non-specific binding properties of the receptors. Finally, this method cannot be utilized to distinguish between adlayer thickness (d_a) and adlayer refractive index (n_a) from the output signal, but it can only be used to determine the adlayer optical thickness (the product of thickness and refractive index; i.e. ($d_a \times n_a$)

In this chapter, an alternative approach for compensation of the background refractive index variations, based on self-referenced spectroscopy, is presented. This approach eliminates the need for the reference channel by increasing the number of linearly independent measurements from the sensing channel - which is possible through an increase in the number of guided modes of the sensor. To this end, several platforms are investigated and their performance is compared with that of the conventional SPR sensor. For each platform, a proper figure of merit (FoM) is employed in order to accurately evaluate the performance of the sensor.

3.2 Self-referenced Spectroscopy

One way to decouple the changes in adlayer (e.g., changes in adlayer or index of refraction) from the variations in the background properties (for example, changes in analyte index of refraction, temperature, or concentration) is to increase the number of independent measurements by increasing the number of independent modes guided by the sensor. From a mathematical point-of-view, this simply means matching the number of measurements (equations) to the number of unknown parameters. We illustrate this point with the following example. Suppose that during a sensing event the output signal is the resonance angle of the first excited mode (θ_1) which is related to the change in background refractive index (Δn_b) and adlayer thickness (Δd_a) according to the following equation[1]

$$\Delta\theta_1 = S_{11}\Delta d_a + S_{12}\Delta n_b$$
$$S_{11} = \frac{\partial\theta_1}{\partial d_a},\; S_{12} = \frac{\partial\theta_1}{\partial n_b} \tag{3.1}$$

It is evident that by only measuring $\Delta\theta_1$ (assuming S_{11} and S_{12}, also known as sensitivities, are known) Eq. (3.1) will not provide a unique value for the change in adlayer thickness (Δd_a) or background refractive index (Δn_b). However, if an additional measurement of the efficiently excited second mode ($\Delta\theta_2$) is performed, then Eq. (3.1) can be rewritten as

$$\Delta\theta_1 = S_{11}\Delta d_a + S_{12}\Delta n_b$$

$$\Delta\theta_2 = S_{21}\Delta d_a + S_{22}\Delta n_b,$$

where it can be solved for Δd_a and Δn_b according to

$$\begin{pmatrix}\Delta d_a\\ \Delta n_b\end{pmatrix} = \begin{pmatrix}S_{11} & S_{12}\\ S_{21} & S_{22}\end{pmatrix}^{-1}\begin{pmatrix}\Delta\theta_1\\ \Delta\theta_2\end{pmatrix}. \tag{3.2}$$

This method is called self-referenced spectroscopy, since it does not require a separate reference channel in order to cancel out the effects of background index variations [3–6]. It goes without saying that for this method to work, the measurements for $\Delta\theta_1$ and $\Delta\theta_2$ should be linearly independent and preferably simultaneous. In the remaining sections of this manuscript several platforms which are suitable for self-referenced spectroscopy are discussed.

3.3 Hybrid Plasmonic Waveguide

The hybrid plasmonic waveguide (HPWG) platform has attracted much interest and has been used for several applications in optical communication systems [7–12]. It consists of a high index medium (e.g. silicon) separated from a gold film by a lower index spacer region (e.g., fluidic channel or silica). By adopting a fluidic channel as the spacer (Fig. 3.1a), the HPWG can also be used for biosensing applications [13, 14]. Figure 3.1b, c illustrate the power density profiles for the transverse magnetic (TM) and transverse electric (TE) modes of this structure with the interface of the various layers marked by dashed white lines. The metal/fluidic-channel interface is parallel to the *y*-*z* plane according to the coordinate system in the figure. The TM mode of this structure is a hybrid mode which is a compromise between the high

[1]Reader may not that (3.1) is an example of total differential of θ_1, where $\theta_1(d_a, n_b)$.

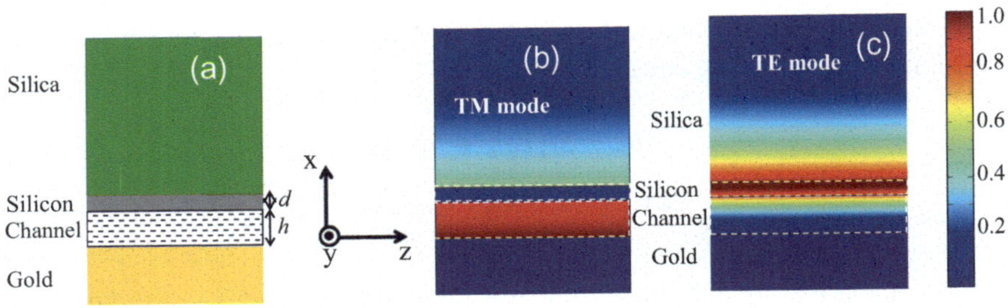

Fig. 3.1 **a** Schematic of the HPWG, coordinate system used in this work is also shown. **b** and **c** guided power densities for the TM and TE modes, respectively. Waveguide dimensions are $h =$ 100 nm, $d = 50$ nm. Wavelength of operation is 1.55 μm [15]

confinement of plasmonic guided mode and the long propagation distance of the dielectric waveguide mode.

The combination of the polarization diversity (i.e., co-existence of the TEand TM modes in different layers) and the high field confinement of the TM mode enables this platform to provide more information about a analyte than possible by using purely plasmonic sensors [15–17]. In addition, the structure allows us to study anisotropic systems with high birefringence (e.g., lipid bilayers) which require the characterization of events parallel and perpendicular to the plane of the biolayer and can be used to find the optical anisotropy (molecular orientation) of anisotropic structures (e.g., proteolipid membranes) [18].

By functionalizing the metal surface for the detection of specific target biomolecules, the HPWG can be used for affinity sensing, where the case for which target molecules are absent in the fluidic channel is shown in Fig. 3.2a. After target molecules are introduced in the channel, they will attach themselves to the metal surface and form an adlayer, as shown in Fig. 3.2b. This modification of the gold surface (i.e. the presence of the adlayer with thickness (d_a) can be detected using different interrogation setup such as Mach–Zehnder interferometer (MZI) or prism coupling. Figure 3.2c shows the HPWG sensor utilized in a MZI configuration. For this case, part of the input light traveling through the HPWG sensor (placed in the "sensor arm") is interfered with light propagating along the "through arm".

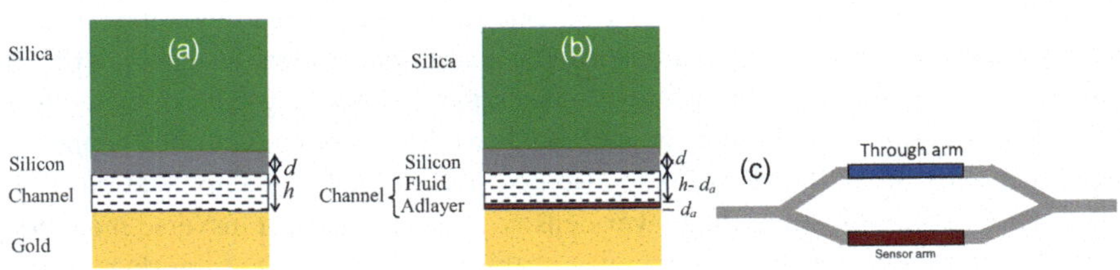

Fig. 3.2 **a** A cross section of the HPWG affinity sensor when the target molecules are absent in the fluidic channel. **b** The same sensor as in (**a**), but now target molecules are present in the channel and have formed an adlayer of thickness a on top of the gold. **c** HPWG sensor in MZI configuration [15]

Table 3.1 Comparing FsoM for HPWG and SPR sensors

Sensor	d (nm)	G^{TM} (nm^{-1})	H^{TM} (RIU^{-1})	G^{TE} (nm^{-1})	H^{TE} (RIU^{-1})	λ_0 (μm)
HPWG	170	0.597	200	0.012	41	1.05
Single interface SPR	–	0.28	339	–	–	0.85

To compare the performance of the HPWG sensor against the conventional single interface SPR sensor, a figure of merit (FoM) for bulk sensing (G) and a FoM for surface sensing (H) from [19] are used. These FoMs are given by

$$G = \left(\frac{\partial n_{\text{eff}}}{\partial d_a}\right) \Big/ k_{\text{eff}} \tag{3.3}$$

$$H = \left(\frac{\partial n_{\text{eff}}}{\partial n_b}\right) \Big/ k_{\text{eff}} \tag{3.4}$$

where n_{eff} and k_{eff} are the real and imaginary parts of the effective mode index. Table 3.1 summarizes the FoMs for the SPR and HPWG sensors for the optimized values of: operating wavelength (λ_0), silicon thickness (d), and for a fixed channel thickness of 100 nm. The permittivity of silica, silicon, and gold are taken from [12].

As is evident from Table 3.1, the TMmode of the HPWG has a larger G value than that of the single interface SPR sensor while the possibility to operate in TE mode[2] enables the HPWG sensor to differentiate surface (Δd_a) from bulk properties (Δn_b) according to

$$\begin{pmatrix} \Delta d_a \\ \Delta n_b \end{pmatrix} = \underbrace{\begin{pmatrix} S_{11} & S_{12} \\ S_{21} & S_{22} \end{pmatrix}^{-1}}_{S} \cdot \begin{pmatrix} \Delta n_{\text{eff}\,1} \\ \Delta n_{\text{eff}\,2} \end{pmatrix}$$

$$S_{11} = \frac{\partial n_{\text{eff}\,1}}{\partial d_a},\; S_{12} = \frac{\partial n_{\text{eff}\,1}}{\partial n_b},\; S_{21} = \frac{\partial n_{\text{eff}\,2}}{\partial d_a},\; S_{22} = \frac{\partial n_{\text{eff}\,2}}{\partial n_b} \tag{3.5}$$

The narrowness of the fluidic channel (approximately 100 nm) and possible difficulty with moving the fluid through such a narrow channel can be elevated if the HPWG in Fig. 3.2 is rotated by 90°, as shown in Fig. 3.3a [17]. Figure 3.3b, c show the power density profiles for the guided modes. The modes are not purely TM and TE in nature as the electric field arrows show. The mode shown in Fig. 3.3b is TM-like (Mode A) and the mode shown in Fig. 3.3c is TE-like (Mode B).

HPWG can also be used in the Kretschmann configuration, as shown in Fig. 3.4a [20, 21]. This configuration eliminates the need for a nano-fluidic channel to deliver the bio-molecules to the sensing region. In this scheme, a silica spacer is located between a thin gold film and a silicon layer (Fig. 3.4a) and a prism coupling, instead

[2]Recall that the conventional SPR sensor only operates in the TM mode.

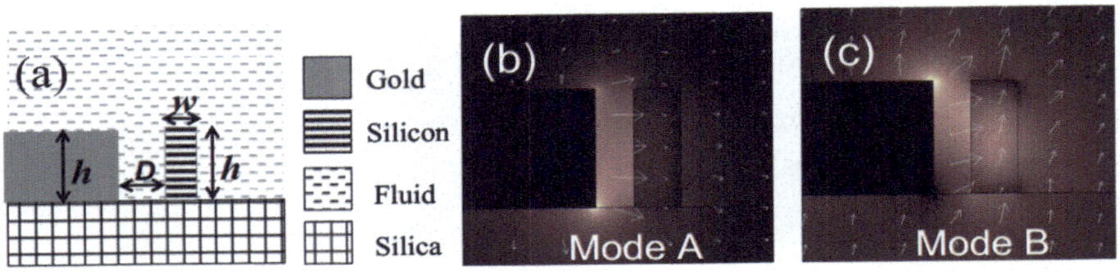

Fig. 3.3 **a** Cross section of the vertical HPWG. **b** and **c** power density profiles of the modes supported by the HPWG for dimensions $h = 340$ nm, $D = 100$ nm, $w = 125$ nm. Wavelength of operation is 1.55 μm [17]

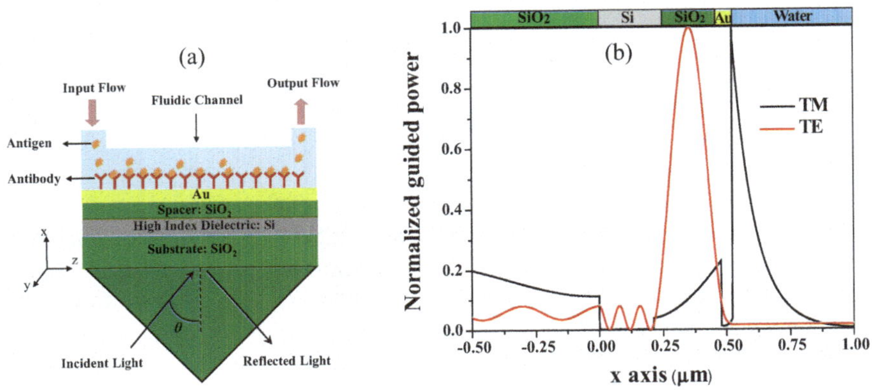

Fig. 3.4 **a** Proposed HPWG structure for biosensing in reflection measurement using Kretschmann configuration. **b** z-component of the Poynting vector for both TM and TE polarizations along the x-axis at the resonance angle. The wavelength of the incident light is 605 nm and the silica, silicon, and gold indices of refraction at this wavelength are taken from [12]. The resonance angles for TM and TE polarizations are 80.76° and 48.27°, respectively [20]

of an MZI, is used to couple the external light to the sensor. The fluidic channel is located on top of the gold film. The binding of target molecules to the bio-receptors immobilized on the gold film [as is shown in Fig. 3.4a] can be monitored by using both and TE polarizations. The mode profiles plotted in Fig. 3.4b belong to the and TE modes of the hybrid structure excited at the resonance angles 80.76° and 48.27°, respectively. Fabrication of such a sensor starts with a silicon-on-glass sample followed by deposition of SiO_2 and a thin gold layer.

To optimize and evaluate the performance of this sensor, the FoM defined in (3.3) and (3.4) are modified for use in angular modulation (when measuring resonance angles). These FoM contain two different components: (1) Sensitivity factor (SF) which is the rate change of the resonance angles with respect to the change in measurand. (2) Sensor merit (SM) which is the ratio of depth of the reflection resonance curve over its full-width-at-half-maximum (FWHM). The product of these two factors is called the combined sensitivity factor (CSF). Depending on whether the quantity of interest during the sensing event is the bulk refractive index (n_b) or the adlayer properties (e.g. adlayer thickness, d_a, or refractive index, (n_a) the SF and CSF are expressed by two different expressions, as the following:

$$\mathrm{CSF}_{\mathrm{surf}} = \mathrm{SF}_{\mathrm{surf}} \cdot \mathrm{SM} = \frac{\partial \theta_{\mathrm{res}}}{\partial d_a} \cdot \frac{R_{\max} - R_{\min}}{\mathrm{FWHM}} \tag{3.6}$$

$$\mathrm{CSF}_{\mathrm{bulk}} = \mathrm{SF}_{\mathrm{bulk}} \cdot \mathrm{SM} = \frac{\partial \theta_{\mathrm{res}}}{\partial n_b} \cdot \frac{R_{\max} - R_{\min}}{\mathrm{FWHM}} \tag{3.7}$$

where $\mathrm{CSF}_{\mathrm{surf}}$ and $\mathrm{CSF}_{\mathrm{bulk}}$ are CSFs for surface and bulk sensing, and $\mathrm{SF}_{\mathrm{surf}}$ and $\mathrm{SF}_{\mathrm{bulk}}$ are the SFs for surface and bulk sensing, respectively. $R_{\max}$ and $R_{\min}$ are the maximum and minimum reflectance intensities close to the resonance angles and θ_{res} is the angle at the resonance. It should be noted that $\mathrm{CSF}_{\mathrm{surf}}$ and $\mathrm{CSF}_{\mathrm{bulk}}$ are inversely related to the limit of detection (LoD) [20] and the resolution of the sensor (minimum detectable in RIU) [22], respectively. The main difference between the CSFs defined in (3.6) and (3.7) and those defined in [19] [i.e., (3.3) and (3.4) of the current manuscript] is the use of FWHM instead of the extinction coefficient (k_{eff}) and inclusion of the resonance depth ($R_{\max} - R_{\min}$). Equations (3.6) and (3.7) are important FoMs which will be used to compare the performances of various sensors when sensors are used in angular modulation (when measuring resonance angles).

Table 3.2 summarizes the optimized sensors FoMs in angular modulation for single interface SPR and HPWG sensors. A genetic algorithm is used to optimize the device dimensions and the operating wavelength. The thicknesses of optimized gold films for the SPR and HPWG sensors are 56 nm and 52 nm, respectively and the optimized silicon and silica (spacer) thicknesses for the HPWG are 25 nm and 30 nm, respectively. As Table 3.2 indicates, the performance of the HPWGin TM mode is comparable to that of the SPR sensor with the additional benefit that unlike the SPR, the HPWG sensor can operate in both TM and TE modes. Therefore, (3.2) can be used to decouple the changes in adlayer thickness (Δd_a) from the background refractive index (Δn_b) by measuring the resonance angles for TM ($\Delta\theta_{\mathrm{TM}}$) and TE modes ($\Delta\theta_{\mathrm{TE}}$). The elements of the matrix $\boldsymbol{S}$ (i.e. the SFs) and its corresponding determinant (to ensure existence of $\boldsymbol{S}^{-1}$) are evaluated according to:

Table 3.2 Comparing the performance of the optimized single interface SPR and HPWG sensors for angular modulation using a prism coupling scheme [15]

Sensor	Prism	λ (nm)	$\mathrm{CSF}^{\mathrm{TM}}_{\mathrm{bulk}}$ (RIU^{-1})	$\mathrm{CSF}^{\mathrm{TM}}_{\mathrm{surf}}$ (nm^{-1})	$\mathrm{CSF}^{\mathrm{TE}}_{\mathrm{bulk}}$ (RIU^{-1})	$\mathrm{CSF}^{\mathrm{TE}}_{\mathrm{surf}}$ (nm^{-1})
Single interface SPR	SiO_2	836	105	0.077	–	–
Hybrid structure	SiO_2	830	100	0.074	0.083	1×10^{-4}

$$\begin{pmatrix} \frac{\partial \theta_{\mathrm{TM}_0}}{\partial d_a} & \frac{\partial \theta_{\mathrm{TM}_0}}{\partial n_b} \\ \frac{\partial \theta_{\mathrm{TE}_0}}{\partial d_a} & \frac{\partial \theta_{\mathrm{TE}_0}}{\partial n_b} \end{pmatrix} = \begin{pmatrix} 0.0006 & 0.5 \\ 0.1 & 135.7 \end{pmatrix} \mapsto \begin{vmatrix} 0.0006 & 0.5 \\ 0.1 & 135.7 \end{vmatrix} = 0.03 \neq 0 \quad (3.8)$$

The non-singularity of the $\boldsymbol{S}$ matrix ensures that the HPWG sensor will allow us to decouple the surface (Δd_a) and bulk effects (Δn_b) using dual polarization spectroscopy in a prism coupling scheme.

3.3.1 Plasmon Waveguide Resonance

In Sect. 3.3 we have presented our simulation results for sensors based on the HPWG platform, originally proposed by our group in 2007 [7–12]. However, hybrid plasmonic sensors are not limited to the aforementioned platform. In this section we will discuss another class of hybrid plasmonic sensors based on plasmon waveguide resonance (PWR) platform and will present our theoretical and experimental results pertaining to this new class of sensors.

The PWR sensor is a promising platform which is composed of a glass substrate coated with a thin gold film and a dielectric layer on top of the gold film, as shown in Fig. 3.5a [23]. The TM mode of this sensor is a hybrid mode resulting from the coupling between the SP wave and the TM mode of the top dielectric layer [the black curve in Fig. 3.5b], while the TE mode is a pure dielectric mode of the top dielectric layer [the red curve in Fig. 3.5b]. The large probing depth of the TM polarized light (up to 9 μm) in the PWR sensor makes it suitable for bulk sensing and detecting large biomaterials such as bacteria, while the small probing depth of the TE polarized light (up to 300 nm) is ideal for surface sensing. The PWR sensor benefits from the small resonance width of the dielectric waveguide modes [Fig. 3.5c] and the high sensitivity of the SP wave.

To better understand the effects of the top dielectric layer on the mode profile, the field profileof TM (Fig. 3.6) and TE modes (Fig. 3.7) are plotted for different

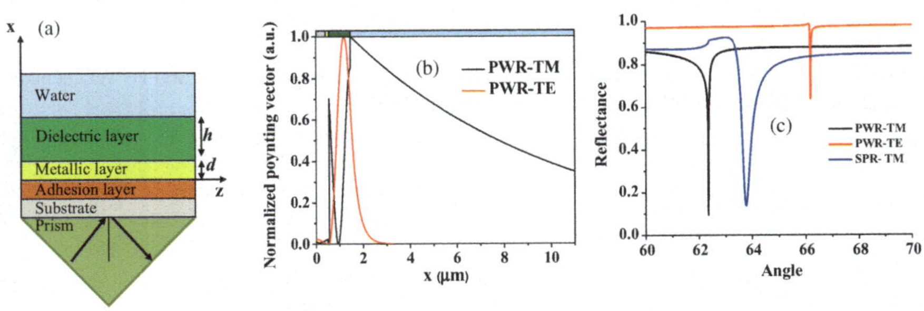

Fig. 3.5 **a** Schematic diagram of the PWR sensor. **b** z-component of the Poynting vector for both TM and TE polarizations in the optimized PWR sensor at the resonance angle of 62.35° and 66.18°, respectively. **c** Reflectance spectrum for the optimized PWR-TM, PWR-TE, and SPR-TM polarizations in black, red, and blue lines, respectively [22]

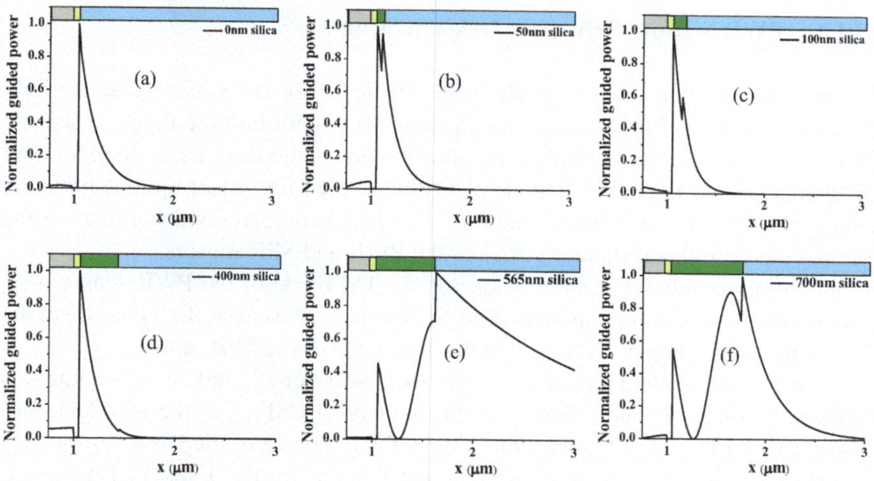

Fig. 3.6 Field profile of the TM mode for different silica layer thicknesses on top of 50 nm gold film at the wavelength of 800 nm. The silica layer thicknesses are: **a** 0 nm, **b** 50 nm, **c** 100 nm, **d** 400 nm, **e** 565 nm, and **f** 700 nm

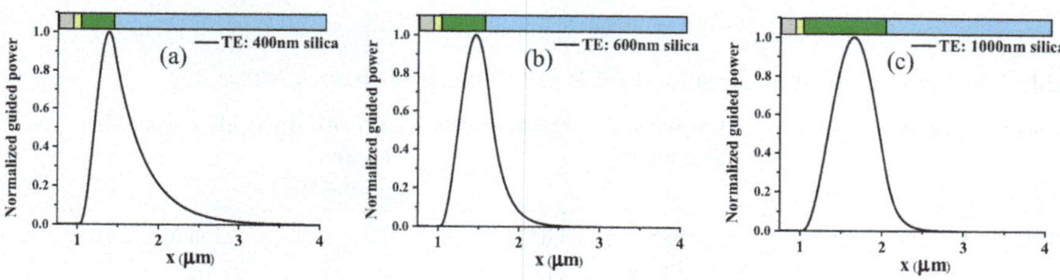

Fig. 3.7 Field profile of the TE mode for different silica layer thicknesses on top of 50 nm gold film at the wavelength of 800 nm. The silica layer thicknesses are: **a** 400 nm, **b** 600 nm, and **c** 1000 nm

thicknesses of the dielectric layer (silica) at the fixed wavelength of 800 nm. For silica thicknesses greater than zero but less than 565 nm (Fig. 3.6a–d), the hybrid TM mode is most intense at the gold-dielectric interface. However, for a silica layer thicker than or equal to 565 nm [as is shown in Fig. 3.6e] the field profile is considerably changed and most of the power is moved to the silica-water interface (instead of the gold-silica interface) with a large penetration into the water. The PWR sensor is mostly operated at this range of top dielectric thicknesses, where the power for the TM mode is the largest at the dielectric-water interface.

The field profile of the TE mode for different thicknesses of silica is shown in Fig. 3.7. The TE mode appears for silica thicknesses larger than 400 nm. As shown in Fig. 3.7a, for silica thickness of 400 nm, the penetration depth of the TE mode into water is the largest, and decreases with increasing silica thickness. Therefore, the TE mode is less sensitive to bulk properties (more sensitive to surface properties) for thicker silica layers.

3.3.1.1 PWR Sensor: Refractive Index Sensing

As mentioned earlier, the TM mode of the PWR sensor is more suitable for bulk refractive index sensing due to its long penetration depth into the fluid. To use the PWR as a refractive index sensor, the dimensions of the device and its operating wavelength are optimized. A genetic algorithm is used for sensor optimization and CSF_{bulk} is adopted as its "fitness function". Table 3.3 compares and summarizes the optimized performance characteristics of the PWR and SPR sensors.

According to Table 3.3, the CSF_{bulk} for the TM mode of the PWR sensor is an order of magnitude larger than that of the SPR sensor. Moreover, the TE mode of the PWR sensor has a larger CSF_{bulk} than the TM mode of the SPR sensor.

The tolerance of the CSF_{bulk} to variations in wavelength and silica thickness is displayed in Fig. 3.8. The SPR sensor has the largest CSF_{bulk} at the wavelength of 1200 nm (see Fig. 3.8a). Figure 3.8b, c show the tolerance of the PWR sensor with respect to the silica thickness at different wavelengths for the TMand TE polarizations, respectively. As compared to the PWR-TM mode, the PWR-TE mode has less sensitivity to the fabrication imperfections. However, for any silica thickness, it is possible to maximize the CSF for the PWR-TM polarization by suitable adjustment of the incident wavelength.

Table 3.3 Comparison of the optimized PWR and SPR sensor characteristics [22]

Sensor	Layers	Optimized parameters	Polarization	Sensitivity factor (deg./RIU)	CSF_{bulk} (RIU^{-1})
PWR	1. BK7 glass 2. Titanium (2 nm) 3. Gold (d) 4. Silica (h)	$h = 585$ nm $d = 48$ nm $\lambda = 830$ nm	TM	78	1716
			TE	13	286
SPR	5. BK7 glass 6. Titanium 7. Gold (d)	$d = 48$ nm $\lambda = 900$ nm	TM	97	97

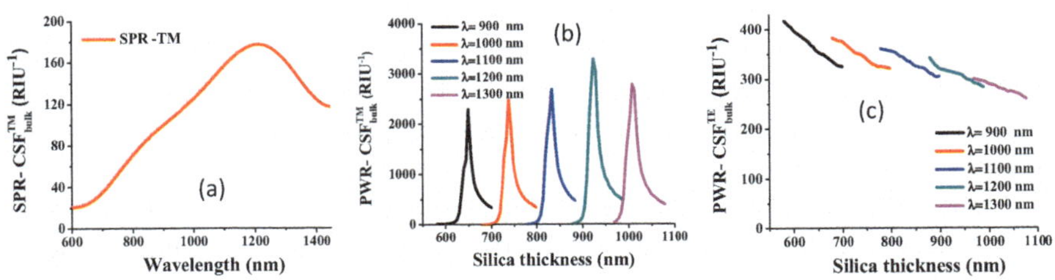

Fig. 3.8 **a** Variations of the CSF for the SPR sensor as a function of incident wavelength. **b** and **c**, variations of the CSF for the PWR-TM and -TE modes as a function of the dielectric (silica) thickness for different incident wavelengths [22]

3.3.1.2 Experimental Setup

BK7 glass substrates (1 × 1 cm) coated with a thin layer of gold (48 ± 1 nm) were purchased from SSENS Ltd. [24]. The samples were initially cleaned with piranha solution at 90 °C for 20 min to remove any organic contaminations and further cleaned in an ultrasonic bath with acetone, isopropanol, and deionized water for 10 min each. Finally, 585 nm of silica was deposited on the SPR samples using a plasma enhanced chemical vapor deposition (PECVD) to fabricate the PWR sensors.

The optical setup used to characterize the PWR and SPR sensors is shown in Fig. 3.9a. To illuminate the sensors, a combination of a super continuum laser (Fianium SC-450), a laser line tunable filter (Photon Etc), and a single mode optical fiber is used. An achromatic lens (*L*1) is used to collimate the light which is then passed through a system composed of a linear polarizer (*P*, oriented at 0°) and a liquid crystal (LC) variable retarder (45° orientations) acting as a polarization switch. A second achromatic lens (*L*2) is used to achieve a converging beam, to cover a desired range of angles. The converging beam is then focused on the sensor through a prism, where a CMOS camera is used to collect the reflected light. The measured and calculated normalized reflectance spectra are shown in Fig. 3.9b.

To evaluate the performance of the PWR and SPR sensors as a refractive index sensor, the resonance angle variation is monitored for different concentrations of solutions flowing over the samples. Three different solutions with 0.5, 1, and 2% of ethanol in deionized water were prepared to create bulk refractive index changes of 0.00014, 0.00028, and 0.00056 RIU, respectively. To create a stable baseline, pure deionized water is first passed over the sample for 5 min. Then an ethanol solution with low concentration of 0.5% is passed for another 5 min. After that, deionized water flows over the sample to recreate the baseline and the same steps are repeated for higher concentrations of ethanol.

Figure 3.10 shows the sensogram for the PWR and SPR sensors at two different wavelengths: (1) He–Ne wavelength (λ = 632 nm), which is a commonly used wavelength for biosensing experiments, and (2) the optimized wavelength which is 830 and 900 nm for the PWR and SPR, respectively. The larger sensitivity of

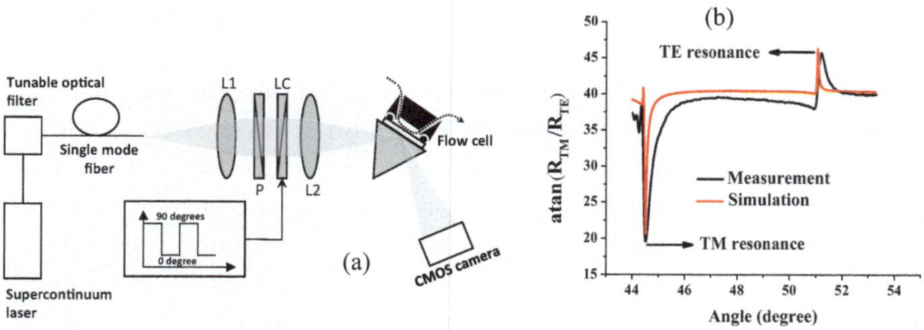

Fig. 3.9 **a** Optical setup used to detect the resonance angle. **b** The experimental (black line) and theoretical (red line) normalized reflectance spectra of the PWR sensor [25]

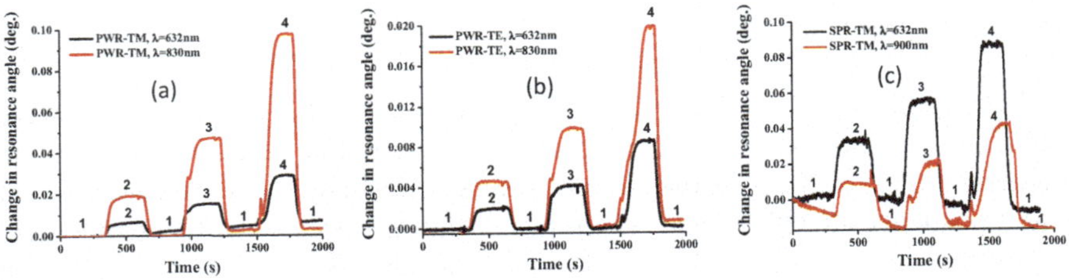

Fig. 3.10 Sensors' responses to the bulk refractive index variations: **a** resonance angle versus time for the TM-polarized PWR sensor at λ = 632 nm and 830 nm. **b** Resonance angle versus time for the TE-polarized PWR sensor at λ = 632 nm and 830 nm. **c** Resonance angle versus time for the TM-polarized SPR sensor at λ = 632 nm, and 900 nm. Solutions are based on: (1) DI water, (2) 0.5% ethanol, (3) 1% ethanol, and (4) 2% ethanol [22]

the PWR-TM mode than that of the PWR-TE mode can be seen by comparing the change in the resonance angles in Fig. 3.10a, b. Figure 3.10c shows the response of the SPR sensor (only TM polarization) to variations of the bulk index. By comparing the baselines of the PWR and SPR sensors in Fig. 3.10, a larger thermal drift of the baseline in the SPR sensor can be observed. The smaller thermal drift of the PWR sensor is due to the thick silica layer with small thermal conductivity (1.4 W m^{-1} K^{-1}) which reduces the heat transfer from the gold film, which has a large thermal conductivity (314 W m^{-1} K^{-1}), to the fluid. Whereas, in the case of the SPR sensor, the metal is in direct contact with the fluid and therefore the heat dissipated by the gold film can easily change the fluid temperature and hence its refractive index. It should be noted that approximately 0.1 °C change in the temperature of water corresponds to 10^{-5} RIU change [26].

The measured results for SPR and PWR sensors at different wavelengths are presented in Table 3.4. Sensor refractive indexresolution (σ_{RI}) is the smallest change in the refractive index that sensor is capable of detecting. This quantity is related to the standard deviation of the sensor output (σ_{SO}) and its sensitivity factor (SF), according to [27]:

$$\sigma_{RI} \equiv \frac{\sigma_{SO}}{SF}. \tag{3.9}$$

Table 3.4 Experimental sensors' characteristics obtained from the sensograms in Fig. 3.10 [22]

Sensor	Polarization	Wavelength (nm)	SF (deg./RIU)	σ_{SO} (deg.)	σ_{RI} (RIU)
PWR	TM	632	43	9.7×10^{-5}	2.3×10^{-6}
		830	164	8.2×10^{-5}	5×10^{-7}
	TE	632	16	4.6×10^{-5}	2.9×10^{-6}
		830	36	5.1×10^{-5}	1.4×10^{-6}
SPR	TM	632	206	6.6×10^{-4}	3.2×10^{-6}
		900	125	3.8×10^{-4}	3×10^{-6}

As can be seen from Table 3.4, the PWR modes (TM and TE) have smaller standard deviation (σ_{SO}) than the SPR-TM mode. This is due to the sharper resonance of the PWR modes, which produces a larger CSF. At longer wavelengths, the standard deviation of the SPR sensor reduces due to its narrower resonance, so the refractive index resolution of the SPR at the wavelength of 900 nm is smaller (better) than that at 632 nm.

According to Table 3.4, the PWR sensor refractive index resolution is 5×10^{-7} RIU and 1.4×10^{-6} RIU for the TM and TE polarizations, respectively; and the refractive index resolution for the SPR sensor is 3×10^{-6} RIU for the TM polarization (the only possible polarization for SPR). The polarization diversity of the PWR sensor (i.e. its ability to operate in both TM and TE modes) can also be used to increase the sensor's functionality as discussed next.

3.3.2 PWR Sensor: Affinity Sensing

The ability of the hybrid plasmonic sensors in general, and PWR in particular, to operate in both TM and TE modes (i.e. polarization diversity) can be utilized in self-referenced and/or multimode spectroscopy measurements. Since the TM and TE modes of the PWR sensor have different penetration depths into the fluid, their response to the change of fluid refractive index (bulk effects) and/or binding to the surface of the sensor (surface effects) are also different. Therefore, by simultaneously monitoring the variations in the resonance angles for both modes ($\Delta\theta_{TM}$, $\Delta\theta_{TE}$) and assuming that the sensitivity of each mode is known (for example, measured during a calibration process or calculated from simulation), two quantities of interest—for example, adlayer thickness (d_a) and bulk index (Δn_b)—can be determined by using (3.2).

To test the ability of the PWR to measure and decouple the bulk and surface effects mentioned above, an experiment based on biotin-streptavidin complex was designed and conducted. The silica surface of the PWR sensor was functionalized with biotin by MicroSurfaces Inc. [28]. To functionalize the SPR sensor with biotin, a solution of biotinylated PEG alkane thiol was passed over the SPR chip for two hours. The functionalized sensors were then fixed between the prism and the flow cell (Fig. 3.9a) using index matching oil. The measured sensogramsof the SPR (only TM mode) and PWR (both TM and TE modes) are shown in Fig. 3.11 [25].

Figure 3.11a shows the response of the SPR sensor, for the TM polarized light, to surface and bulk refractive index variations. Figure 3.11b, on the other hand, shows the response of the TE (red curve) and TM (black curve) modes of the PWR sensor also to the surface and bulk variations. Consistent with our predictions, the TE mode of the PWR experiences a larger change with respect to the surface variation as

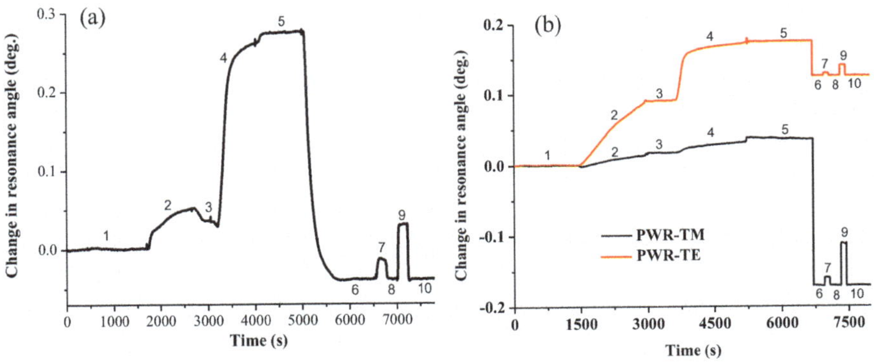

Fig. 3.11 **a** Angular positions of the resonance dip versus time for the SPR sensor. **b** Angular positions of the resonance dip versus time for the PWR sensor. Solutions are (1) PBS, (2) 1 μg/mL Streptavidin, (3) PBS, (4) 10 μg/mL Streptavidin, (5) PBS, (6) DI water, (7) 0.01 M salted water, (8) DI water, (9) 1% ethanol, and (10) DI water [25]

Table 3.5 Experimental sensor characteristics obtained from the sensograms shown in Fig. 3.11 [25]

Sensor	Polarization	Surface sensitivity (deg./nm)	Bulk sensitivity (deg./RIU)	LoD (ng/ml)	Resolution (RIU)
PWR	TM	0.008	112	55	2.6×10^{-6}
	TE	0.04	37	9	7.4×10^{-6}
SPR	TM	0.06	154	63	5×10^{-6}

compared to the TM [steps 1 to 5 in Fig. 3.11b], while the TM mode is more affected by the change in the bulk refractive index [steps 6 to 10 in Fig. 3.11b].[3]

Table 3.5 summarizes and compares the PWRPWR and SPR properties and performances. In this table, surface sensitivity (deg./nm) designates the sensitivity to the change in adlayer thickness; whereas bulk sensitivity (deg./RIU) is the sensitivity to the change in bulk (fluid) refractive index. As the entries in the table indicate, the LoD for the TE mode of the PWR sensor is smaller than that of the TM mode of the PWR and the TM mode of the SPR. Therefore, the TE mode of the PWR is an excellent choice for sensing surface binding. Moreover, the resolution of the TM mode of the PWR is smaller than that of the TEmode of PWR and TM mode of the SPR. Hence, the TM mode of the PWR is an excellent choice for sensing bulk refractive index.

Aside from the superior performance of the PWR as compared to SPR sensor, the simultaneous presence of the TMand TE modes in the PWR allows us to decouple the surface and bulk properties (an option not available for SPR sensor.)

[3]Recall that TE mode of the PWR is strongly bounded to the surface of the sensor with a penetration depth of a few hundreds of nanometers; whereas the TM mode penetrates largely into the fluid with a penetration depth of a few micrometers.

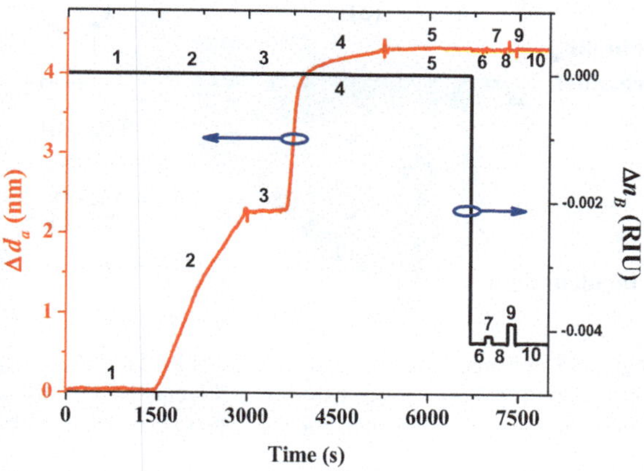

Fig. 3.12 Surface binding thickness (adlayer thickness) and bulk refractive index changes obtained from Fig. 3.11b and (3.2). Solutions are (1) PBS, (2) 1 μg/mL Streptavidin, (3) PBS, (4) 10 μg/mL Streptavidin, (5) PBS, (6) DI water, (7) 0.01 M salted water, (8) DI water, (9) 1% ethanol, and (10) DI water [25]

Figure 3.12 shows the result of decoupling the adlayer thickness changes from the bulk index variations by using the linear model given in (3.2). The black line shows the bulk (also called background or buffer) index change as a function of time—which is almost constant during the streptavidin attachment (steps 1 to 5) and changes when the bulk (buffer) refractive index variations are introduced (steps 6 to 10). The red line corresponds to changes in the adlayer thickness, which increases with attachment of the streptavidin to biotin (steps 1 to 5) and remains constant when the buffer refractive index changes (step 6 to 10). Therefore, by using simultaneous dual polarization spectroscopy at a single wavelength and a single sensing channel, the changes in adlayer thickness are completely decoupled from the variations in bulk refractive index. This ability to separate the bulk and surface effects, by using polarization diversityof the PWR, is an important feature of the hybrid plasmonic sensors, leading to their utilization as a platform for multi-mode spectroscopy.

3.4 Dielectric Grating SPR Sensor

The dielectric grating SPR sensor (DGSPR) is another platform which can be used for self-referenced spectroscopy [29]. DGSPR is very similar to the PWR sensor, where the main difference is replacing the top dielectric layer (in the PWR sensor) with a dielectric grating (see Fig. 3.13a). This enables the DGSPR to be used for studying the surface bindings on the gold surface (between the dielectric strips), instead of the silica surface of PWR; i.e. one can functionalize the gold surface instead of the silica surface.

The DGSPR sensor can guide a pure SPR [TM2 in Fig. 3.14a] in addition to two PWR modes [TM1 and TE1 in Fig. 3.14a]. Therefore, by careful adjustment of the dimensions of the grating and operating wavelength, the three modes of the DGSPR sensor can be utilized to decouple three different unknown quantities, and hence obtain more information from the analyte. Figure 3.14b shows the field profile

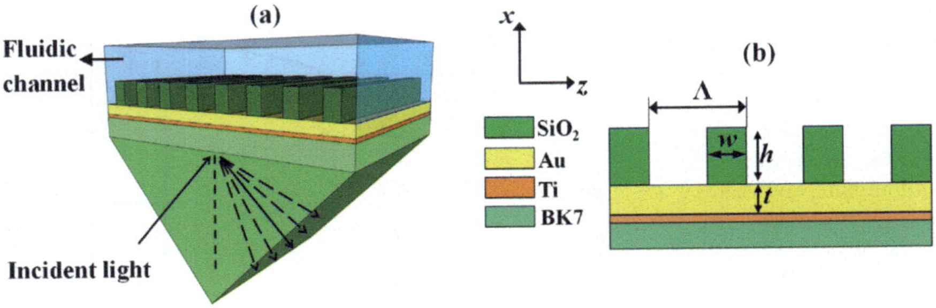

Fig. 3.13 **a** Three dimensional schematic of the DGSPR sensor. **b** Two dimensional cross section of the DGSPR sensor. Silica nanowires of rectangular profile are assumed to be infinite in length (along the *y*-axis) with periodicity of Λ, thickness of *h*, and filling factor of *w*/Λ [29]

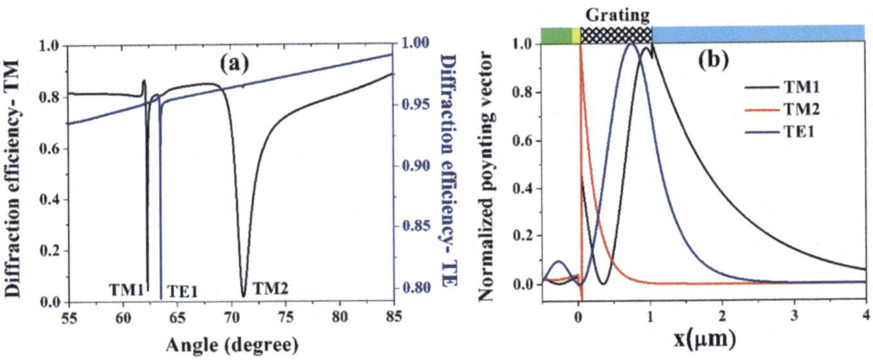

Fig. 3.14 **a** Reflectance spectrum for the optimized DGSPR sensor. **b** z-component of the Poynting vector for both TM and TE polarizations at resonance angles of 62.28, 71.11, and 63.5° for TM1, TM2, and TE1 modes, respectively [29]

of all three modes at their respective resonance angles. According to the figure, the TM1 mode (black line) is a hybrid mode generated by the coupling between the SP wave propagating on top of the gold surface and the waveguide mode of the dielectric grating. The TM2 mode (red line) is a pure SP wave propagating at the metal-dielectric grating interface. The TE1 mode (blue line) is the dielectric mode mostly confined inside the grating region.

To simulate the DGSPR response, an in-house code based on the rigorous coupled wave analysis (RCWA) is developed. The RCWA is an exact solution of the Maxwell's equations for electromagnetic fields which can calculate the diffraction efficiencies of the reflected or transmitted light intensity from gratings [30]. To evaluate the performance of the DGSPR sensor in decoupling surface and bulk parameters, a FoM based on Eq. (3.10) is defined:

$$\mathrm{FoM} = \mathrm{CSF}^{\mathrm{TM1}}_{\mathrm{surf\ thick}} \cdot \mathrm{CSF}^{\mathrm{TM2}}_{\mathrm{surf\ thick}} \cdot \mathrm{CSF}^{\mathrm{TE1}}_{\mathrm{bulk\ index}}. \tag{3.10}$$

Table 3.6 The optimized dimensions of DGSPR sensor

Wavelength (nm)	Gold thickness (t)	Grating thickness (h)	Periodicity (Λ)	Fill factor (w/Λ)
825	49 nm	992 nm	890 nm	0.4

where, $CSF^{TM1}_{surf\ thick}$ and $CSF^{TM2}_{surf\ thick}$ correspond to the adlayer (surface) thickness variations for TM1 and TM2 modes, respectively, and $CSF^{TE1}_{bulk\ index}$ is the CSF corresponding to the bulk refractive index changes for the TE polarization. A genetic algorithm is used to optimize the sensor with the above FoM as its "fitness function". Table 3.6 shows the optimized dimensions of the DGSPR sensor.

Table 3.7, summarizes the performance of the optimized DGSPR sensor with respect to variations in the surface and bulk parameters for each mode. In this table, in order to facilitate the comparison between different modes supported by the sensors, we have used the following notations: (1) The performance of each mode as a result of variations in adlayer thickness and adlayer index are indicated by CSF^{surf}_{thick} and CSF^{surf}_{index} respectively. (2) The performance of each mode as a consequence of changes in the bulk refractive index is designated by CSF^{bulk}_{index}. As entries in the table show the CSF^{surf}_{thick} and CSF^{surf}_{index} of the TM modes are in the same range as the optimized SPR sensor. On the other hand, the CSF^{bulk}_{index} of the TM1 is an order of magnitude larger than that of the SPR sensor, due to its longer penetration into the buffer [black line in Fig. 3.14b]. The TEmode of the DGSPR sensor has a smaller CSF^{surf}_{thick} and CSF^{surf}_{index} than its TM modes, but has a larger CSF^{bulk}_{index} than the SPR which makes it ideal for detecting bulk index variations with less cross sensitivity to surface effects.

Similar to the discussion in Sect. 3.2, during a sensing event, the three modes supported by the DGSPR can be used to calculate and decouple three unknown parameters. As an example, suppose that variations in adlayer thickness (Δd_a), adlayer refractive index (Δn_a), and bulk refractive index (Δn_b) are the quantities of interest. Then, these can be decoupled by using:

Table 3.7 Performance characteristics of the optimized SPR and DGSPR sensors [29]

Sensor		TM1	TM2	TE1
DGSPR	CSF^{surf}_{thick}	0.031 nm^{-1}	0.031 nm^{-1}	0.0003 nm^{-1}
	CSF^{surf}_{index}	0.51 RIU^{-1}	0.76 RIU^{-1}	0.01 RIU^{-1}
	CSF^{bulk}_{index}	619 RIU^{-1}	40 RIU^{-1}	377 RIU^{-1}
SPR	CSF^{surf}_{thick}	0.057 nm^{-1}	–	–
	CSF^{surf}_{index}	0.76 RIU^{-1}	–	–
	CSF^{bulk}_{index}	96 RIU^{-1}	–	–

$$\begin{pmatrix} \Delta\theta_{\mathrm{TM1}} \\ \Delta\theta_{\mathrm{TM2}} \\ \Delta\theta_{\mathrm{TE1}} \end{pmatrix} = \underbrace{\begin{pmatrix} \frac{\partial\theta_{\mathrm{TM1}}}{\partial d_a} & \frac{\partial\theta_{\mathrm{TM1}}}{\partial n_a} & \frac{\partial\theta_{\mathrm{TM1}}}{\partial n_b} \\ \frac{\partial\theta_{\mathrm{TM2}}}{\partial d_a} & \frac{\partial\theta_{\mathrm{TM2}}}{\partial n_a} & \frac{\partial\theta_{\mathrm{TM2}}}{\partial n_b} \\ \frac{\partial\theta_{\mathrm{TE1}}}{\partial d_a} & \frac{\partial\theta_{\mathrm{TE1}}}{\partial n_a} & \frac{\partial\theta_{\mathrm{TE1}}}{\partial n_b} \end{pmatrix}}_{S} \cdot \begin{pmatrix} \Delta d_a \\ \Delta n_a \\ \Delta n_b \end{pmatrix} \tag{3.11}$$

In (3.11), the S matrix contains the sensitivity factors of each mode to the surface and bulk parameters (d_a, n_a, and n_b). During a sensing event, by measuring the resonance angle variations ($\Delta\theta_{\mathrm{TM1}}$, $\Delta\theta_{\mathrm{TM2}}$, $\Delta\theta_{\mathrm{TE1}}$) and measuring (or calculating) the sensitivities factors, the changes in the quantities of interest can be found according to:

$$\begin{pmatrix} \Delta d_a \\ \Delta n_a \\ \Delta n_b \end{pmatrix} = S^{-1} \times \begin{pmatrix} \Delta\theta_{\mathrm{TM1}} \\ \Delta\theta_{\mathrm{TM2}} \\ \Delta\theta_{\mathrm{TE1}} \end{pmatrix}, S = \begin{pmatrix} 0.003 & 0.1 & 72 \\ 0.065 & 1.1 & 81 \\ 0.0001 & 0.0015 & 61 \end{pmatrix} \tag{3.12}$$

For the DGSPR of Table 3.6, the determinant of the S matrix is calculated to be 0.2, which confirms that S is not a singular matrix and hence invertible. Therefore, the three-mode spectroscopy using the DGSPR can provide enough information to distinguish between three unknown quantities (Δd_a, Δn_a and Δn_b), a task that is not possible by using a conventional SPR sensor that relies on a single polarization measurement.

3.5 Metallic Grating SPR Sensor

A metallic grating SPR (MGSPR) sensor is another platform that can be used for three-mode spectroscopy [31, 32]. The sensor is a metallic grating loaded on top of a SPR substrate, as shown in Fig. 3.15. The metallic grating is used to split the incoming light into several diffraction orders. Assuming that the dispersion properties of the

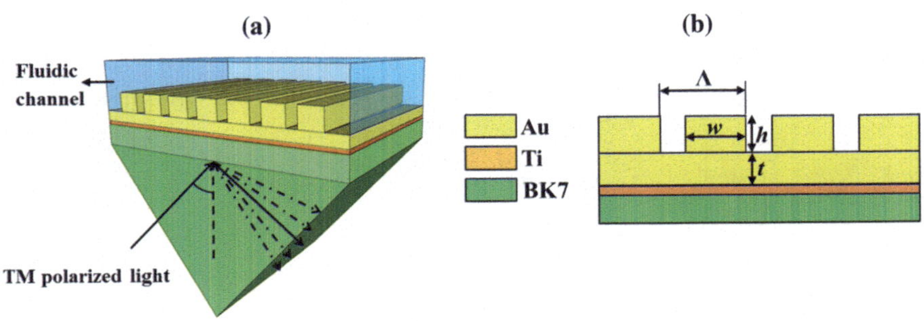

Fig. 3.15 **a** Three dimensional schematic of a MGSPR sensor. **b** Two dimensional cross section of the MGSPR sensor. Gold nanowires of rectangular profile are assumed to be infinite in length with periodicity of Λ, thickness of h, and filling factor of w/Λ [32]

SP are not affected by the grating, the momentum conservation between the free space light and the SP wave, when using prism coupling, can be expressed as [33]:

$$\pm\sqrt{\frac{\varepsilon_m \varepsilon_b}{\varepsilon_m + \varepsilon_b}} = n_p \sin(\theta_{SPR}) + m\frac{\lambda}{\Lambda}. \quad (3.13)$$

Here, θ_{SPR},ε_m, ε_b, n_p, m, λ and Λ are: angle of incidence at the resonance frequency, metal relative permittivity, buffer (fluid) relative permittivity, prism index of refraction, diffraction orders, wavelength of the incident light in free space, and period of the grating. The grating generates a series of beams that are diffracted away from the corrugated surface (see Fig. 3.15a); where the '+' and '−' signs in Eq. (3.13) correspond to the positive ($m > 0$) and negative ($m < 0$) diffraction orders. The MGSPR provides us with the possibility of efficient coupling between the diffraction orders of the grating and the SP wave, which makes the MGSPR a good candidate for multi-mode spectroscopy.

To study the effects of a grating on the performance of the sensor, we have used the RCWA code mentioned earlier. A metallic grating was optimized so that three SP waves at two different wavelengths were excited efficiently. Thus, variations in the adlayer properties (Δd_a and Δn_a) can be differentiated from the bulk index variations (Δn_b) via dual wavelength measurements. Table 3.8 summarizes the optimized dimensions and wavelengths for the MGSPRand SPR sensors.

The diffraction efficiency of the reflected light for the optimized MGSPR at a wavelength of 845 nm is shown in Fig. 3.16a. There are two distinct resonances in

Table 3.8 The optimized dimensions of MGSPR and SPR sensors [32]

	Wavelength (nm)	Gold thickness (t) (nm)	Grating thickness(h)	Periodicity (Λ)	Fill factor (w/Λ)
MGSPR	845	26	241 nm	325 nm	0.7
SPR	915	50	–	–	–

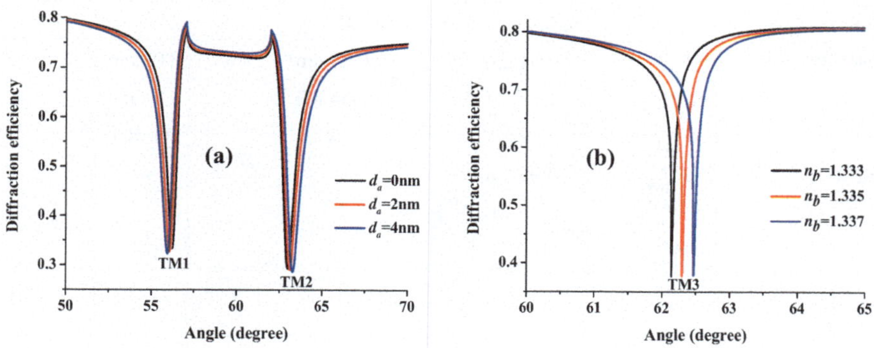

Fig. 3.16 Diffraction efficiencies of the reflected light with TM polarization for the optimized PGSPR sensor at the incident wavelengths of **a** 845 nm, and **b** 970 nm [32]

the spectrum corresponding to two counter-propagating SP waves: (1) The first TM mode (TM1), with smaller angle of resonance, is excited by the first negative diffraction order ($m = -1$); (2) the second TM mode (TM2), with larger resonance angle, is excited by the 0th diffraction order ($m = 0$). These two modes are highly sensitive to the variations of adlayer properties (n_a, d_a). For three-mode spectroscopy, a third mode (TM3) is also optimized at a second wavelength (970 nm) with high sensitivity to the bulk refractive index variations (Fig. 3.16b). The mode (TM3) is also excited by the 0th diffraction order of the MGSPR sensor.

Table 3.9, summarizes and compares the response of MGSPR sensor to variations in the surface and bulk parameters, with that of the SPR sensor. Although, the CSF^{surf}_{thick} and CSF^{surf}_{index} of the SPR are slightly larger than the corresponding values for the MGSPR, in the case of SPR these quantities cannot be decoupled. However, the MGSPR has two modes (TM1, TM2), each sensitive to the surface binding, and a third mode (TM3) with large sensitivity to the bulk refractive index variations. These modes together can be used to decouple the surface and bulk effects.

To determine the quantities of interest from the variations in the three resonance angles ($\Delta\theta_{TM1}$, $\Delta\theta_{TM2}$, $\Delta\theta_{TM3}$) of the MGSPR sensor, (3.12) is modified for dual-wavelength spectroscopy according to:

$$\begin{pmatrix} \Delta d_a \\ \Delta n_a \\ \Delta n_b \end{pmatrix} = S^{-1} \times \begin{pmatrix} \Delta\theta_{M1}(\lambda_1) \\ \Delta\theta_{M2}(\lambda_1) \\ \Delta\theta_{M3}(\lambda_2) \end{pmatrix}, \quad S = \begin{pmatrix} \frac{\partial\theta_{TM1}(\lambda_1)}{\partial d_a} & \frac{\partial\theta_{TM1}(\lambda_1)}{\partial n_a} & \frac{\partial\theta_{TM1}(\lambda_1)}{\partial nb} \\ \frac{\partial\theta_{TM2}(\lambda_1)}{\partial d_a} & \frac{\partial\theta_{TM2}(\lambda_1)}{\partial n_a} & \frac{\partial\theta_{TM2}(\lambda_1)}{\partial nb} \\ \frac{\partial\theta_{TM3}(\lambda_2)}{\partial d_a} & \frac{\partial\theta_{TM3}(\lambda_2)}{\partial n_a} & \frac{\partial\theta_{TM3}(\lambda_2)}{\partial n_b} \end{pmatrix} = \begin{pmatrix} 0.065 & 1.026 & 71.84 \\ 0.075 & 1.066 & 86.85 \\ 0.005 & 0.057 & 82.89 \end{pmatrix} \tag{3.14}$$

The determinant of the S matrix is equal to 0.56, which confirms the non-singularity of the S matrix. Therefore, (3.14) leads to a unique solution for Δd_a, Δn_a, and Δn_b. Thus the MGSPR sensor can provide enough information, via dual-wavelength spectroscopy, to distinguish between the quantities of interest.

Table 3.9 The optimized performance characteristics for MGSPR and single interface SPR sensors [32]

Sensor		TM1	TM2	TM3
MGSPR	CSF^{surf}_{thick}	0.04 nm^{-1}	0.04 nm^{-1}	0.02 nm^{-1}
	CSF^{surf}_{index}	0.6 RIU^{-1}	0.54 RIU^{-1}	0.29 RIU^{-1}
	CSF^{bulk}_{index}	42 RIU^{-1}	44 RIU^{-1}	369 RIU^{-1}
SPR	CSF^{surf}_{thick}	0.057 nm^{-1}	–	–
	CSF^{surf}_{index}	0.76 RIU^{-1}	–	–
	CSF^{bulk}_{index}	96 RIU^{-1}	–	–

3.5.1 *Summary*

Self-reference spectroscopy was investigated in different platforms in order to determine and differentiate the parameters that can affect the sensor output signal during a sensing event. Two planar structures were studied for two-mode spectroscopy. The first structure was a hybrid plasmonic waveguide (HPWG) which enabled spectroscopy with two orthogonal polarization modes, while each mode had different sensitivities to the variations in bulk and/or surface properties. The optimization of the HPWG in Mach–Zehnder interferometer (MZI) and prism coupling configurations was studied and the appropriate mechanism to decouple surface and bulk properties was discussed. The second hybrid structure was a plasmonic waveguide resonance (PWR) which similar to the HPWG supports two orthogonal polarizations, but overcomes the limitations associated with a narrow fluidic channelin the HPWG sensor. The PWR sensor guides a TM mode with high sensitivity to the bulk parameters and a TE mode with high sensitivity to the surface index variations. The utilization of this sensor for refractive index sensing and thin film measurement was experimentally demonstrated. It was shown that the TM mode of the PWR sensor has smaller refractive index resolution (2.6×10^{-6} RIU) than that of the SPR sensor (5×10^{-6} RIU), while the TE mode has a smaller LoD (9 ng/ml) than that of the TM mode of the SPR sensor (63 ng/ml). Finally, efficient decoupling of the surface reactions from the bulk index variations was demonstrated experimentally using the PWR sensor.

Lastly, two additional sensors were proposed in order to decouple three measurands using three-mode spectroscopy. These sensors were: a dielectric grating surface plasmon resonance (DGSPR) sensor and a metallic grating surface plasmon resonance (MGSPR) sensor. The DGSPR sensor, which contains a dielectric grating, was designed and optimized for dual polarization measurements at a single wavelength where the sensor guides two TM modes and one TE mode. The MGSPR sensor, which contains a metallic grating, utilizes dual-wavelength measurements, where it excites three TM modes at two different wavelengths. Therefore, grating-based sensors can provide the user with more information from the sample under test—and this information can be used to decouple different quantities of interest.

References

1. J. Homola, Present and future of surface plasmon resonance biosensors. Anal. Bioanal. Chem. **377**, 528–539 (2003)
2. G.G. Nenninger, J.B. Clendenning, C.E. Furlong, S.S. Yee, Reference-compensated biosensing using a dual-channel surface plasmon resonance sensor system based on a planar lightpipe configuration. Sens. Actuators, B: Chem. **51**, 38–45 (1998)
3. R. Slavík, J. Homola, H. Vaisocherová, Advanced biosensing using simultaneous excitation of short and long range surface plasmons. Meas. Sci. Technol. **17**, 932–938 (2006)

4. J.T. Hastings, J. Guo, P.D. Keathley, P.B. Kumaresh, Y. Wei, S. Law et al., Optimal self-referenced sensing using long—and short—range surface plasmons. Opt. Express **15**, 17661–17672 (2007)
5. J. Guo, P.D. Keathley, J.T. Hastings, Dual-mode surface-plasmon-resonance sensors using angular interrogation. Opt. Lett. **33**, 512–514 (2008)
6. Y. Wan, Z. Zheng, Z. Lu, J. Liu, J. Zhu, Self-referenced sensing based on a waveguide-coupled surface plasmon resonance structure for background-free detection. Sens. Actuators, B: Chem. **162**, 35–42 (2012)
7. M. Z. Alam, J. Meier, J. S. Aitchison, M. Mojahedi, Super mode propagation in low index medium, in CLEO: 2007 (Optical Society of Americam Washington, DC), paper JThD112, 2007
8. R.F. Oulton, V.J. Sorger, D.A. Genov, D.F.P. Pile, X. Zhang, A hybrid plasmonic waveguide for subwavelength confinement and long-range propagation. Nat. Photonics **2**, 496–500 (2008)
9. M.Z. Alam, J. Meier, J.S. Aitchison, M. Mojahedi, Propagation characteristics of hybrid modes supported by metal-low-high index waveguides and bends. Opt. Express **18**, 12971–12979 (2010)
10. M. Z. Alam, J. Stewart Aitchison, M. Mojahedi, Compact and silicon-on-insulator-compatible hybrid plasmonic TE-pass polarizer. Opt. Lett. **37**, 55–57 (2012)
11. M.Z. Alam, J.S. Aitchison, M. Mojahedi, Polarization-independent hybrid plasmonic coupler for a silicon on insulator platform. Opt. Lett. **37**, 3417–3419 (2012)
12. E. D. Palik, *Handbook of Optical Constants of Solids* (Academic Press, Inc, 1985)
13. M.S. Kwon, Theoretical investigation of an interferometer-type plasmonic biosensor using a metal-insulator-silicon waveguide. Plasmonics **5**, 347–354 (2010)
14. L. Zhou, X. Sun, X. Li, J. Chen, Miniature microring resonator sensor based on a hybrid plasmonic waveguide. Sensors **11**, 6856–6867 (2011)
15. M.Z. Alam, F. Bahrami, J.S. Aitchison, M. Mojahedi, Analysis and optimization of hybrid plasmonic waveguide as a platform for biosensing. Photonics J., IEEE **6**, 1–10 (2014)
16. M. Z. Alam, F. Bahrami, J. S. Aitchison, M. Mojahedi, "A hybrid waveguide sensor for highly sensitive biosensing," in CLEO: 2011 (Optical Society of Americam Washington, DC), paper JWA111, 2011
17. M. Z. Alam, J. S. Aitchison, M. Mojahedi, Vertical wall affinity sensor with polarization diversity. Optical Sensors (Optical Society of Americam Washington, DC) paper SMB3, 2011
18. Z. Salamon, M.F. Brown, G. Tollin, Plasmon resonance spectroscopy: probing molecular interactions within membranes. Trends Biochem. Sci. **24**, 213–219 (1999)
19. P. Berini, Bulk and surface sensitivities of surface plasmon waveguides. New J. Phys. **10** (2008)
20. F. Bahrami, M.Z. Alam, J.S. Aitchison, M. Mojahedi, Dual polarization measurements in the hybrid plasmonic biosensors. Plasmonics **8**, 465–473 (2013)
21. F. Bahrami, M.Z. Alam, J.S. Aitchison, M. Mojahedi, "A plasmonic affinity biosensor with dual polarization based on hybrid plasmonic platform," in CLEO: 2012 (Optical Society of Americam Washington, DC), paper JTh2A, 2012
22. F. Bahrami, M. Maisonneuve, M. Meunier, J. Stewart Aitchison, M. Mojahedi, An improved refractive index sensor based on genetic optimization of plasmon waveguide resonance. Opt. Express **21**, 20863–20872 (2013)
23. F. Bahrami, J.S. Aitchison, M. Mojahedi, "A highly optimized plasmon waveguide resonance biosensor," IEEE Phonics Conference, paper TuB3, pp. 208–209 (2012)
24. S. Patskovsky, E. Bergeron, M. Meunier, Hyperspectral darkfield microscopy of PEGylated gold nanoparticles targeting CD44-expressing cancer cells. J. Biophotonics (2013)
25. F. Bahrami, M. Maisonneuve, M. Meunier, J.S. Aitchison, M. Mojahedi, Self-referenced spectroscopy using plasmon waveguide resonance biosensor. Biomed. Opt. Express **5**, 2481–2487 (2014)
26. R.B.M. Schasfoort, A.J. Tudos, *Handbook of Surface Plasmon Resonance* (Royal Society of Chemistry, 2008)
27. M. Piliarik, J. Homola, Surface plasmon resonance (SPR) sensors: approaching their limits? Opt. Express **17**, 16505–16517 (2009)

28. M. Inc. http://www.proteinslides.com/
29. F. Bahrami, J.S. Aitchison, M. Mojahedi, Multimode spectroscopy using dielectric grating coupled to a surface plasmon resonance sensor. Opt. Lett. **39**, 3946–3949 (2014)
30. M.G. Moharam, D.A. Pommet, E.B. Grann, T.K. Gaylord, Stable implementation of the rigorous coupled-wave analysis for surface-relief gratings: enhanced transmittance matrix approach. J. Opt. Soc. Am. A: Opt. Image Sci., Vision **12**, 1077–1086 (1995)
31. F. Bahrami, M. Maisonneuve, M. Meunier, J.S. Aitchison, M. Mojahedi, Multimode spectroscopy with plasmonic sensors," in CLEO: 2012 (Optical Society of Americam Washington, DC), paper JTu4A, 2013
32. F. Bahrami, J.S. Aitchison, M. Mojahedi, Dual-wavelength spectroscopy of a metallic-grating-coupled surface plasmon resonance biosensor. Photonics J., IEEE **7**, 1–7 (2015)
33. J. Homola, I. Koudela, S.S. Yee, Surface plasmon resonance sensors based on diffraction gratings and prism couplers: sensitivity comparison. Sens. Actuators, B: Chem. **54**, 16–24 (1999)

Chapter 4
Refractive Index Sensing with Anisotropic Hyperbolic Metamaterials

Nikolaos Vasilantonakis, Gregory A. Wurtz, and Anatoly V. Zayats

Abstract Metamaterials with hyperbolic dispersion based on metallic nanorod arrays provide a flexible platform for the design of bio- and chemical sensors and active nanophotonic devices that allow the incorporation of refractive-index-dependent materials into or on top of the metamaterial. In this Chapter, we describe sensing geometries with anisotropic materials and overview the refractive index sensitivity of the resonances of the metamaterial, in different regimes. The examples of various sensing applications for label-free analyte recognition, hydrogen detection and ultrasound monitoring will be described. A general strategy for maximising the refractive index sensitivity of nanorod-based hyperbolic metamaterials will be discussed and can be used as the basis for the design of new ultrasensitive chemo- and bio-sensors that outperform both surface-plasmon polaritons and localised surface-plasmon based transducers.

4.1 Introduction to Sensing with Plasmonic Metamaterials

The refractive index sensitivity of plasmonic and waveguide resonances forms the basis of commercial and newly emerging optical sensing techniques for label-free biosensing and chemical identification [1–4], as well as active nanophotonic components [5, 6]. In the former class of applications, the presence of an analyte substance modifies the eigen-modes of the nanostructure, a change that can be detected by a

N. Vasilantonakis
Institute of Electronic Structure and Laser Foundation for Research and Technology Hellas (FORTH), P.O. Box 1385, 71110 Heraklion, Crete, Greece

G. A. Wurtz
Department of Physics, University of North Florida, Building 50/Office 1532 1 UNF Drive, Jacksonville, FL 32224, USA

A. V. Zayats (✉)
Department of Physics and London Centre for Nanotechnology, King's College London, Strand, London WC2R 2LS, United Kingdom
e-mail: a.zayats@kcl.ac.uk

R. De La Rue et al. (eds.), *Biomedical Optical Sensors*,
Biological and Medical Physics, Biomedical Engineering,
https://doi.org/10.1007/978-3-030-48387-6_4

shift of the resonant wavelengths of the structure or by changes of transmitted or reflected light intensity. In the latter, the refractive index changes are induced by external stimuli, such as temperature, acoustic pressure, external static electric or magnetic fields, or indeed optical fields, via nonlinear effects in the surrounding dielectric or metal [6]. The strong modification of the optical response in plasmonic nanostructures arises from the strong confinement of the electromagnetic field near the metal/dielectric interface. Both sensing and active nanophotonic devices can make use of macroscopic thin metal films or nanostructured surfaces where surface electromagnetic waves called surface plasmon polaritons (SPPs) propagate—or use nanoparticles and their assemblies that support localised surface plasmons (LSPs) [2, 3, 7].

Surface plasmon resonance (SPR) biosensors use SPP waves for the detection of binding events, lifetime measurements or molecular concentration, based on the attenuated total internal reflection (ATR) configuration [1]. Due to the strong field confinement of SPPs, sensing limits are greatly enhanced, possibly exceeding 3,000 nm per refractive index unit (RIU) [8], and these limits can be even further boosted via phase-sensitive interferometry [9]. Nonetheless, SPP-based techniques have restrictions in detecting small-molecule analytes, typically ones that are smaller than 500 Da, making them problematic for modern nanoscale chemical and biochemical tasks [10]. An alternative route is to use ***LSP*** localized surface plasmon (LSP) modes on plasmonic nanoparticles that provide an even stronger field confinement and are, therefore, more sensitive to smaller-size molecules [2, 11]. However, the overall sensitivity provided by LSPs is typically orders of magnitude smaller than for SPPs, not exceeding 100 to 300 nm/RIU [2, 12].

Recently, plasmonic metamaterials have been demonstrated to provide record refractive index sensitivity for biosensing, ultrasound detection and large effective Kerr-type nonlinearities [13]. In particular, the class of anisotropic metamaterials based on arrays of strongly interacting, aligned plasmonic nanorods exhibits hyperbolic dispersion, with one negative (ε_z) and two positive effective permittivity tensor components $\left(\varepsilon_{x,y}\right)$, leading to a metamaterial with hyperbola-shaped isofrequency contours. This unique isofrequency surface enables a plethora of applications, from guiding and imaging beyond the diffraction limit [14, 15] to enhanced nonlinearities [5, 16] and chemo- and bio-sensing [5, 6, 17]. Additionally, selected resonances of plasmonic nanorod metamaterials have been shown to exhibit a strong sensitivity to the thickness change of a dielectric load [18], ultrasensitive detection of ultrasound [19]—and ultrafast (sub-ps) response times due to optical nonlocality [16].

4.2 Differences Between Isotropic and Anisotropic Materials for Sensing

We now compare the sensing capabilities of isotropic and anisotropic media and show the advantages of the latter, based on some simple paradigms. We first inspect the dispersion relations for the half-space media and compare the derived expressions for the isotropic and anisotropic cases. We then focus our attention on the sensitivity of the dispersion for each scenario and explain why sensing is important, from the applied point of view, and why anisotropy plays a crucial role.

4.2.1 Comparison of Dispersion

Let us consider two half-space media with permittivities $\{\varepsilon_1, \varepsilon_2\}$ and permeabilities $\{\mu_1 = \mu_2 = \mu_0\}$. The material ε_1 is taken to be gold (Au) and follow the Drude-Sommerfeld model, while the material ε_2 can either be isotropic or anisotropic. In this scenario, the SPP dispersion will have the form:

$$k_{\text{spp}} = k_0\sqrt{\frac{\varepsilon_1\varepsilon_2}{\varepsilon_1+\varepsilon_2}} \quad \text{(isotropic)} \tag{4.1}$$

$$k_{\text{spp}} = k_0\sqrt{\frac{\varepsilon_{\text{zz}}\varepsilon_1(\varepsilon_{\text{xx}}-\varepsilon_1)}{\varepsilon_{\text{xx}}\varepsilon_{\text{zz}}-\varepsilon_1^2}} \quad \text{(anisotropic)} \tag{4.2}$$

where k_0 is the wavevector in vacuum and $\varepsilon_{\text{xx}} = \varepsilon_{\text{yy}} \neq \varepsilon_{\text{zz}}$ are the diagonal components of the permittivity tensor. (The off-diagonal components are zero). To examine these two cases, we can keep the permittivity component along one direction fixed and vary the other component. Figure 4.1a, b demonstrate the scenario where $\varepsilon_2^{iso} = \varepsilon_{xx}^{ani} = 4$ and ε_{zz}^{ani} vary from 1 to 9. An immediate observation is that when $\varepsilon_{zz}^{ani} < \varepsilon_2^{iso}$the SPP dispersion shifts to higher energies and vice versa. A similar shift is also true for the imaginary part of the dispersion, as shown in Fig. 4.1b. Therefore, by tuning the permittivity along a given direction, the SPP resonance can be changed—which is not possible with isotropic materials since, by definition, the optical response for a given polarization remains unaffected along different axes. A more realistic case where ε_1 follows the experimental values from [20] is depicted in Fig. 4.1c, d. Here the SPP dispersion shift along the energy axis is still valid, but is not clearly distinct. The reason for the decreased energy shift is the influence of the interband transitions in Au that occur close to a photon energy of 2.4 eV $\left(\omega/\omega_p = 0.23\right)$. What is more pronounced is the change of the resonance energy along the ordinary axis. Note however that, although the SPP wavevector can be enhanced by choosing the appropriate anisotropic permittivity, the imaginary part is also increased and therefore losses become significant. The main conclusion is that, regardless of the theoretical or experimental approximation, the SPP dispersion can be tuned by carefully choosing the appropriate set of $\left\{\varepsilon_{\text{XX}}^{\text{ani}}, \varepsilon_{\text{ZZ}}^{\text{ani}}\right\}$—a feature that is absent in isotropic media, since they have no directionality.

To obtain a broader view of the validity of equations derived so far, we can examine the electromagnetic field distribution for the SPP excitation. Since the wavevectors are known, the fields in the medium ε_1 will be:

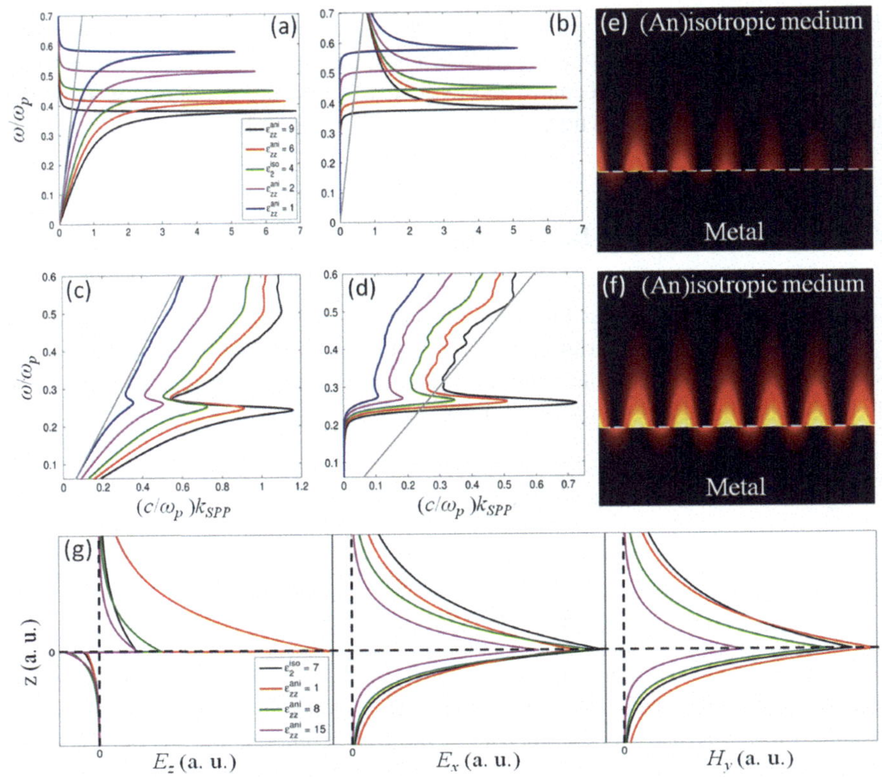

Fig. 4.1 Comparison of SPP dispersion between isotropic and anisotropic media. Real (**a**) and imaginary part (**b**) of dispersion. Material ε_1 follows the Drude-Sommerfeld model for Au ($\omega_p = 1.367 \times 10^{16}$ Hz). Material ε_2 has $\varepsilon_2^{\text{iso}} = 4$ for the isotropic case, while the anisotropic case has $\varepsilon_{\text{xx}}^{\text{ani}} = \varepsilon_2^{\text{iso}}$—and $\varepsilon_{\text{zz}}^{\text{ani}}$ varies from 1 to 9. **c**, **d** The same as (a,b) but using the experimental values for Au taken from reference [20]. Note how the dispersion can be tuned by changing the anisotropic permittivity. **e**, **f** The norm of the electric field showing the excitation of a SPP in a metal-(an)isotropic interface, with and without losses respectively. **g** Typical cross-sections of the electromagnetic field components at that interface for various $\varepsilon_{\text{zz}}^{\text{ani}}$. Notice that both E_x and H_y are continuous at the interface, while E_z is not—because of surface charge accumulation

$$\left.\begin{aligned} \boldsymbol{H}_1(\boldsymbol{r},t) &= (0, H_0, 0)^{\mathrm{T}} e^{i(k_x x - k_{z1} z - \omega t)} \\ \boldsymbol{E}_1(\boldsymbol{r},t) &= \left(-\frac{cH_0 k_{z1}}{\omega\varepsilon_1}, 0, -\frac{cH_0 k_x}{\omega\varepsilon_1}\right)^{\mathrm{T}} e^{i(k_x x - k_{z1} z - \omega t)} \end{aligned}\right\} \tag{4.3}$$

and for the material ε_2 :

$$\left.\begin{aligned} \boldsymbol{H}_2(\boldsymbol{r},t) &= (0, H_0, 0)^{\mathrm{T}} e^{i(k_x x + k_{z2} z - \omega t)} \\ \boldsymbol{E}_2(\boldsymbol{r},t) &= \left(-\frac{cH_0 k_{z2}}{\omega\varepsilon_{\text{xx}}^{\text{ani}}}, 0, -\frac{cH_0 k_x}{\omega\varepsilon_{\text{zz}}^{\text{ani}}}\right)^{\mathrm{T}} e^{i(k_x x + k_{z2} z - \omega t)} \end{aligned}\right\} \tag{4.4}$$

where H_0 is a constant magnetic field amplitude and $k_x = k_{SPP}$. Note that for the case where $\varepsilon_{xx}^{ani} = \varepsilon_{zz}^{ani} = \varepsilon_2^{iso}$ it is simplified to the well-known isotropic expression. Based on the aforementioned equations, the field distributions can be plotted. Figure 4.1e, f show the magnitude of the electric field for a SPP travelling along a typical metallo-dielectric interface (either iso- or anisotropic) with and without losses, respectively. Note that, for the lossless case, the SPP can travel infinitely along the surface—while, for the lossy case, the propagation length is finite—due to the small, but non negligible, imaginary part of the permittivity. Figure 4.1g shows several cross-sections of the electromagnetic field components for different permittivity values. It is clearly shown that, when ε_{zz}^{ani} increases, the field components are more localized to the surface—leading to strong optical confinement—regardless of the fact that the permittivity along the x-direction is the same as in the isotropic case. The E_z component has a discontinuity at the interface, due to surface charge accumulation. Note however that the permittivities for either an isotropic or an anisotropic medium cannot change arbitrarily with respect to the metallic medium; they both need to keep k_x real and k_z imaginary at all times.

4.2.2 *Comparison of Sensitivity*

Examination of the sensitivity, i.e. the change of the ordinary wavevector with respect to the permittivity, of the SPP dispersion is another important factor that plays a critical role in applications that depend on small changes (typically 10^{-3}–10^{-4}) of the refractive index such as biosensors [2, 21–23]. As a consequence, having the ability to manipulate the sensitivity of the dispersion is a key factor. Based on (4.1) and (4.2), we can measure the derivative of the SPP dispersion with respect to the permittivity, for the anisotropic and isotropic scenarios, respectively. To compare these two cases, we follow the recipe discussed previously and keep the permittivity component along one direction fixed, while varying the other. Using ε_1 data for Au taken from Johnson and Christy [20], we set $\varepsilon_2^{iso} = \varepsilon_{xx}^{ani}$ and plot it for different values of ε_{zz}^{ani}. The results are shown in Fig. 4.2 for a typical wavelength of 600 nm ($\varepsilon_1 = -9.43 + 1.50i$).

Inspecting the figures, we clearly observe that both anisotropic sensitivities, $\partial k_x^{ani}/\partial \varepsilon_{xx}^{ani}$ and $\partial k_x^{ani}/\partial \varepsilon_{zz}^{ani}$ can be tuned in order to achieve resonances for a different ε_{xx}^{ani} range. Additionally, when $\varepsilon_{zz}^{ani} < \varepsilon_{xx}^{ani}$, the sensitivity with respect to ε_{zz}^{ani} has greater values than the one with respect to ε_{xx}^{ani} (Fig. 4.2a) while the opposite occurs when $\varepsilon_{zz}^{ani} > \varepsilon_{xx}^{ani}$ (Fig. 4.2d). The sensitivity can therefore be altered by using the appropriate set of permittivity components along different directions. The characteristic form of all the sensitivities is due to the surface plasmon resonance. Close to this resonance, there is a drastic increase of the slope, which explains the positive sensitivity peak, and then it sharply decreases for even higher permittivity values, affirming the negative slope. Lastly, the anisotropic sensitivity along a certain permittivity axis can be tuned via the appropriate permittivity components. For instance, Fig. 4.2d shows that the sensitivity with respect to ε_{xx}^{ani} is superior to the one with respect to

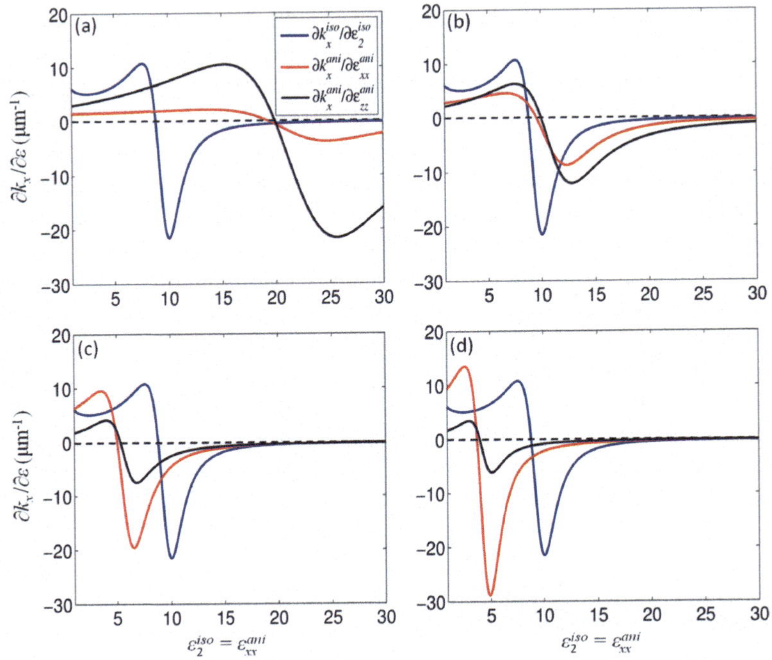

Fig. 4.2 The sensitivity of SPP dispersion for a semi-infinite metallodielectric pair. The metal considered is Au with properties taken from reference [20], while the dielectric is either isotropic or anisotropic. Figures (**a**) to (**d**) depict the dispersion sensitivity for a broad range of $\varepsilon_2^{\mathrm{iso}} = \varepsilon_{\mathrm{xx}}^{\mathrm{ani}}$ when $\varepsilon_{\mathrm{zz}}^{\mathrm{ani}}$ is 4, 8, 15 and 20, respectively. Note how the anisotropic sensitivity can be tuned along the permittivity plane, in contrast to the isotropic one, which remains static—since it is independent of $\varepsilon_{\mathrm{zz}}^{\mathrm{ani}}$. For all of the graphs, the wavelength was 600 nm, giving $\varepsilon_1 = -9.43 + 1.50i$

$\varepsilon_{\mathrm{zz}}^{\mathrm{ani}}$. This behaviour can be reversed by having $\varepsilon_{\mathrm{xx}}^{\mathrm{ani}} = 4$ and letting $\varepsilon_{\mathrm{zz}}^{\mathrm{ani}}$ vary (not shown). The main outcome is that the anisotropy that is inserted through the tensorial form of the permittivity can create various sensing dispersions—and at the same time control the position where the sensing is maximized through careful inspection of permittivity components at a given wavelength. Conversely, the isotropic sensitivity can only create a single sensitivity curve and remains static for variations of $\varepsilon_{\mathrm{zz}}^{\mathrm{ani}}$, since it has no directionality.

To investigate sensitivity in a more thorough way, we can examine the dispersion along the anisotropic permittivity plane. Keeping the same configuration as before, Fig. 4.3 demonstrates both sensitivity dispersions, $\partial k_x^{\mathrm{ani}}/\partial \varepsilon_{\mathrm{xx}}^{\mathrm{ani}}$ and $\partial k_x^{\mathrm{ani}}/\partial \varepsilon_{\mathrm{zz}}^{\mathrm{ani}}$, for two typical wavelengths at 600 and 700 nm. A universal trend that is valid in all figures is the sign flip of the sensitivity close to the resonance (red and blue curves), which has already been discussed and is connected with the surface plasmon resonance. The characteristic bending is due to the change of the resonance along the permittivity plane. Comparing Fig. 4.3a, b with 4.3c, d, we observe that, as the wavelength increases, both anisotropic sensitivities need a higher set of $\left\{\varepsilon_{\mathrm{xx}}^{\mathrm{ani}}, \varepsilon_{\mathrm{zz}}^{\mathrm{ani}}\right\}$ to satisfy the resonance condition. This means that sensing can be tuned either through wavelength

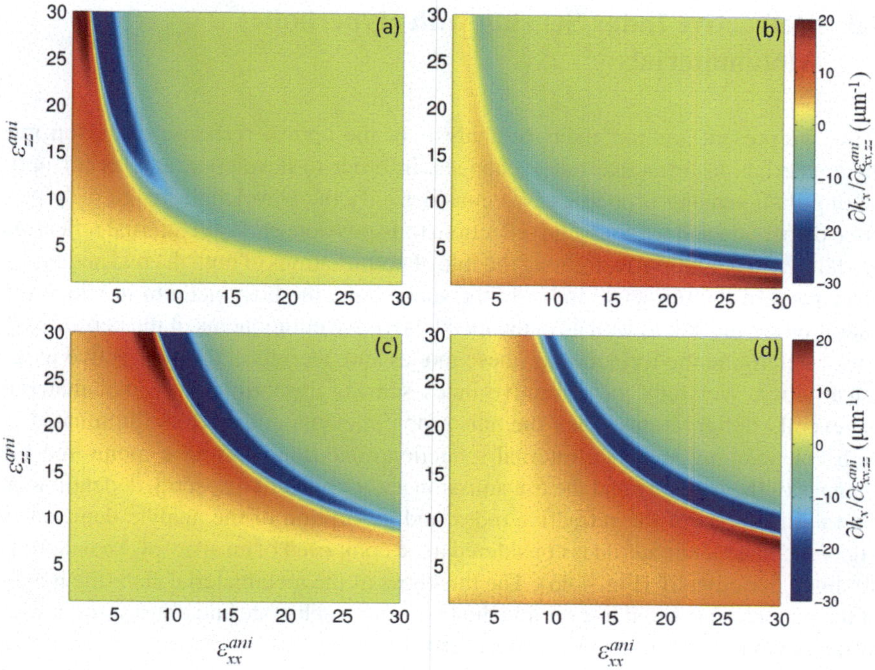

Fig. 4.3 Anisotropic sensitivity dispersion with respect to $\varepsilon_{\rm xx}^{\rm ani}$ (**a**) and $\varepsilon_{\rm zz}^{\rm ani}$ (**b**) for a broad range of permittivities at a wavelength of 600 nm. **c** and **d**: as before but at a wavelength of 700 nm. Note the increase in the sensitivity close to the surface plasmon resonance and its shift to higher ($\varepsilon_{\rm xx}^{\rm ani}$, $\varepsilon_{\rm zz}^{\rm ani}$) as the wavelength increases

or permittivity optimization. Although it is always easier to change the wavelength, there are situations where the frequency range is limited (for example applications that function only in the visible spectrum) and therefore permittivity optimization is the only approach. Here anisotropy has a crucial role since it allows us to change permittivity along different directions, making it more flexible in comparison with the isotropic counterpart. Lastly, the two anisotropic sensitivities exhibit complementary behavior close to the resonance. This means that, for a given wavelength, if we need to achieve high sensitivity with respect to $\varepsilon_{\rm xx}^{\rm ani}$ then $\varepsilon_{\rm zz}^{\rm ani}$ should be increased while $\varepsilon_{\rm xx}^{\rm ani}$ decreases—and vice versa.

So, based on some simple paradigms, we have shown how anisotropy allows a more advanced manipulation of the SPP dispersion and at the same time provides a more flexible sensing configuration with respect to the isotropic scenario. However, the anisotropic permittivities were varied without any specific limitation, which is of course not possible in the real world. Consequently, in the following sections we investigate the refractive index sensitivity of metallic nanorod hyperbolic metamaterials and their impact in various applications.

4.3 Refractive Index Sensing with Hyperbolic Metamaterials

Here, we present a comprehensive analysis of the optical response of anisotropic metamaterials to refractive index changes, in order to develop strategies for optimizing their sensing properties. We consider only the wavelength range of hyperbolic dispersion and we study the reflection, transmission, and total internal reflection modalities of operation, examining the role of perturbations of both the real and imaginary parts of the refractive index in the superstrate, the host medium in which the nanorods are placed, as well as in the metal itself. All components of the permittivity tensor determine the formation of these modes and, therefore, their refractive index sensitivities. We study the refractive index sensing capabilities of a metamaterial in several configurations when the anisotropic metamaterial slab is illuminated in both conventional and total internal reflection conditions and have monitored the changes of the intensity and the transmission and/or reflection spectra, depending on the modifications of the refractive index and absorption of the analyte depicted in Fig. 4.4a. The metamaterial is considered to be composed of an array of Au nanorods arranged periodically (Fig. 4.4a). The thickness of the metamaterial slab (the height of the nanorods) is l, and it is sandwiched between a substrate and a superstrate, with refractive indices of n_{sub} and n_{sup}, respectively.

The Maxwell Garnett (MG) approximation was followed to derive the tensor of the effective permittivity of the anisotropic metamaterial [24]. Considering an array of rods in the xy plane, the effective permittivities for ordinary and extraordinary axes take the form:

$$\varepsilon_{x,y}^{\mathrm{eff}} = \varepsilon_h \frac{(1+p)\varepsilon_{\mathrm{Au}} + (1-p)\varepsilon_h}{(1-p)\varepsilon_{\mathrm{Au}} + (1+p)\varepsilon_h} \tag{4.5}$$

$$\varepsilon_z^{\mathrm{eff}} = p\varepsilon_{\mathrm{Au}} + (1-p)\varepsilon_h \tag{4.6}$$

where $p = \pi(r/d)^2$ defines the nanorod filling factor, with d being the period of the array and r is the nanorod radius, ε_{Au} and ε_h are the permittivities of Au [20] and host medium, respectively. Note that the same period is considered in both x and y directions—and therefore $\varepsilon_x^{\mathrm{eff}} = \varepsilon_y^{\mathrm{eff}}$. The MG approximation breaks down for wavevectors close to the Brillouin zone boundary. However, for the typical period of 100 nm considered here, the first Brillouin zone boundary is close to $k_x \approx 31\ \mu\mathrm{m}^{-1}$, which is far from the regime investigated $\left(k_x < 17\ \mu\mathrm{m}^{-1}\right)$. Additionally, the wavelength range where $\mathrm{Re}\left(\varepsilon_z^{\mathrm{eff}}\right)$ vanishes (the ENZ regime) requires special considerations—and the ENZ and elliptic dispersion regimes will not be considered in the present work [25, 26].

In order to examine the refractive index sensitivity of the optical properties of the metamaterial, the effective permittivity has been designed to achieve hyperbolic dispersion throughout the visible spectral range. Figure 4.4b shows the real and imaginary parts of the components of the effective permittivity tensor for typical

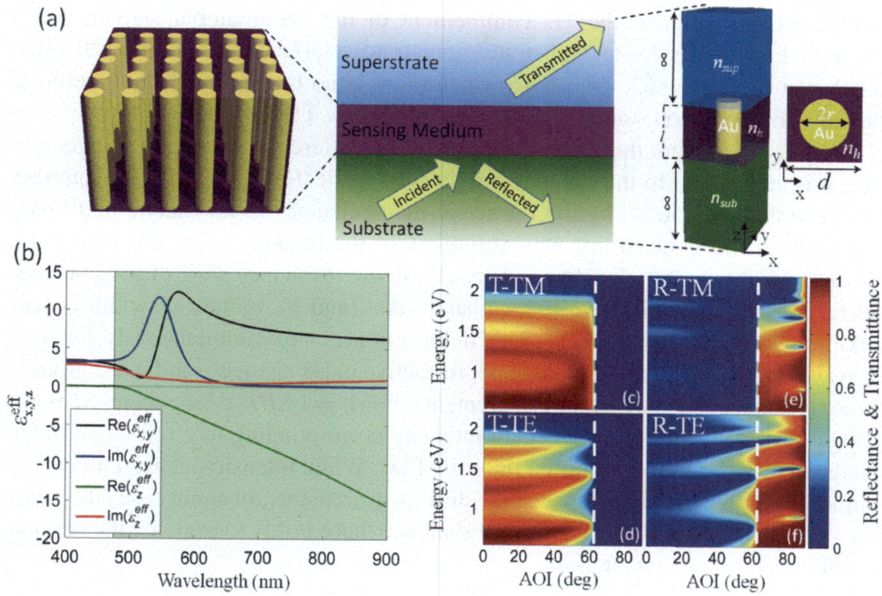

Fig. 4.4 **a** Left: Schematic of the metamaterial transducer made of an array of Au nanorods embedded in a host environment (analyte); Middle: Schematic of the typical experimental realization of refractive index sensing experiments in the reflection or transmission geometry; Right: Schematics of the unit cell of the metamaterial. **b** Effective permittivities of the metamaterial in a water-like analyte ($n_h = 1.33$) with the nanorod period $d = 100$ nm and radius $r = 40$ nm. The green area shows the hyperbolic dispersion regime where $\varepsilon_z < 0$. **c–f** Transmittance and reflectance dispersions for **c**, **e** TM- and **d**, **f** TE-polarized light. Geometry is the same as in (**b**). The substrate is glass ($n_{\text{sub}} = 1.5$) and superstrate is water ($n_{\text{sup}} = 1.33$). The height of the nanorods is $l = 400$ nm. In all of the dispersions, the TIR occurs at an angle of incidence of 62.46°, as indicated by the dashed line

nanorod array parameters and a water-like permittivity for the host medium. For this geometry, $\varepsilon_z^{\text{eff}}$, the permittivity component for light polarized along the nanorod axis, is negative for wavelengths longer than approximately 475 nm (Fig. 4.4b), while the transverse permittivity $\varepsilon_{x,y}^{\text{eff}}$, for light polarized perpendicular to the nanorod axes, is always positive—exhibiting resonant behavior near a wavelength of 540 nm, where the imaginary part of the permittivity reaches a maximum. For wavelengths above 540 nm, the transverse component of the imaginary part of the permittivity is small relative to its real counterpart and does not exhibit any resonances.

Figure 4.4c–f show the transmittance and reflectance dispersions for both TM and TE-polarizations, respectively [27]. Unbound modes that leak into both the substrate and the superstrate exist below the critical angle—and θ_c appear as maxima (minima) in the transmittance (reflectance)—while 'leaky' modes, homogeneous in the substrate, are only present above θ_c and their effect is most pronounced in the reflectance of the structure. These modes are quantized solutions of the wavevector (k_z), due to the 1D confinement of the metamaterial slab in the z-direction. For the glass/water (substrate/superstrate) interface, total internal reflection (TIR) occurs at $\theta_c = 62.46°$, above which there is no transmission, and therefore transmission-detection sensing is possible only below TIR angles.

In order to compare the sensing capabilities of different transducers and geometries, it is convenient to introduce two figures of merit (FoMs). FoM_λ characterizes the spectral shift induced by the refractive index changes of an analyte and FoM_I characterizes the induced intensity variations of the transmitted or reflected light. We define $\text{FoM}_\lambda = (\Delta\lambda/\delta\lambda)/\Delta n$, where $\Delta\lambda$ is the resonance shift of a metamaterial resonance for a refractive index change Δn—and $\delta\lambda$ is the full-width at half maximum of the resonance [28]. This definition accounts simultaneously for both the wavelength shift of a given mode per refractive index change—and the sharpness of the resonance. For intensity measurements, $\text{FoM}_I = (\Delta I/I_0)/\Delta n$, where ΔI is the change in the transmitted or reflected intensity corresponding to a refractive index change of Δn—and I_0 is the initial intensity [28]. While intensity measurements are simpler to implement than the spectral shift measurements, in some cases the latter may provide better sensitivity [17]. We shall use these FoMs to evaluate the sensing capabilities of the metamaterial.

4.3.1 *Effective Permittivity Sensitivity to Geometry and Refractive Index Variations*

We begin by examining the sensitivity of the effective permittivities $\varepsilon_{x,y}^{\text{eff}}$ and $\varepsilon_z^{\text{eff}}$, which determine the optical response of the metamaterial, on the refractive index changes of the analyte $n_h = \sqrt{\varepsilon_h' + i\varepsilon_h''}$,—which is considered to be a host medium where nanorod arrays are placed. To evaluate this sensitivity, we study the partial derivative of (4.5) and (4.6) with regard to the permittivity components of each constituent, obtaining:

$$\frac{\partial \varepsilon_{x,y}^{\text{eff}}}{\partial \varepsilon_h'} = -i\frac{\partial \varepsilon_{x,y}^{\text{eff}}}{\partial \varepsilon_h''} = \frac{P\varepsilon_{Au} + 2\varepsilon_h}{P\varepsilon_h + \varepsilon_{Au}} - P\varepsilon_h \frac{P\varepsilon_{Au} + \varepsilon_h}{(P\varepsilon_h + \varepsilon_{Au})^2} \tag{4.7}$$

$$\frac{\partial \varepsilon_z^{\text{eff}}}{\partial \varepsilon_h'} = -i\frac{\partial \varepsilon_z^{\text{eff}}}{\partial \varepsilon_h''} = 1 - p \tag{4.8}$$

where $1 < P = (1 + p) = (1 - p) \leq 8.32$. The above expressions allow us to track selectively the response of the effective permittivities when either the real or the imaginary part of the permittivities of the constituents changes. Figure 4.5 illustrates this behavior for the same metamaterial as in Fig. 4.4b. The response of $\varepsilon_z^{\text{eff}}$ to the refractive index modifications is extremely broadband and constant across the spectral range considered, for both the real and imaginary parts (Fig. 4.5a, b), and is determined solely by the nanorod filling factor via p (4.8). On the other hand,

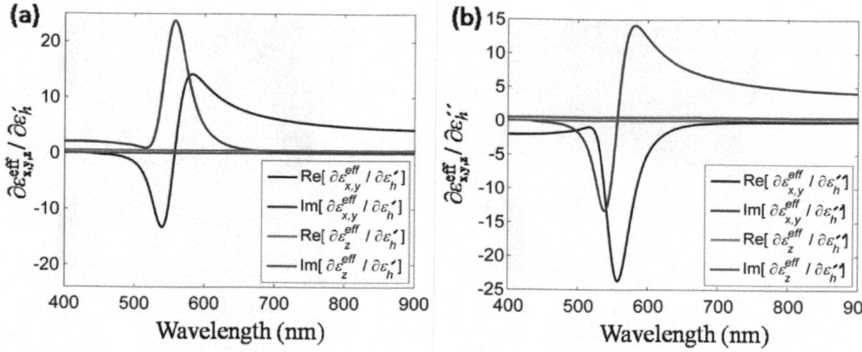

Fig. 4.5 The spectral dependence of the derivative of the effective permittivity with respect to (**a**) ε_h', and (**b**) ε_h'' modifications, for a filling factor of $p = 0.5$

$\varepsilon_{x,y}^{\text{eff}}$ exhibits the strongest changes close to its resonance at 540 nm (Fig. 4.4b) and strongly decreases for lower nanorod filling factors. Beyond the resonance at longer wavelengths, inspecting the dependence of $\varepsilon_{x,y}^{\text{eff}}$ on the host medium refractive index, one notices that it always exceeds the sensitivity of $\varepsilon_z^{\text{eff}}$ (Fig. 4.5a, b), while both are similar at low nanorod filling factors.

To examine the impact of the filling factor (p) more extensively, the dispersion diagrams of the derivative of $\varepsilon_{x,y}^{\text{eff}}$ with respect to ε_h' and ε_h'' are presented in Fig. 4.6. All the refractive index sensitivity dispersions are dominated by a dispersive-shape resonance, which red-shifts and increases in magnitude with increasing nanorod filling factor. Both the dispersive behavior and the spectral shift observed in Fig. 4.6 can be linked to the spectral sensitivity of the resonance in $\varepsilon_{x,y}^{\text{eff}}$ shown in Fig. 4.4b, which red-shifts with both an increase in ε_h and the nanorod filling factor. It is important to note that Fig. 4.6a, c are exactly the same; this is a direct consequence of the first equality in (4.7).

4.3.2 *Mode Frequency Dependence on the Refractive Index of the Analyte*

All the components of the effective permittivity tensor are important for defining the behavior of the TM modes of the metamaterial and, therefore, their sensing capabilities. At the same time, only the $\varepsilon_{x,y}^{\text{eff}}$ components define the TE modes. To obtain better insight into the sensing properties of the unbounded, leaky, and waveguided modes of the metamaterial transducer, we consider the same analytic model for the case of an anisotropic metamaterial layer as in [15]. Neglecting phase shifts at the boundaries of the metamaterial, we obtain a simple analytical approximation for the wavevector component k_x of the TE and TM modes supported by the metamaterial slab as:

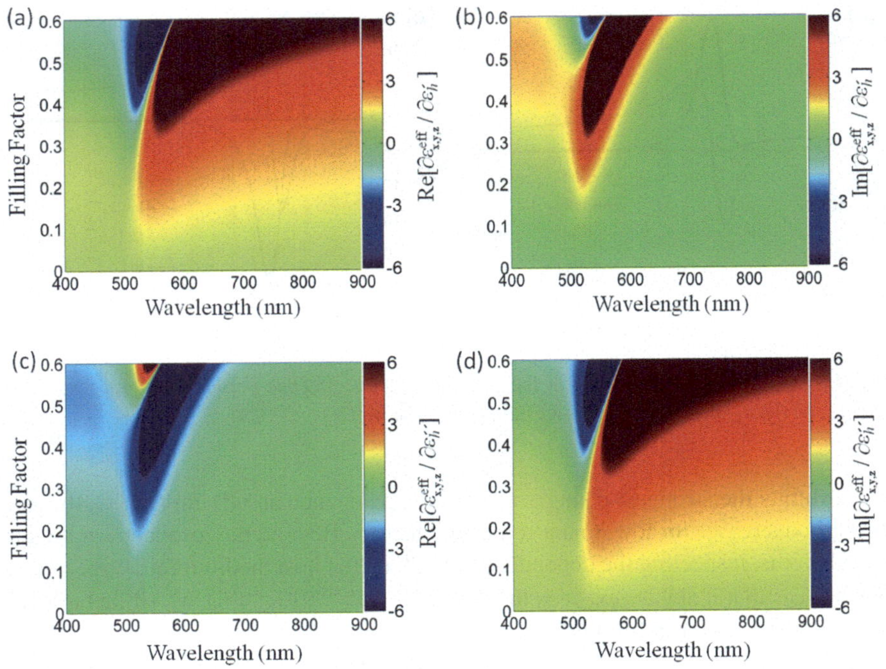

Fig. 4.6 Real (**a**), (**c**) and imaginary (**b**), (**d**) derivatives of the effective permittivity maps with respect to (**a**), (**b**) $\varepsilon_h^{'}$ and (**c**), (**d**) $\varepsilon_h^{''}$. Notice that (**a**) and (**d**) are the same, which is a consequence of (4.7)

$$\left(k_x^{\mathrm{TE}}\right)^2 = \varepsilon_{x,y}^{eff} k_0^2 - \left(\frac{q\pi}{l}\right)^2 \tag{4.9}$$

$$\left(k_x^{\mathrm{TM}}\right)^2 = \varepsilon_z^{\mathrm{eff}} k_0^2 - \left(\frac{q\pi}{l}\right)^2 \frac{\varepsilon_z^{\mathrm{eff}}}{\varepsilon_{x,y}^{\mathrm{eff}}} \tag{4.10}$$

The quantization $q\pi/l$ emerges from the transversal confinement of the wavevector component perpendicular to the interfaces of the metamaterial, with the integer $q > 0$ referring to the mode number. Unbounded, leaky, and waveguide modes then satisfy the condition $k_x^{\mathrm{TE,TM}} < n_{\mathrm{sup}} k_0 n_{\mathrm{sup}} k_0 \leq k_x^{\mathrm{TE,TM}} < n_{\mathrm{sub}} k_0$ and $k_x^{\mathrm{TE,TM}} \geq n_{\mathrm{sub}} k_0$ respectively, with the mode frequency:

$$\omega_q^{\mathrm{TE}} = \frac{c_0}{\sqrt{\varepsilon_{x,y}^{\mathrm{eff}}}} \sqrt{\left(k_x^{\mathrm{TE}}\right)^2 + \left(\frac{q\pi}{l}\right)^2} \tag{4.11}$$

$$\omega_q^{\mathrm{TM}} = c_0 \sqrt{\frac{\left(k_x^{\mathrm{TM}}\right)^2}{\varepsilon_z^{\mathrm{eff}}} + \left(\frac{q\pi}{l}\right)^2 \frac{1}{\varepsilon_{x,y}^{\mathrm{eff}}}} \tag{4.12}$$

where the number of supported modes is determined by the sign of the permittivity components. For higher-order modes, it may be necessary to take into account deviations from the EMT due to spatial dispersion effects [25]. The sensitivity of the mode frequency $\omega_q^{\mathrm{TE,TM}}$ to variations in ε_h, for a given q and $k_x^{\mathrm{TE,TM}}$, immediately follows as:

$$\frac{\partial\omega_q^{\mathrm{TE}}}{\partial\varepsilon_h^{'}} = -i\frac{\partial\omega_q^{\mathrm{TE}}}{\partial\varepsilon_h^{''}} = -\frac{\omega_q^{\mathrm{TE}}}{2\varepsilon_{x,y}^{\mathrm{eff}}}\frac{\partial\varepsilon_{x,y}^{\mathrm{eff}}}{\partial\varepsilon_h^{'}} = i\frac{\omega_q^{\mathrm{TE}}}{2\varepsilon_{x,y}^{\mathrm{eff}}}\frac{\partial\varepsilon_{x,y}^{\mathrm{eff}}}{\partial\varepsilon_h^{''}} \tag{4.13}$$

$$\begin{aligned}\frac{\partial\omega_q^{\mathrm{TM}}}{\partial\varepsilon_h^{'}} = -i\frac{\partial\omega_q^{\mathrm{TM}}}{\partial\varepsilon_h^{''}} &= -\frac{c_0^2}{2\omega_q^{\mathrm{TM}}}\left[\left(\frac{k_x^{\mathrm{TM}}}{\varepsilon_z^{\mathrm{eff}}}\right)^2\frac{\partial\varepsilon_z^{\mathrm{eff}}}{\partial\varepsilon_h^{'}} + \left(\frac{q\pi}{l\varepsilon_{x,y}^{\mathrm{eff}}}\right)^2\frac{\partial\varepsilon_{x,y}^{\mathrm{eff}}}{\partial\varepsilon_h^{'}}\right] \\ &= -i\frac{c_0^2}{2\omega_q^{\mathrm{TM}}}\left[\left(\frac{k_x^{\mathrm{TM}}}{\varepsilon_z^{\mathrm{eff}}}\right)^2\frac{\partial\varepsilon_z^{\mathrm{eff}}}{\partial\varepsilon_h^{''}} + \left(\frac{q\pi}{l\varepsilon_{x,y}^{\mathrm{eff}}}\right)^2\frac{\partial\varepsilon_{x,y}^{\mathrm{eff}}}{\partial\varepsilon_h^{''}}\right]\end{aligned} \tag{4.14}$$

Note the connection of the mode frequency with the real and imaginary parts of the permittivity—which is a direct consequence of (4.7), (4.8). Equations (4.13) and (4.14), together with equations (4.5) and (4.6), indicate the important interplay between the absolute values and variations in $\varepsilon_{x,y}^{\mathrm{eff}}$ and $\varepsilon_z^{\mathrm{eff}}$ for determining shifts in the mode frequencies, e.g. when one of the permittivity components becomes small. Their relative contributions also depend on the nanorod filling factor and wavelength.

The TM-mode position dependence with respect to the real part of the constituent permittivities is plotted in Fig. 4.7 within the 1st Brillouin zone, in order to ensure the validity of EMT. As the mode number increases from $q = 1$ to 5, there is an order of magnitude increase in the sensitivity of the modal frequency to variations in the refractive index of the host medium (Fig. 4.7a). The superior refractive index

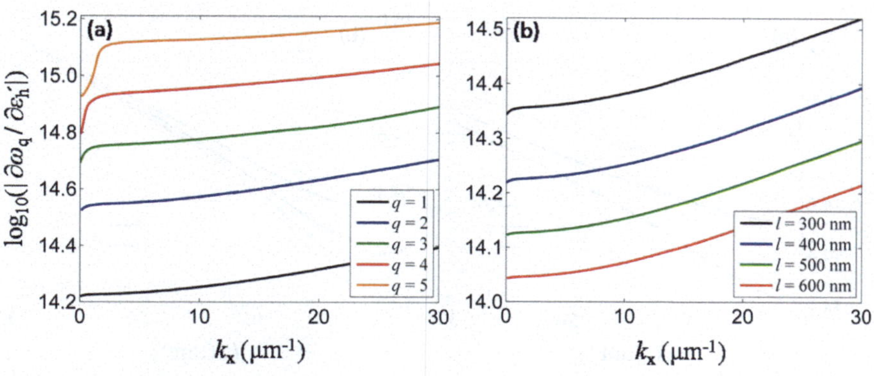

Fig. 4.7 The TM-mode frequency shift with $\varepsilon_h^{'}$ variation: **a** the first five modes ($q = 1$–5) for $l =$ 400 nm and **b** the fundamental mode ($q = 1$) for various transducer thicknesses. The geometry is the same as in Fig. 4.4b

sensitivity of high-order modes is a consequence of their spectral position close to the resonance in $\varepsilon_{x,y}^{\mathrm{eff}}$ (Fig. 4.4)—and the increased field gradients inside the metamaterial for higher-order modes. These gradients are determined by the spatial frequency of the mode $q\pi/l$ increasing with increasing q value or decreased sensor thickness l. Interestingly, this behavior leads to an increase in the modal sensitivity as the mode shifts to shorter wavelengths, a trend that is opposite to that observed with conventional SPR or LSPR transducers. In particular, shifting the resonance frequency of the fundamental mode ($q = 1$) from 0.5 to 1.5 eV, by decreasing the metamaterial layer thickness from 500 nm to 130 nm, results in a 500% increase in sensitivity as monitored in both transmittance and reflectance at normal incidence. An increase in the nanorod filling factor p, all other parameters being kept constant, leads to a decrease in the sensitivity of the modal position. Again, this behavior is reminiscent of the increased modal delocalization within the metamaterial with decreasing frequency.

The modal frequency sensitivity with respect to ε_h'' was also examined, giving a similar trend to Fig. 4.7a, but with a one-to-two orders of magnitude smaller sensitivity to the variations in ε_h'. This behavior indicates that the real parts of the refractive index of the host medium will affect sensing more than their imaginary counterparts. Monitoring the refractive index changes in the analyte at normal incidence provides the lowest sensitivity with regards to the spectral shift of the mode (Fig. 4.7).

The same analysis, but in the situation where the TE-mode frequency varies, is presented in Fig. 4.8. The modal frequency range in Fig. 4.8a is approximately the same as in the TM-case (Fig. 4.7a), meaning that the sensing capabilities between these two polarizations are comparable $\left(\partial\omega_q^{\mathrm{TM}}/\partial\varepsilon_h' \sim \partial\omega_q^{\mathrm{TE}}/\partial\varepsilon_h'\right)$. Only the fundamental mode ($q = 1$) has a different behavior, with a modal frequency variation that is distinctively smaller at low wavevectors—which is due to the absence of a cut-off frequency for the fundamental mode in TE-polarization. It is important to note that the TE-mode frequency variation is increasing faster at higher wave-vectors,

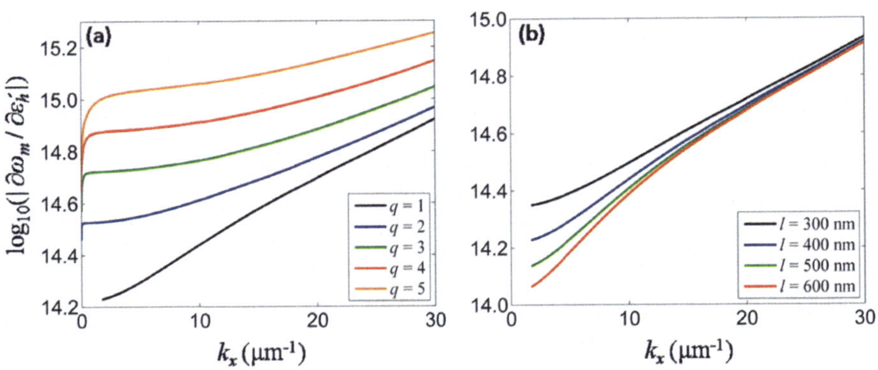

Fig. 4.8 The TE-mode frequency shift with ε_h' variation: **a** the first five modes ($q = 1$–5) for $l = 400$ nm and **b** the fundamental mode ($q = 1$) for various transducer thicknesses. The geometry is the same as in Fig. 4.4b

compared with its TM counterpart—and eventually all the modes converge at very high wave-vectors. This phenomenon can be understood as being due to the positive group velocity of the TE waveguide modes, as explained in [15]. Similarly to the TM scenario, the modal frequency sensitivity with respect to ε_h'' was also examined, giving one-to-two orders of magnitude smaller sensitivity in comparison with ε_h'.

4.3.3 *Sensitivity Dependence on the Thickness of the Metamaterial Transducer*

As a next step, the impact of the thickness of the metamaterial transducer, l, was examined. Figure 4.7b shows the $q = 1$ mode frequency sensitivity with respect to ε_h', as l varies from 300 to 600 nm. In both cases, as the waveguide thickness increases, the sensitivity drops, in agreement with the modal position shift to lower frequencies. A similar behavior is observed for higher-order modes. This behavior can be clarified more explicitly by examining the normal incidence behavior of the modal resonance frequency, simplifying (4.14) to $\partial\omega_q^{\mathrm{TM}}/\partial\varepsilon_h' = -[c_0\pi(\varepsilon_{x,y}^{\mathrm{eff}})^{-3/2}/2l]\partial\varepsilon_{x,y}^{\mathrm{eff}}/\partial\varepsilon_h'$ and clearly showing the inverse proportionality of the sensitivity of the modal resonance to the metamaterial thickness l for unbound modes. Therefore, for a given mode number, a thinner transducer will be more sensitive. This result has been verified using numerical simulations as long as the mode considered had a frequency that did not exceed the frequency corresponding to a free-space wavelength of 650 nm, or beyond the range of high losses due to $\mathrm{Im}(\varepsilon_{x,y}^{\mathrm{eff}})$ (Fig. 4.4b).

In the case of TE-modes (Fig. 4.8b), the modal frequency variation has comparable values, in comparison with the TM scenario, for low wavevectors $(k_x^{\mathrm{TE,TM}} < 15\,\mu\mathrm{m}^{-1})$—but becomes significantly higher as the wave-vector increases. A striking difference is the minor dependence of the TE-mode frequency change as l increases from 300 to 600 nm (Fig. 4.8b).

4.3.4 *Sensing Variations of the Real Part of the Refractive Index*

We start with the non-absorbing case, corresponding to the situation of an analyte consisting of non-resonant molecules. In this situation, changes in the refractive index may be induced through the Kerr nonlinearities of the host medium or superstrate, electro-optical, thermo-optical or pressure effects. Figure 4.9a–c show the reflectance sensitivity (FoM_I) for different wavelengths and angles of incidence for TM-polarization, when the changes in the refractive index originate from either the superstrate $(\Delta n_{\mathrm{sup}})$, the host medium (Δn_h)—or both simultaneously (Δn_b). The FoM_I has a dispersive behavior, changing sign in the vicinity of the modal resonances (Fig. 4.9b, c). This behavior is observed as a result of a simultaneous change in both

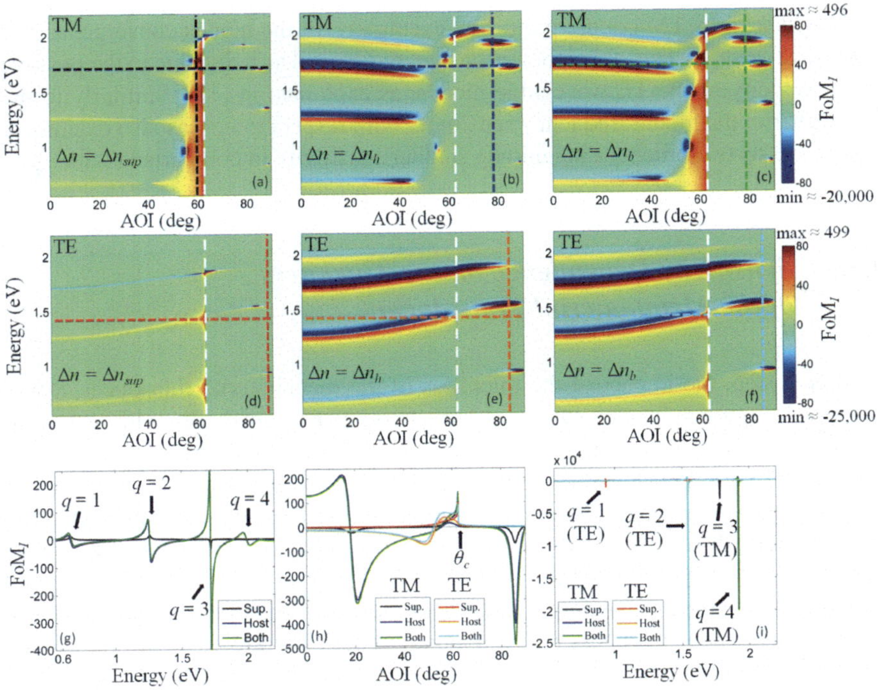

Fig. 4.9 Spectral and angular dependences of the intensity figure of merit (FoM_I) for the changes of the refractive index ($\Delta n = 0.002$) of the superstrate (**a**), (**d**), host medium (**b**), (**e**) and both (**c**), (**f**)—for **a–c** TM-polarization and **d–f** TE-polarization. The superstrate light line is also shown (white dashed line). The color scale is the same, in order to enable comparison. **g** Cross-sections of (**a**)–(**f**) at normal incidence. **h** Cross-sections of (**a**)–(**f**) tracking the $q = 3$ mode (horizontal dashed lines) at 1.7 eV for TM-polarization and $q = 2$ mode at 1.4 eV for TE-polarization. **i** Cross-sections of (**a**)–(**f**) at angles where the highest sensitivity is observed (vertical dashed lines). All images were acquired from use of the EMT approximation

the intensity (ΔI) and wavelength ($\Delta\lambda$) of the metamaterial modes. However, when ΔI has a dominant role, the FoM_I retains its sign (for example, in Fig. 4.9a for the angular range between 0 and 50°). From Fig. 4.9a, it is clear that the strongest variations of the optical response are observed in the vicinity of the modes of the metamaterial transducer, but with a maximum sensitivity that is near the critical angle θ_c. The reason for this is that when the refractive index of the superstrate is changed by Δn_{sup} the critical angle itself will be affected—leading to a strong sensitivity along the superstrate light line, the so-called near cut-off regime. This behavior was also captured with 3D numerical analysis and showed an order of magnitude higher sensing close to θ_c than at lower angles.

Significantly different behavior is observed for the variations of the refractive index of the host medium Δn_h (Fig. 4.9b). In this case, the sensitivity of the leaky modes is higher than that of the unbound modes. When both the refractive index of the host medium and the superstrate change, a combination of the individual cases above is observed, although not just as a simple addition of the individual sensitivities, since modes from the metamaterial layer are penetrating into the superstrate medium (Fig. 4.9c). Not surprisingly, the results show that a small change in the refractive index of the superstrate affects the sensitivity only at angles very close to θ_c, while for any other incidence angle the host medium has a dominant role. Both TM (Fig. 4.9a–c) and TE (Fig. 4.9d–f) modes of the metamaterial transducer have comparable sensitivities if the analyte is incorporated between the rods. Fullwave simulations using the commercial software COMSOL have confirmed this trend, but with TE modes exhibiting slightly lower sensitivity. (The figure is not shown). This is not surprising, given the nature of the anisotropic waveguide modes determined by all components of the effective permittivity tensor. For low filling factors, the role of the TM modes increases, when the $\varepsilon_{x,y}^{\text{eff}}$ sensitivity becomes smaller (not shown).

Figure 4.9g–i show the sensitivity for selected wavelengths and angles of incidence. The spectral dependence of FoM_I at normal incidence (Fig. 4.9g) reveals that the mode with the highest sensitivity ($q = 3$) reaches an FoM_I value that exceeds 400, with the host medium being mainly responsible for this enhanced sensitivity. As the mode number increases from $q = 1$ to 3, the FoM_I also increases—in agreement with analytical examination (Fig. 4.7a). However, the $q = 4$ mode, located close to 2 eV, shows a decreased sensing capability that is contrary to the analytical scenario—which is due to the increased absorption of $\varepsilon_{x,y}^{\text{eff}}$ (Fig. 4.4b) at wavelengths below 650 nm (>1.9 eV), which are not tracked with the analytical model discussed in Sect. 4.4. Figure 4.9h examines the angular dependence of the sensitivity for the modes with the highest sensitivities: $q = 3$ at 1.7 eV for TM-polarization and $q = 2$ at 1.4 eV for TE-polarization. These plots show a typical dispersive behavior of the modal position, with high sensitivities observed on both sides of the resonance. As expected, the highest FoM_I value is observed above θ_c for leaky modes. This behavior is better observed in Fig. 4.9i, which depicts the global maximum FoM_I value for TM-polarization, for the $q = 4$ mode located at a wavelength of about 645 nm and for an angle of incidence of about 79°, above the critical angle. Similarly for the TE-case, the highest FoM_I value is observed for the $q = 2$ mode for a wavelength close to 870 nm and at 84°. Only variations in the superstrate refractive index lead to high sensitivity—which lies just below θ_c, at 58.8°, for reasons already explained above. In the following sections, we shall concentrate only on sensing based on the TM modes—pointing out that, in the case of variation, sensing based on the TE-modes also has comparable but slightly lower FoM_I.

Regardless of the polarization chosen, the $q = 1$ mode exhibits the highest resonance shift—reaching almost 4,000 nm/RIU, while the highest $\text{FoM}_\lambda \approx 167$ is observed for the $q = 3$ mode, outperforming both SPR and LSPR sensors which have typical FoM_λ values around 23 [2] and 8 [2, 29], respectively. The origin of such a high FoM_λ value is in the influence of the analyte—not only directly on the electromagnetic mode properties, as is usual for conventional sensors, but also on the effective permittivity of the metamaterial, which is connected to the host medium refractive index (Figs. 4.5 and 4.6), on the microscopic level, which is the result of plasmon-plasmon interactions within the nanorod array [18, 30, 31]. The intensity figure of merit (FoM_I) is proportional to $\left(\Delta R / R_0\right) / \Delta n$ and can be directly observed in Fig. 4.9. Depending on the wavelength and AOI, FoM_I can be as high as 25,000, significantly higher that any LSP- and SPP-based [28] sensors (Fig. 4.9i). Both FoM_I and FoM_λ can be further tailored and enhanced via the geometrical parameters of the transducer, with FoM_λ reaching as high as 300 being demonstrated experimentally [16].

4.3.5 Sensing via Variations in the Imaginary Part of the Refractive Index

Turning to the impact of variations in the absorption on the metamaterial transducer response, Fig. 4.10a–c show the reflectance variations for different wavelengths and angles of incidence, for TM-polarized incident light, with changes in the imaginary part of the refractive index of either the superstrate (Δn_{sup}), the host medium (Δn_h), or both simultaneously (Δn_b). As was the case with the changes in the refractive index (Fig. 4.9a–c), the sensitivity of the transducer originates from the modal dispersion of the metamaterial slab. Interestingly, an increase in the absorption leads to an increased reflection for some angles, while for other angles the behavior is opposite. Again, increased sensitivity is observed for higher order modes—going from $q = 1$ to 3, with decreased sensitivity for the $q = 4$ mode due to losses (Fig. 4.10d).

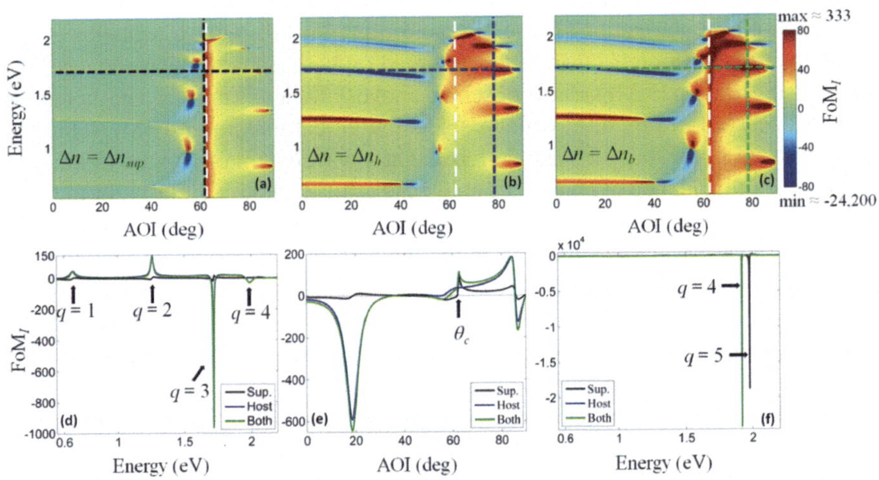

Fig. 4.10 Spectral and angular dependencies of the intensity figure of merit (FoM_I) for the changes of the absorption ($\Delta n = 0.002i$) of the superstrate: (**a**), host medium (**b**) and both (**c**), for TM-polarization. The superstrate light line is also shown (white dashed line). The color scale is the same, in order to enable comparison. **d** Cross-sections of (**a**)–(**c**) at normal incidence. **e** Cross-sections of (**a**)–(**c**) tracking the $q = 3$ mode (horizontal dashed lines) at 1.7 eV. **f** Cross-sections of (**a**)–(**c**) at angles where the highest sensitivity is observed (vertical dashed lines). All images were acquired from use of the EMT approximation

In a similar manner to that for the nonabsorbing case, the mode with the highest FoM_I value ($q = 3$) depicted in Fig. 4.10d showed a resonance shift of more than 700 nm/RIU, while the $q = 1$ mode exhibits the highest resonant shift—reaching values close to 4,000 nm/RIU. Tracking the $q = 3$ mode at various AOI (Fig. 4.10e), its sensitivity is enhanced around angles of both 20° and 85°, where the mode is leaky. The FoM_λ of the $q = 3$ mode in Fig. 4.10d is close to 167, similar to the non-absorbing analytes discussed in Sect. 4.5.1. However, its FoM_I reaches a value of almost 1,000, or at least two times higher than the corresponding one in Fig. 4.9g for the same mode ($q = 3$), since the influence of the imaginary part of the permittivity on reflection is stronger than the real part. The highest FoM_I in Fig. 4.10(f) exceeds 24,000—which is, significantly, 25% higher than the TM- FoM_I for sensing based on non-absorbing analytes (Fig. 4.9i).

4.4 Examples of Metamaterial Sensors

Refractive index sensing is a basic functionality that is suitable for a wide range of sensing scenarios where the refractive index changes in one or another way - e.g. for either bio- or chemical sensing, where the presence of certain molecules modifies the refractive index or the absorption of an analyte [1–4]; for acoustic processes that lead to material compression and, therefore, related refractive index variations [19, 32]; or indeed refractive index variations that are introduced by external stimuli, such as nonlinear optical [5, 16, 33], electro-optical [34–36] and thermo-optical processes [37, 38]. Here, we present several examples of the applications of nanorod metamaterials for biosensing [23], ultrasound detection [19] and hydrogen-gas detection [17].

4.4.1 Label-Free Biosensing

The first experimental evaluation of the sensing characteristics of the plasmonic nanorod metamaterial transducer was carried out in a standard streptavidin-biotin affinity model [23]. Biotin is a prominent example of an analyte (a vitamin) with very small molecular weight (244 Da). The response of the system to this analyte can show the potential of the metamaterial for the detection of small compounds (drugs, vitamins, hormones, etc.), for which conventional SPR devices are not sufficiently sensitive. Because of the small size of the molecule, the detection of biotin by conventional SPR requires relatively high concentrations—of 100 μM or higher [39]. After several functionalization steps, including immobilization of a streptavidin-melamide complex as a receptor [23], biotin of different concentrations was injected into the flow cell (Fig. 4.11). The biotin-streptavidin binding led to an immediate increase in the signal up to its saturation, corresponding to a dynamic equilibrium of the binding process for the injected concentration of the biotin. Taking into account the noise

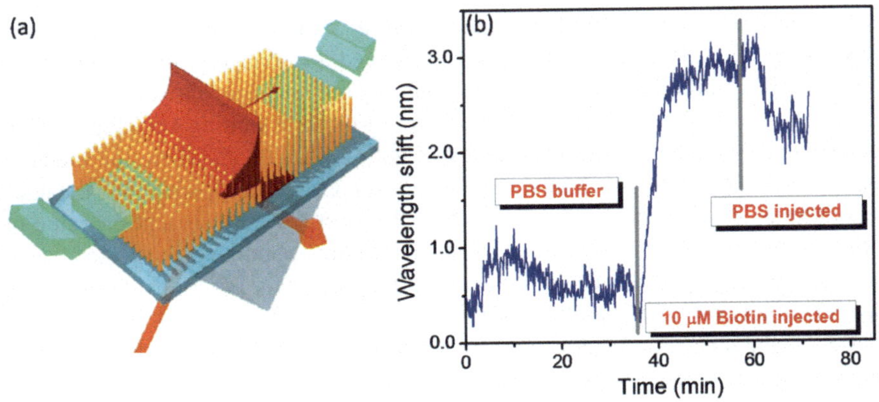

Fig. 4.11 **a** Schematic of the biosensing experiment. **b** Response of the metamaterial-based biosensor to the reaction of biotin-streptavidin binding recorded in the ATR geometry at a wavelength of 1230 nm. The nanorod array parameters are: length = 380 ± 20 nm, diameter = 25 ± 5 nm and spacing = 60 ± 10 nm between the rods (center-to-center). Adapted from [23]

level in the system after averaging over the 0.05 nm wavelength spectral-range, the detection limit of the metamaterial-based sensor to biotin can be estimated as being below 300 nm—which is more than 2 orders of magnitude better than for conventional SPR sensors that use continuous films. The detection limit for large analytes, such as streptavidin with a molecular weight of 60,000 Da, can be in the pM range. Such a gain in sensitivity is due to a combination of the higher sensitivity of metamaterial to change in the bulk refractive index and the mass accumulation effect that results from the much larger surface area of the nanoporous matrix. It is important that the sensing scheme makes it possible to follow the course of interactions in real time and to resolve all the kinetic constants. A routine ultra-sensitive determination of these constants is one of the major advantages of label-free plasmonics sensing technologies over fluorescent labeling methods.

The parameters of the waveguide mode used for these measurements lead to a sensitivity of 32,000 nm/RIU and an FoM of 330 [23], making this metamaterial an extremely sensitive transducer that significantly outperforms conventional SPP- and LSP- based biosensors and is therefore promising for improvement in the detection limit of low molecular weight analytes.

4.4.2 *Ultrasound Sensing*

The high sensitivity of plasmonic metamaterials has been used to develop a transducer for ultrasound based detection [19]. The metamaterial transducer exhibits a dramatic improvement in sensitivity to acoustic pressure compared with standard surface plasmon polariton or piezoelectric material based sensors. This ultra-sensitive response to the pressure induced changes of the refractive index is provided by a modification of the optical properties of the metamaterial that arise from a change in the interaction between the plasmonic nanorods—as well as by modification of the plasmonic resonances of the nanorod. A high sensitivity and broadband acoustic frequency response, together with non-resonant optical readout, makes this transducer an attractive alternative to existing methods for the detection of high-frequency ultrasound waves.

When subjected to the pressure of an ultrasound pulse, the reflection from the metamaterial slab changes in response to the periodic modification of the refractive index produced by the acoustic wave propagation through the metamaterial slab. The response curves for metamaterials with different parameters, and therefore different supported modes, show a similar linear signal-pressure dependence when used as a sensing channel (Fig. 4.12). Due to the non-resonant nature of the sensing mechanism, there are no stringent requirements on the wavelength of the probe light, but rather it should be within the resonance band ($\Delta\lambda \sim 100$ nm). For comparison, the transducer

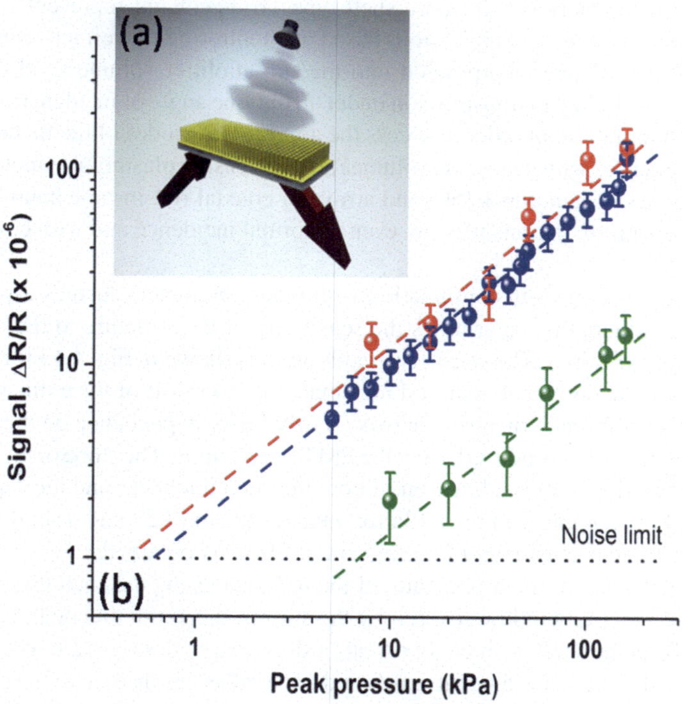

Fig. 4.12 **a** Schematic of the ultrasound detection experiment. **b** The signal dependence on the ultrasound power for two different metamaterial transducers (red and blue lines) and a reference SPR-based sensor (black). The metamaterial transducers operate in a non-resonant regime under illumination in the total internal reflection configuration, at an angle of incidence of 45°. The SPR transducer is based on a 50 nm thick Au film and operates in the resonant regime at the SPR angle (illuminated through a glass prism at an angle of incidence of 64°). The operating wavelength is 600 nm. Adapted from [19]

based on the resonant excitation of surface plasmon polaritons, while also exhibiting a linear dependence in the range of studied parameters, shows more than an order of magnitude lower sensitivity under the same conditions.

The upper detection limit of the transducer is determined by the noise in the detection channel, which corresponds to a sensitivity level, in the measured configuration, of less than 500 Pa—and is at least an order of magnitude better than the corresponding detection limit for SPR-based ultrasound sensors [40–44], as well as commercially available high-frequency ultrasound transducers based on piezoelements [45, 46]. Moreover, it is approaching the sensitivity of the state-of-the-art micro-ring resonator transducers [47, 48], without compromising the sensitivity at the higher frequencies for which microring-based sensors are not applicable.

4.4.3 Core-Shell Refractive Index Structures

Plasmonic metamaterials with core-shell-type arrangements represent a suitable nanostructured system for applications based on monitoring the refractive index (RI) changes of a substance incorporated into the sub-attolitter volume shell [49]. The transducer was studied in transmission under an oblique angle of incidence and under TM light illumination, in order to access the appropriate modes of the metamaterial. Similar geometries with nanoscale volumes around sensing plasmonic nanostructures based on arrays of nanotubes [50] and arrays of coaxial rod-in-tube nanostructures [51] allow operation in transmission, even at normal incidence and with comparable performance.

The shell thickness—in combination with other parameters such as aspect ratio, separation, and length—determines the sensitivity of the structure to the refractive index changes within it. The sensitivity calibration is shown in Fig. 4.13 for different parameters of the core-shell nanorod metamaterial. The shift of the extinction resonance for the different structures follows a fairly linear dependence on the shell RI. The behavior is well reproduced using the EMT simulations. The slope of these curves determines FoM_{λ}—and is a function of both the shell thickness and the aspect ratio of the rod. For example, in Fig. 4.13a for an aspect ratio of 27 and a shell thickness of 20 nm, a typical sensitivity of around 100 nm/RIU is obtained.

For a fixed value of the aspect ratio of the rod, increasing the shell thickness will increase the spectral window covered by the shift of the extinction peak for a given change of RI in the shell. A linear fit reveals a slope that scales as ~12 *s*, where *s* is the thickness of the shell, a value that was found to be rather insensitive to the rod aspect ratio. The significant bandwidth of the extinction resonance provides simplicity in the experimental implementation, the metamaterial considered in Fig. 4.13, and provides sensing of RI changes in the range from 1.0 to 1.6, with a sensitivity on the order of 2×10^{-2}, when using a wavelength within the 80 nm bandwidth. An improved FoM can be achieved using material with lower losses, albeit at the price of reduced bandwidth. The main advantage of the structure considered is however in its flexible geometry and the possibility of engineering the optical properties. In particular, it

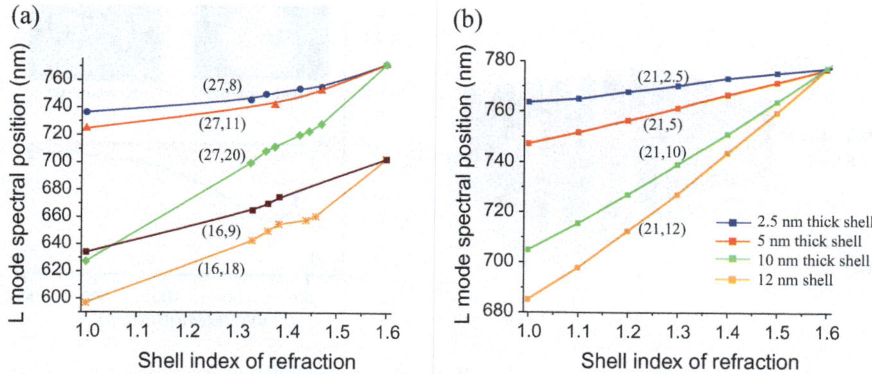

Fig. 4.13 Spectral position of the extinction resonance for an array of Au-core/shell nanorods embedded in an alumina matrix, plotted for different parameters (α, β), where α is the rod aspect ratio and β is the shell thickness, as a function of the refractive index of the shell. **a** Experimental results. The values of $n = 1.6$ and $n = 1$ correspond to rods surrounded by a homogeneous environment (no shell) made of alumina and air, respectively. **b** Theoretical dependences from the effective medium model calculations, exemplifying the experimental behavior. Adapted from [49]

allows the volume of the sensed substance to be controlled with a resolution on the order of a fraction of an attoliter (10^{-18} L) through control of the thickness of the nanoshell—and independently enables tuning of the spectral range of operation over the visible and infrared spectral ranges by varying the aspect ratio of the plasmonic core.

4.4.4 Hydrogen Gas Sensors

Efficient and accurate detection of hydrogen is extremely important, due to the potential hazards associated even with very low concentrations of hydrogen. Large-area core-shell nanorod metamaterials, synthesized using highly-ordered porous alumina templates, provide extremely strong, easily seen by the eye, modification of transmission and reflection of light produced by the presence of hydrogen gas, due to the modification of the modes of the metamaterial [17]. More than 30% change in both the reflection and transmission of the metamaterial layer is observed for 2% hydrogen mixed with nitrogen gas (Fig. 4.14). The performance of these metamaterial sensors is far superior to conventional palladium sensors and can provide visual hydrogen detection in practical environments, without the need for sophisticated optical or electronic detection systems.

In order to achieve sensitivity to hydrogen, hybrid Au-Pd bi-metallic metamaterials based on Au-core and Pd-shell nanorods (Fig. 4.14a) or composite Au-Pd-Au nanorods have been used. Palladium is used as the transducer in many hydrogen

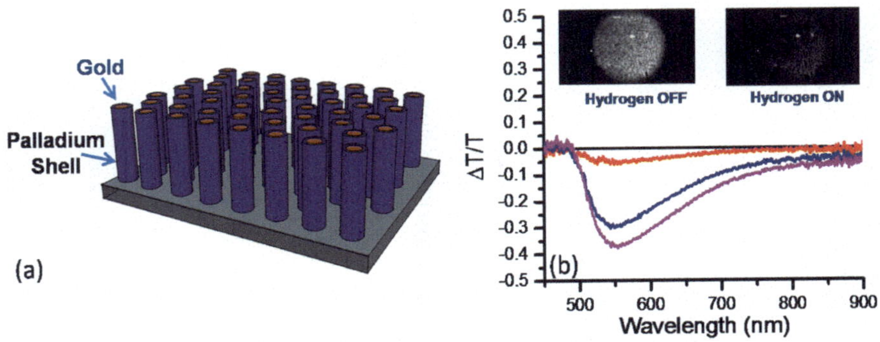

Fig. 4.14 **a** Schematics of the Au-core/Pd-shell metamaterial for hydrogen gas sensing. **b** Variation of the metamaterial transmission at the angle of incidence of 40° for *p*-polarised light, on exposure to 2% hydrogen gas in nitrogen, with increasing exposure time. The insert shows images of the sensor, both in the absence and presence of hydrogen: the brightness is significantly reduced on exposure to H_2. The metamaterial parameters are: Au nanorod diameter 70 nm, separation 100 nm, length 195 nm, and Pd shell thickness 3 nm. Adapted from [17]

sensing technologies because Pd-hydride formation is reversible and thermodynamically stable under ambient conditions - and leads to changes in both the DC conductivity and optical properties, on exposure to hydrogen, due to electronic and structural modifications [52]. The variations in the refractive index can be detected optically with very high sensitivity by incorporation of Pd in a plasmonic metamaterial.

When the core nanoshell array is illuminated by *p*-polarized light in the transmission geometry, the transmission of the metamaterial slab mode decreases by almost 40% at 550 nm wavelength (Fig. 4.14b), when hydrogen gas is introduced into a flow cell. At the same time, the reflection also dramatically increases by about 40% at a wavelength of 600 nm, corresponding to a leaky waveguide mode. The high sensitivity of this optical system to changes in hydrogen concentration results from a combination of the physical changes (expansion) and the corresponding change in the complex permittivity of the palladium shell. Both of these changes result in a change in the plasmonic properties of each individual nanorod and the modification of the permittivity of the shell also strongly modifies the optical coupling between them. It reacts chemically, forming palladium hydride reversibly, and results in volume expansion of the lattice constant, which is accompanied by a shift in the Fermi level, resulting in a change in both the real and imaginary parts of the refractive index of the palladium. The properties of such a metamaterial are based on these pronounced optical interactions. The optical interactions and the optical resonance produce extraordinary changes in the transmission from the material caused by a shift in the resonant wavelength and a change in overall extinction. The sensitivity of segmented Au-Pd-Au nanorods is smaller [17].

4.5 Summary

We have presented an overview of the sensing capabilities and practical examples of sensing applications based on anisotropic plasmonic materials and, in particular, hyperbolic metamaterials based on nanorods. The analysis carried out shows the advantages of the anisotropic materials in general - and hyperbolic metamaterials in particular - for refractive index sensing applications. These capabilities of the nanorod-based metamaterials stem from their strong sensitivity to the plasmon-mediated interactions between the nanorods in the array that form the metamaterial, providing unprecendented sensitivity when the analyte is incorporated between the nanorods—and modifies the interaction. The analysis that we have presented provides the design principles for the nanorod metamaterial based transducers with increased refractive index sensitivities that are expected for the higher-order modes of the metamaterial slab and its smaller thickness. The metamaterial transducer is shown to provide enhanced sensing performance by comparison with both the SPP and LSP-based geometries presented in the literature to date, in terms of both the figures of merit $FoM_{\lambda,I}$ and the nm/RIU sensitivity characteristics achievable. Hyperbolic metamaterials can be used as a design strategy to enhance the flexibility of ultra-sensitive transducers for bio-sensing, chemical sensing and nonlinear nanophotonic devices.

References

1. J. Homola (ed.) *Surface Plasmon Resonance Based Sensors* (Springer, 2006)
2. J.N. Anker, W.P. Hall, O. Lyandres, N.C. Shah, J. Zhao, R.P. Van Duyne, Biosensing with plasmonic nanosensors. Nat. Mat. **7**, 442–453 (2008)
3. M.E. Stewart, C.R. Anderton, L.B. Thompson, J. Maria, S.K. Gray, J.A. Rogers, R.G. Nuzzo, Nanostructured plasmonic sensors. Chem. Rev. **108**, 494–521 (2008)
4. B. Liedberg, C. Nylander, I. Lundstrum, Surface-plasmon resonance for gas-detection and biosensing. Sens. Actuators B **4**, 299–304 (1983)
5. M. Kauranen, A.V. Zayats, Nonlinear plasmonics. Nat. Phot. **6**, 737–748 (2012)
6. A.V. Zayats, S. Maier (eds.), *Active Plasmonics and Tunable Plasmonic Metamaterials* (Wiley, 2014)
7. S. Lal, S. Link, N.J. Halas, Nano-optics from sensing to waveguiding. Nat. Phot. **1**, 641–648 (2007)
8. M. Svedendahl, S. Chen, A. Dimitriev, M. Kall, Refractometric sensing using propagating versus localized surface plasmons: a direct comparison. Nano Lett. **9**, 4428–4433 (2009)
9. A.V. Kabashin, P.I. Nikitin, Surface plasmon resonance interferometer for bio- and chemical-sensors. Opt. Commun. **150**, 5–8 (1998)
10. P.N. Prasad (ed.), *Introduction to Biophotonics* (Wiley-Interscience, 2003)
11. S. Nie, S.R. Emory, Probing single molecules and single nanoparticles by surface-enhanced Raman scattering. Science **275**, 1102–1106 (1997)
12. A.J. Haes, R.P. Van Duyne, A unified view of propagating and localized surface plasmon resonance biosensors. Anal. Bioanal. Chem. **379**, 920–930 (2004)
13. P. Wang, M.E. Nasir, A.V. Krasavin, W. Dickson, Y. Jiang, A.V. Zayats, Plasmonic metamaterials for nanochemistry and sensing. Acc. Chem. Res. **52**, 3018–3028 (2019)

14. J. Yao, Z. Liu, Y. Liu, Y. Wang, C. Sun, G. Bartal, A.M. Stacy, X. Zhang, Optical negative refraction in bulk metamaterials of nanowires. Science **321**, 930 (2008)
15. N. Vasilantonakis, M.E. Nasir, W. Dickson, G.A. Wurtz, A.V. Zayats, Bulk plasmon-polaritons in hyperbolic nanorod metamaterial waveguides. Laser Photonics Rev. **9**, 345–353 (2015)
16. G.A. Wurtz, R. Pollard, W. Hendren, G.P. Wiederrecht, D.J. Gosztola, V.A. Podolskiy, A.V. Zayats, Designed ultrafast optical nonlinearity in a plasmonic nanorod metamaterials enhanced by nonlocality. Nat. Nanotech. **6**, 107–111 (2011)
17. M.E. Nasir, W. Dickson, G.A. Wurtz, W.P. Wardley, A.V. Zayats, Hydrogen detected by the naked eye: optical hydrogen gas sensors based on core/shell plasmonic nanorod metamaterials. Adv. Mat. **26**, 3532–3537 (2014)
18. W. Dickson, G.A. Wurtz, P. Evans, D. O'Connor, R. Atkinson, R. Pollard, A.V. Zayats, Dielectric-loaded plasmonic nanoantenna arrays: a metamaterial with tuneable optical properties. Phys. Rev. B **76**, 115411 (2007)
19. V.V. Yakolev, W. Dickson, A. Murphy, J. McPhillips, R.J. Pollard, V.A. Podolskiy, A.V. Zayats, Ultrasensitive non-resonant detection of ultrasound with plasmonic metamaterials. Adv. Mat. **25**, 2351–2356 (2013)
20. P.B. Johnson, R.W. Christy, Optical constants of the noble metals. Phys. Rev. B **6**, 4370–4379 (1972)
21. T. Sannomiya, O. Scholder, K. Jefimovs, C. Hafner, A.B. Dahlin, Investigation of plasmon resonances in metal films with nanohole arrays for biosensing applications. Small **7**, 1653–1663 (2011)
22. H.N. Daghestani, B.W. Day, Theory and applications of surface plasmon resonance, resonant mirror, resonant waveguide grating, and dual polarization interferometry biosensors. Sensors **10**, 9630–9646 (2010)
23. A.V. Kabashin, P. Evans, S. Pastkovsky, W. Hendren, G.A. Wurtz, R. Atkinson, R. Pollard, V.A. Podolskiy, A.V. Zayats, Plasmonic nanorod metamaterials for biosensing. Nat. Mat. **8**, 867–871 (2009)
24. J. Elser, R. Wangberg, V.A. Podolskiy, E.E. Narimanov, Nanowire metamaterials with extreme optical anisotropy. Appl. Phys. Lett. **88**, 261102 (2006)
25. B.M. Wells, A.V. Zayats, V.A. Podolskiy, Nonlocal optics of plasmonic nanowire metamaterials. Phys. Rev. B **89**, 035111 (2014)
26. K.-T. Tsai, G.A. Wurtz, J.-Y. Chu, T.-Y. Cheng, H.-H. Wang, A.V. Krasavin, J.-H. He, B.M. Wells, V.A. Podolskiy, J.-K. Wang, Y.-L. Wang, A.V. Zayats, Looking into meta-atoms of plasmonic metamaterial. Nano Lett. **2014**, 4971–4976 (2014)
27. N. Vasilantonakis, G.A. Wurtz, V.A. Podolskiy, A.V. Zayats, Refractive index sensing with hyperbolic metamaterials: strategies for biosensing and nonlinearity enhancement. Opt. Express **23**, 14329–14343 (2015)
28. J. Feng, V.S. Siu, A. Roelke, V. Mehta, S.Y. Rhieu, G.T.R. Palmore, D. Pacifici, Nanoscale plasmonic interferometers for multispectral, high-throughput biochemical sensing. Nano Lett. **2012**, 602–609 (2012)
29. H. Liao, C.L. Nehl, J.H. Hafner, Biomedical applications of plasmon resonant metal nanoparticles. Nanomedicine **1**, 201–208 (2006)
30. G.A. Wurtz, W. Dickson, D. O'Connor, R. Atkinson, W. Hendren, P. Evans, R. Pollard, A.V. Zayats, Guided plasmonic modes in nanorod assemblies: Strong electromagnetic coupling regime. Opt. Express **16**, 7460–7470 (2008)
31. G.A. Wurtz, P.R. Evans, W. Hendren, R. Atkinson, W. Dickson, R.J. Pollard, A.V. Zayats, Molecular plasmonics with tuneable exciton- plasmon coupling strength in j-aggregate hybridized Au nanorod assemblies. Nano Lett. **7**, 1297–1303 (2007)
32. O. Leroy, M. A. Breazeale, Eds., Physical Acoustics: Fundamentals and Applications, Springer, (1991)
33. A. Neira, N. Olivier, M.E. Nasir, W. Dickson, G.A. Wurtz, A.V. Zayats, Eliminating material constraints for nonlinearity with plasmonic metamaterials. Nat. Comm. **6**, 7757 (2015)
34. P.R. Evans, G.A. Wurtz, W.R. Hendren, R. Atkinson, W. Dickson, A.V. Zayats, Electrically switchable non-reciprocal transmission of plasmonic nanorods with liquid crystal. Appl. Phys. Lett. **91**, 043101 (2007)

35. W. Dickson, G.A. Wurtz, P.R. Evans, R.J. Pollard, A.V. Zayats, Electronically-controlled surface plasmon dispersion and optical transmission through hole arrays in metal films with liquid crystal. Nano Lett. **8**, 281–286 (2008)
36. G. Si, Y. Zhao, E.S.P. Leong, Y.J. Liu, Liquid-crystal-enabled active plasmonics: a review. Materials **7**, 1296–1317 (2014)
37. J.-S.G. Bouillard, W. Dickson, D.P. O'Connor, G.A. Wurtz, A.V. Zayats, Low-temperature plasmonics of metallic nanostructures. Nano Lett. **12**, 1561–1565 (2012)
38. J. Gosciniak, S.I. Bozhevolnyi, T.B. Andersen, V.S. Volkov, J. Kjelstrup-Hansen, L. Markey, A. Dereux, Thermo-optic control of dielectric-loaded plasmonic waveguide components. Opt. Express **18**, 1207–1216 (2010)
39. http://www.thermo.com/eThermo/CMA/PDFs/Articles/articlesFile_2449.pdf
40. G. Paltauf, H. Schmidt-Kloiber, K.P. Kostli, M. Frenz, Optical method for two-dimensional ultrasonic detection. Appl. Phys. Lett. **75**, 1048–1050 (1999)
41. J. Boneberg, S. Briaudeau, Z. Demirplak, V. Dobler, P. Leiderer, Two-dimensional pressure measurements with nanosecond time resolution. Appl. Phys. Lett. **A69**, S557–S560 (1999)
42. R. Nuster, G. Paltauf, P. Burgholzer, Comparison of surface plasmon resonance devices for acoustic wave detection in liquid. Opt. Express **15**, 6087–6095 (2007)
43. A. Schilling, O. Yavas, J. Bischof, J. Boneberg, P. Leiderer, Absolute pressure measurements on a nanosecond time scale using surface plasmons. Appl. Phys. Lett. **69**, 4159–4161 (1996)
44. C.L. Wong, H.P. Ho, K.S. Chan, S.Y. Wu, C. Lin, Application of spectral surface plasmon resonance to gas pressure sensing. Opt. Eng. **44**, 124403 (2005)
45. P.C. Beard, A. Hurrell, T.N. Mills, Characterization of a polymer film optical fiber hydrophone for use in the range 1 to 20 MHz: a comparison with PVDF needle and membrane hydrophones. IEEE Trans. Ultrasound Ferroelectr. Freq. Control **47**, 256–264 (2000)
46. S. Tadigadapa, K. Mateti, Piezoelectric MEMS sensors: state-of-the-art and perspectives. Meas. Sci. Tech. **20**, 092001 (2009)
47. S.L. Chen, S.W. Huang, T. Ling, S. Ashkenazi, L.J. Guo, Polymer microring resonators for high-sensitivity and wideband photoacoustic imaging. IEEE Trans. Ultrasound Ferroelectr. Freq. Control **56**, 2482–2491 (2009)
48. T. Ling, S.-L. Chen, L.J. Guo, Fabrication and characterization of high Q polymer micro-ring resonator and its application as a sensitive ultrasonic detector. Opt. Express **19**, 861–869 (2011)
49. P.R. Evans, G.A. Wurtz, R. Atkinson, W. Hendren, D. O'Connor, W. Dickson, R.J. Pollard, A.V. Zayats, Plasmonic core/shell nanorod arrays: subattoliter controlled geometry and tunable optical properties. J. Phys. Chem. C **111**, 12522–12527 (2007)
50. J. McPhillips, A. Murphy, M.P. Jonsson, W.R. Hendren, R. Atkinson, F. Hook, A.V. Zayats, R.J. Pollard, High-performance biosensing using arrays of plasmonic nanotubes. ACS Nano **4**, 2210–2216 (2010)
51. A. Murphy, Y. Sonnefraud, A.V. Krasavin, P. Ginzburg, F. Morgan, J. McPhillips, G. Wurtz, S.A. Maier, A.V. Zayats, R. Pollard, Fabrication and optical properties of large-scale arrays of gold nanocavities based on rod-in-a-tube coaxials. Appl. Phys. Lett. **102**, 103103 (2013)
52. T. Shegai, P. Johansson, C. Langhammer, M. Kall, Directional scattering and hydrogen sensing by bimetallic Pd-Au nanoantennas. Nano Lett. **12**, 2464–2469 (2012)

Chapter 5
Photonic Crystal Biosensors

Martina Gerken and Richard De La Rue

Abstract This chapter is concerned with biosensor devices that exploit photonic crystal (PhC) principles in their operation. We have chosen to use a definition of the term 'photonic crystal' that is fairly broad. In surveying the already extensive literature, we have accepted that the author's or authors' choice of the words 'photonic crystal' in a description of their work makes it appropriate for consideration in the present chapter.

5.1 Introduction

This chapter is concerned with biosensor devices that exploit photonic crystal (PhC) principles in their operation. We have chosen to use a definition of the term 'photonic crystal' that is fairly broad. In surveying the already extensive literature, we have accepted that the author's or authors' choice of the words 'photonic crystal' in a description of their work makes it appropriate for consideration in the present chapter.

The guidelines that we have just stated are consistent with describing and analysing the operation of a particular set of photonic devices that are based on physically solid structures that exhibit a spatially periodic variation in their optical properties. In this chapter the physical structures that are at the core of the photonic crystal biosensor devices, sub-systems and systems of interest may exhibit spatial periodicity in one, two or three dimensions. In the case of one-dimensionally periodic structures, the words 'diffraction grating' have been used routinely by optical physicists and engineers for the structures used, for example, in spectroscopy to produce a wavelength-dependent (and therefore frequency-dependent) change in the propagation direction of a collimated beam of light. Where the collimated light-beam has

M. Gerken (✉)
Faculty of Engineering, Kiel University, 24143 Kiel, Germany
e-mail: mge@tf.uni-kiel.de

R. De La Rue
School of Engineering, University of Glasgow, Glasgow G12 8QQ, UK

R. De La Rue et al. (eds.), *Biomedical Optical Sensors*,
Biological and Medical Physics, Biomedical Engineering,
https://doi.org/10.1007/978-3-030-48387-6_5

power in it that is distributed over a range of wavelengths, i.e. a spectrum, the diffraction grating *disperses* the propagating light over a corresponding range of angles, thereby enabling measurement of the wavelength via a suitable calibrated detection system.

One-dimensionally periodic structures that are characterised as 'photonic crystal' structures typically diffract light in much the same way as does a conventional diffraction grating—and so they can be considered as particular manifestations of diffraction gratings. Extension of the spatial periodicity to two or even three dimensions arguably justifies the use of the term 'photonic crystal' more strongly, particularly when the close analogy is recognised between the interaction of photonic crystal structures with light and the Bragg diffraction process for X-rays by classical crystalline solid materials, due to the periodic arrangement of their constituent atoms. A characteristic feature of the Bragg diffraction process for X-rays interacting with a monocrystalline solid, provided that the illuminating radiation is sufficiently close to monochromatic and it is well-collimated, is the production, on a suitably located detector plate, of a regular pattern of diffraction 'spots'. The positions of the diffraction spots are determined by the alignment of the various sets of lattice planes that make up the crystal. Measurement of the corresponding angles can be used to determine the symmetry and the lattice positions of the atoms/molecules that make up the crystal. An intrinsic assumption that is normally made is that the scattering process, for each individual crystal lattice plane, is weak—so that multiple scattering effects can be neglected.

At the other extreme, the one-dimensionally periodic grating structure routinely used in distributed feedback (DFB) and distributed Bragg reflector (DBR) lasers involves, in first-order operation, only one guided-wave mode in each of the forward and backward directions within the periodic grating—with Bragg scattering coupling the two modes to produce feedback and/or reflection.

Spatial periodicity in the optical properties is a characteristic feature of photonic crystal structures. An obvious generalisation of this statement is that optical waves, for our purposes, are simply tangible manifestations of Maxwell's equations—and that the frequency or wavelength plays a directly scalable role in relation to the characteristic periodicity of the environment within which the light (or electromagnetic waves more generally) is propagating. Bandgap properties and the associated behaviour are characteristic features of the propagation of waves in periodic media, whether the waves of interest are matter waves, electromagnetic waves, electronic waves, acoustic waves or some other form of wave.

The term *photonic crystal* was coined by E. Yablonovitch and by S. John as an appropriate term for the description of periodic optical media—by way of analogy with the 'electronic' properties of, in particular, single-crystal semiconducting materials [1, 2]. The transport of information and energy in a photonic crystal is controlled by the aggregate *group velocity* of the Bloch modes at the specified frequency of the electromagnetic wave (i.e. light at optical frequencies) that encounters the photonic crystal medium. The velocity of a Bloch mode in a periodic medium is determined by coherent multiple (Bragg) scattering by the regular arrangement of the photonic 'atoms' of the photonic crystal.

Device functionality that arises, for example, from the formation of resonant micro-cavities within a photonic crystal 'environment' typically involves controlled deviation from periodicity. Relevant deviations from periodicity include the insertion of small or large 'defects' of one kind or another into the interior of an otherwise (in principle) perfectly periodic photonic crystal—and the relative displacement of entire finite or notionally infinite blocks of periodic photonic crystal regions with respect to one another. As we shall describe later in this chapter, microcavities may be created in various ways in both one-dimensional and two-dimensional photonic crystal structures. Detailed design determines the magnitude, frequency and sharpness (i.e. the quality-factor or *Q-factor*) of the characteristic resonance(s) of the microcavity. In the context of biomedical sensing, photonic crystal based microcavity structures have the key characteristic features of very small resonant electromagnetic modal volumes that are precisely located in position on a nanometre scale—and accurately defined and very sharp resonance frequencies that are determined by the detailed geometry of the microcavity. The resonance frequencies are also readily perturbed by small changes in local refractive index. The various different examples of photonic crystal microcavities that have now been demonstrated have the common feature, in most cases, of being fabricated using the planar technologies of microelectronics and nano-scale engineering.

Significant deviations from periodicity in photonic crystal structures also include the possibility of forming, for example, chirped structures where there is local (quasi-) periodicity—but also progressive and systematic variation in the periodicity with spatial position. More-or-less simple variations in periodicity can be defined algorithmically and exploited to modify the response of a system. An obvious enough example is the use of controlled chirping of the PhC periodicity to produce a focusing action in space and/or time.

One of the authors of this chapter has previously been responsible for the production of an introductory review [3], with some useful references, on the application of photonic crystal principles, structures and devices in biomedical sensing. Several more substantial reviews are available in the literature [4, 5]. B. Troia et al. [4] have covered the application of various photonic crystal structures in a range of sensing applications. Integrated photonic crystal sensors based on refractive index change are considered in some detail, with a strong emphasis on device structures and processes that are relevant to biomedical sensing. In all the cases cited, the devices are based on planar fabrication processes, most notably ones based on silicon-on-insulator (SOI) and the structures are, in many cases, microcavity resonators. Sensors based on interferometric photonic crystal channel guide structures and on slow light behaviour near photonic band edges also feature. Refractive index change, in the context of biomedical sensing, is particularly relevant to label-free sensing, together with specific binding processes that intrinsically modify the effective (guided-wave) refractive index and waveguide modal distribution. The review also extends to several useful definitions of sensitivity that are closely relevant to biomedical sensing. The interested reader should note that the word 'sensible' that appears, for instance, in Fig. 12 of B. Troia et al. [4] should be understood to mean 'sensitive', as elsewhere in that article.

The review by B. Troia et al. [4] also covers sensors based on three-dimensionally periodic photonic crystal structures and the important topic of sensing based on photonic crystal (more generally 'microstructured') fibers. The latter topic is also covered in the chapter in the present book by Ding and Brambilla (Chap. 6). Another review by an overlapping group of authors [6] deals with the more restricted topic of chemical sensors realised in the more general form of 'photonic structures'.

A further review article of considerable value is that by M. C. Estevez et al. [5]. It covers, in a comprehensive manner, (integrated) optical devices that are characteristically realised using planar technology and then specifically applied to biosensing. The essential processes of biofunctionalisation and immobilization of bio-material are explained before a range of different waveguide-based planar device structures are reviewed, with devices based on photonic-crystal principles receiving much attention and thirty seven references being cited.

D. Threm et al. [7] have reviewed the application of various types of photonic crystal structures for use as biosensors. The structures are categorised under the headings: 1D photonic crystals (periodic deposited thin-film multilayer structures), photonic crystal slabs, photonic crystal waveguides and photonic crystal microcavities. All four categories fit to the description of 'on-chip integration'.

X. Fan et al. [8] have reviewed optical biosensors that exploit label-free principles, i.e. typically changes in local refractive index associated with specific binding of target bio-material, rather than, for instance, fluorescent labelling. Photonic crystal structures are cited as providing one category of optical biosensors, with sub-categories that include waveguides, passive microcavities, optically pumped microcavity lasers and photonic crystal fibers.

R. V. Nair and R. Vijaya [9] have addressed the topic of sensors based on photonic crystal structures, over a wide range of measurands. In particular, they have considered both photonic crystal based biosensors and photonic crystal based refractive index sensors—which may also be relevant in bio-sensing. Likewise, measurement, for instance, of glucose levels in human beings can be considered as biomedical sensing—while the sensing of oils and the sensing of gases, which are also considered, may be relevant as forms of environmental sensing.

A quite general feature that applies to many photonic crystal based biosensors, but also to other bio-sensing structures (e.g. biosensors based on meta-surfaces), is the use of planar substrates on which the sensing 'surfaces' have been organised. The sensing surface of such bio-sensors is often realised by using planar fabrication processes that include thin-film deposition, lithographic patterning and etching processes. One interesting planar fabrication process in this context is the use of a colloidal silica microsphere monolayer formed and deposited on a glass substrate to pattern the deposition of an evaporated metal film into a structured meta-surface on the glass substrate [10]. Similar processes have been used to realise arrays of nano-pillars and nano-holes [11, 12], using both silica and polymeric microspheres.

One characteristic approach to constructing a bio-sensor is to create a more-or-less uniform and substantial area of photonic crystal patterning on the surface of a planar substrate. As a reasonably typical example, a thousand period grating with a period of 0.5 μm will occupy a square area of only 0.5 mm $\times$ 0.5 mm (i.e. 0.25 mm^2).

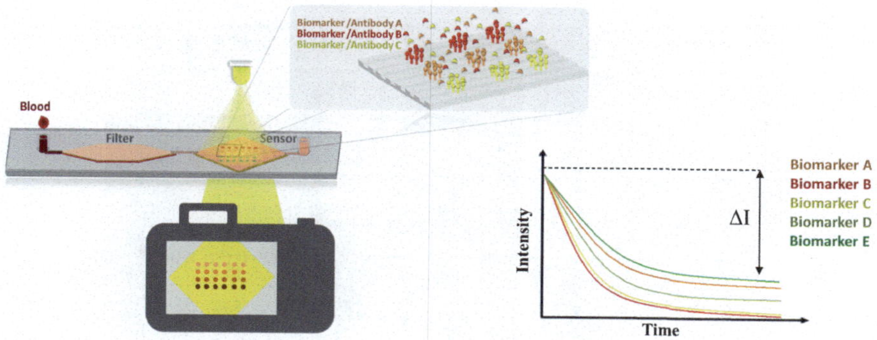

Fig. 5.1 Schematic of sensor with photonic crystal slab as transducer for biomolecular binding and out-of-plane illumination. Reprinted with permission from [13]. Copyright (2015) Optical Society of America

Such photonic crystal patterning can take the form of one-dimensional (1D) or two-dimensional (2D) grating structures. The resulting *photonic crystal slab* can be used for multiparametric sensing as well as enhancement purposes in a photonic crystal microscope.

Figure 5.1 depicts as an example the schematic of multiparametric biomarker detection using out-of-plane illumination and imaging. Such a microfluidic chip with photonic crystal detection units allows for label-free and real-time protein detection in small reagent volumes [13]. In the case of the use of a photonic crystal slab in a photonic-crystal enhanced microscope, the entire sensor surface area may be imaged, for example, by taking spectra line-by-line and computer-based image processing to obtain the hyperspectral image.

Alternatively, the sensor surface may be covered partially by localised detection areas in the form of photonic crystal microcavities. Consider a situation in which light is distributed, for example, to 10,000 individual microcavities organised in a square array where each microcavity is addressed by a feeder waveguide that continues further to feed other microcavities. For such an arrangement, an area of only 1 mm^2 would be required. This calculation assumes that each microcavity, together with a section of feeder and interconnection waveguide, occupies an area of 10 μm $\times$ 10 μm, i.e. 100 μm^2. This, even if the numbers involved are somewhat optimistic, serves to emphasize the scope of what is possible—in terms of sensor compactness, mass-scale production of sensor chips and the generation of large amounts of data to be analysed. For this alternative case—where a 2D array of microcavities is fed by means of an array of waveguides, which in turn is scanned at its output by using an electronically scanned 1D photo-detector array—some form of optical spatial multiplexing or distribution system will be required.

Figure 5.2 is a scanning electron micrograph (SEM) image that shows a specific example of a photonic crystal micro-cavity array that has been evaluated in a prototype biosensing role [14]. In this case—realised using silicon photonics technology, i.e. silicon-on-insulator (SOI) wafer sections—the PhC microcavities were based on

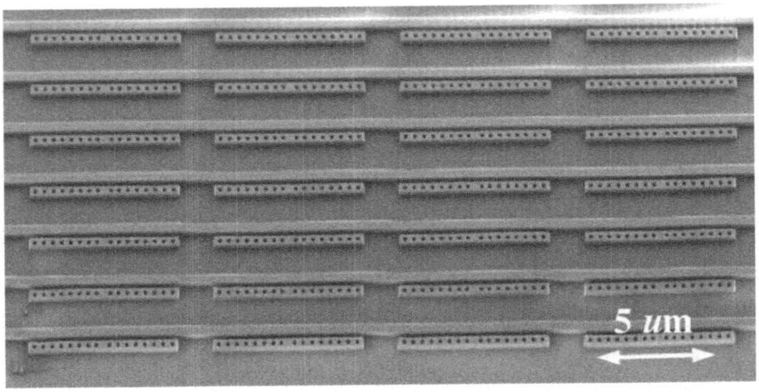

Fig. 5.2 SEM image of photonic crystal micro-cavity. Reprinted with permission from [14]. Copyright (2008) Optical Society of America

one-dimensional photonic crystal (Bragg) mirrors, in the form of periodically spaced holes embedded in photonic wire waveguides. Relatively simple cavity structure design was sufficient to achieve useful sensitivity values with bio-material deposited from solution and delivered through microfluidic channels.

Figure 5.3 shows SEM images of three-dimensional (3D) photonic crystal structures. Figure 5.3b depicts a synthetic opal structure assembled from spherical 'atoms' in a face-centred cubic (fcc) crystal lattice [15]. For the purposes of this chapter, it is interesting that the naturally occurring photonic crystal material opal—and its synthetic form—are both of direct relevance to the present context of photonic crystal based biosensing. The intrinsic porosity of synthetic opal and inverse opal structures can be exploited in biosensing through permeation of biomaterial in solution into the photonic crystal [16, 17]. Furthermore, a single monolayer of colloidal silica spheres deposited on a substrate is already sufficient for the realisation of biosensor functionality [18].

This chapter is structured as follows. Section 5.2 treats biosensing with photonic crystal slabs using out-of-plane illumination. Section 5.3 considers photonic crystal micro-cavities and slow light effects. Section 5.4 gives an overview on other types of photonic crystal biosensors such as 3D photonic crystal biosensors. Conclusions are drawn in Sect. 5.5.

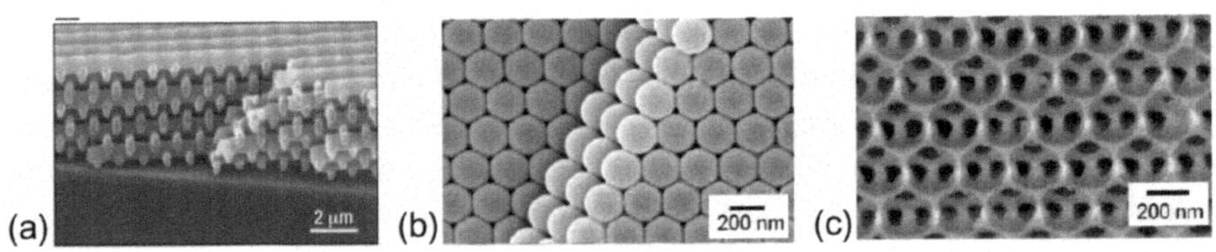

Fig. 5.3 SEM images of 3D photonic crystals fabricated by **a** layer-by-layer direct laser writing. Reprinted with permission from [19]. Copyright (2004) Springer. **b** Self assembly of 220-nm polystyrene and **c** inversion of polystyrene structure by silica. **b**, **c** Reprinted with permission from [15]. Copyright (2007) Optical Society of America

5.2 Photonic Crystal Slab Sensing Using Out-of-Plane Illumination

Nanostructured waveguides with a periodic structure at or close to the surface are of primary interest for photonic crystal biosensors that use out-of-plane illumination. In the first place, one-dimensionally (1D) periodic structures, as shown in the example of Fig. 5.4a, have been employed. In other words, the periodic structure is a specific type of grating. Furthermore, the interaction of light with such a grating leads to diffraction, provided that the grating period is appropriately related to the wavelength of the light that illuminates the grating. A commonly used name for such a waveguide with a 1D periodic nanostructure is 'resonant waveguide grating' (RWG). We prefer to label this grating as a 'photonic crystal slab' (PCS), since this term includes the more general case of a waveguide with a two-dimensionally

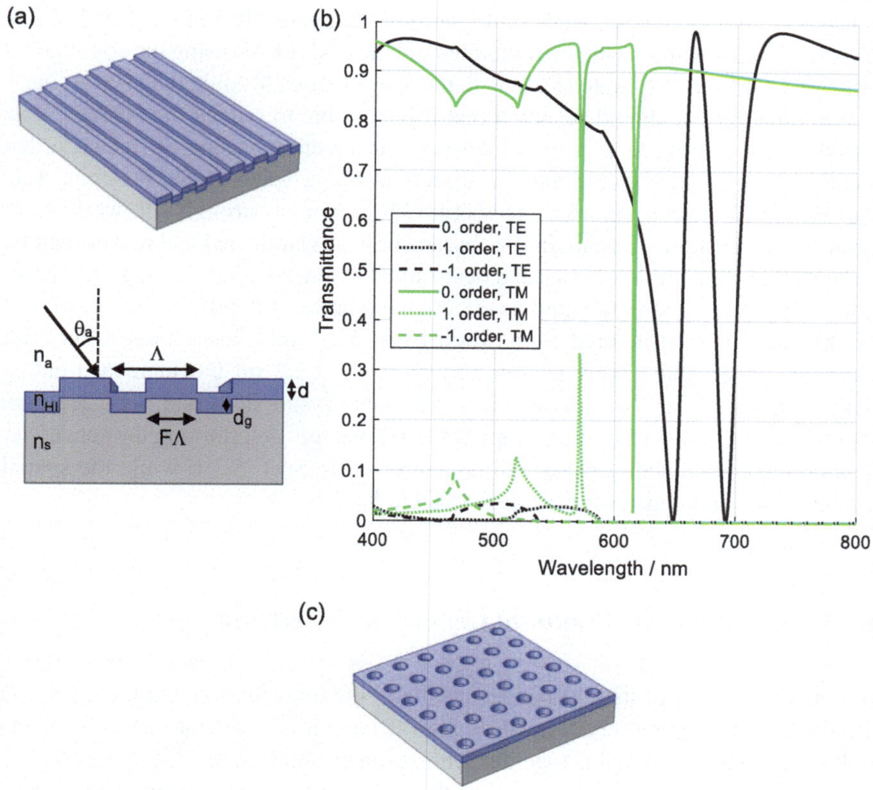

Fig. 5.4 **a** Schematic of a one-dimensional (1D) photonic crystal slab (PCS) (also named resonant waveguide grating, RWG). **b** Simulated transmission spectrum for TE- and TM-polarization of different diffraction orders at $\theta_a = 4°$ with air as analyte ($\Lambda = 370$ nm, $F\Lambda = 220$ nm, $n_a = 1$, $n_{HI} = 2.44$, $n_s = 1.52$, $d = 85$ nm, $d_g = 60$ nm). **c** Schematic of two-dimensional (2D) PCS with square symmetry

nanostructured surface—as depicted, for example, in Fig. 5.4c. The name 'photonic crystal' is arguably appropriate, since it can be organised so that only a narrow range of wavelengths are selected for strong reflection or, more generally, for large changes in reflection or transmission. In the absence of significant losses due to scattering or absorption, reflection and transmission are complementary.

For a general diffraction situation, light is scattered into multiple diffraction orders and the total amount of light (i.e. the power in the light beam) that emerges in the various diffraction orders (including the transmitted or reflected 'zeroth order' light) must equal the amount of light that is incident on the photonic crystal slab, for the case of negligible absorption. While the most 'obvious' situation for diffraction, particularly in the context of microscopy, is one where the light is incident normally on the surface of the photonic crystal slab—the more general situation of incidence at an angle with respect to the normal to the surface, and from either 'above' or 'below', has been used advantageously in different system layouts, as detailed in the following subsections. Figure 5.4b gives an example of the transmission into different diffraction orders for an incidence angle of 4°. The simulation results in Figs. 5.4, 5.5, 5.6, 5.7 and 5.8 are obtained with the finite-element method (FEM) using the commercial software COMSOL Multiphysics® with the Wave Optics Module (COMSOL Inc.).

For generality, it should be noted that it is possible to reduce the characteristic period of the grating to a point where no light emerges in any diffracted order, because there is no possible angular direction where wave-vector matching (i.e. photon momentum conservation) is possible [23]. When this strongly sub-wavelength situation occurs, light is (again in the absence of losses) reflected and transmitted by amounts that are complementary and determined by the weighted average refractive index of the grating structure and its effective thickness or depth.

This section is structured as follows: Sect. 5.2.1 introduces biosensing with photonic crystal slabs. This is followed by Sect. 5.2.2 on biochemical binding analyses and cellular assays employing photonic crystal slabs and Sect. 5.2.3 on photonic crystal enhanced microscopy (PCEM). Factors determining the sensitivity of photonic crystal slab sensors will be addressed in Sect. 5.2.1, while the spatial resolution for imaging sensors is considered in Sect. 5.2.3.

5.2.1 Principles of Photonic Crystal Slab Sensing

The physical origin of the resonances visible in the transmission and the reflection spectrum of a photonic crystal slab is the diffraction of light into and out of the high-index waveguide layer, together with resonant interference [20, 24–26]. The basic ray schematic in Fig. 5.5a depicts the different components involved. Part of an incident beam of light ($\boldsymbol{E}_{\mathrm{inc}}$) is directly transmitted ($\boldsymbol{E}_{\mathrm{t}}$) and part is reflected ($\boldsymbol{E}_{\mathrm{r}}$) by the waveguide layer. Additionally, higher order diffraction may cause part of the beam to couple into a waveguide mode ($\boldsymbol{E}_{\mathrm{d},1}$). This quasi-guided mode is diffracted out of the waveguide by the grating structure. For constructive interference of the reflected parts of the beam ($\boldsymbol{E}_{\mathrm{r}}, \boldsymbol{E}_{\mathrm{r}}', \boldsymbol{E}_{\mathrm{r}}'', \ldots$) a resonance peak is obtained in reflection.

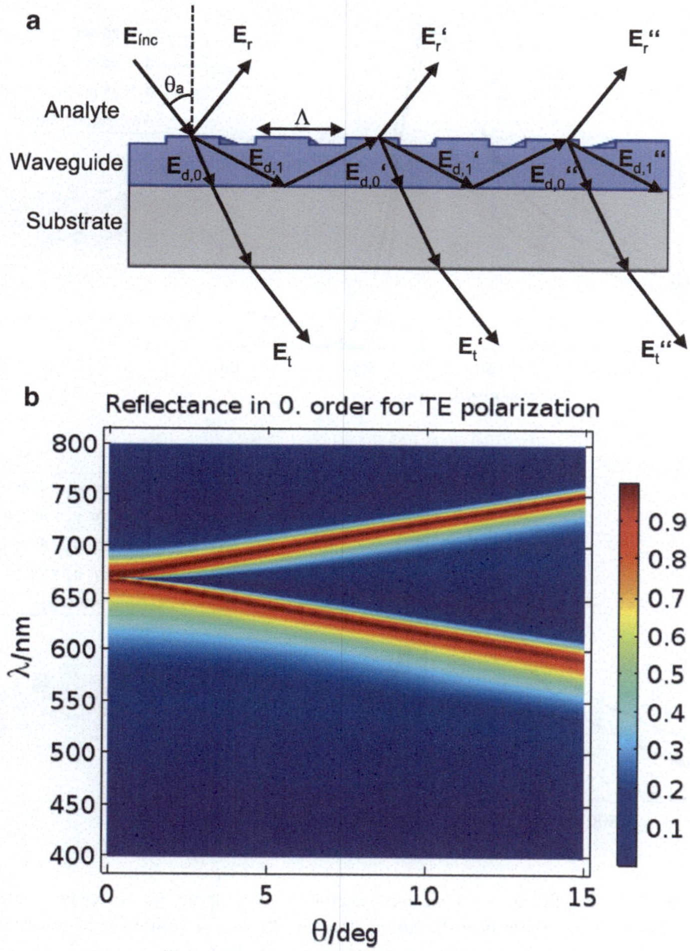

Fig. 5.5 **a** Light ray propagation in a nanostructured waveguide (adapted from [20]). **b** Simulated reflectance as a function of incidence angle θ and wavelength λ for the square grating structure in Fig. 5.4b for TE polarization

This peak is called a guided-mode resonance. Simultaneously, a dip is observed in the transmission spectrum. Figure 5.5b gives an example of the reflectance as a function of incidence angle θ and wavelength λ for the one-dimensional grating structure in Fig. 5.4a. The resonance angle θ_B and wavelength λ_B are related by the Bragg equation obtained from momentum conservation [27]:

$$\sin(\theta_B) = \pm n_{eff} \pm m\frac{\lambda_B}{\Lambda}; \quad m = 1, 2, \ldots \tag{5.1}$$

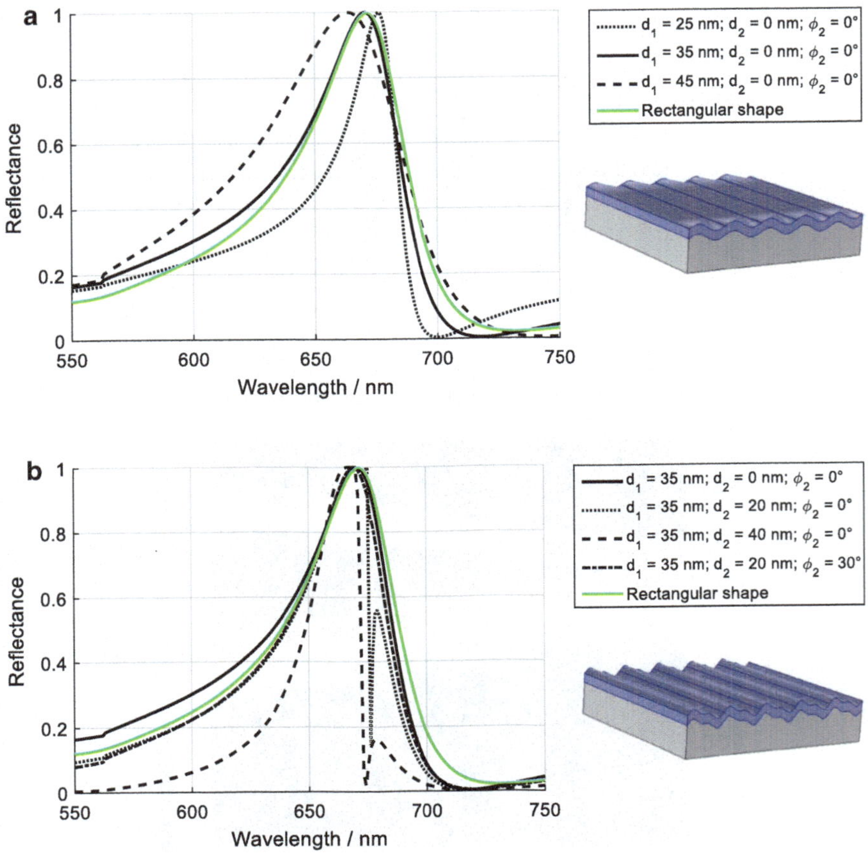

Fig. 5.6 Simulated TE reflectance for sinusoidal gratings with upper and lower waveguide surface described by $s(x)$ and a waveguide thickness of $d = 85$ nm. **a** Influence of amplitude of the fundamental spatial Fourier component. Schematic depicts PCS with $d_1 = 35$ nm, $d_2 = 0$ nm, $\phi_2 = 0°$. **b** Influence of amplitude of second-harmonic spatial Fourier component and relative phase of Fourier components. Schematic depicts PCS with $d_1 = 35$ nm, $d_2 = 20$ nm, $\phi_2 = 0°$

Here, n_{eff} is the effective refractive index of the waveguide mode, Λ is the period of the grating and m is an integer. At normal incidence ($\theta = 0°$) a set of counter-propagating waveguide modes is excited. The splitting of the two modes, i.e. the width of the bandgap between the modes, depends on the grating parameters. D. Rosenblatt et al. [20] give analytical expressions for the case of a 'thin' grating ($d_g \ll \lambda$) obtained from a perturbation approach. It is demonstrated that the fundamental spatial Fourier component of the grating profile determines the Bragg scattering strength, while the second-harmonic spatial Fourier component causes the mode splitting as it couples the counter-propagating waveguide modes.

For dielectric photonic crystal slabs with low absorption, Bragg scattering out of the waveguide dominates the waveguide mode losses. Thus, strong scattering leads

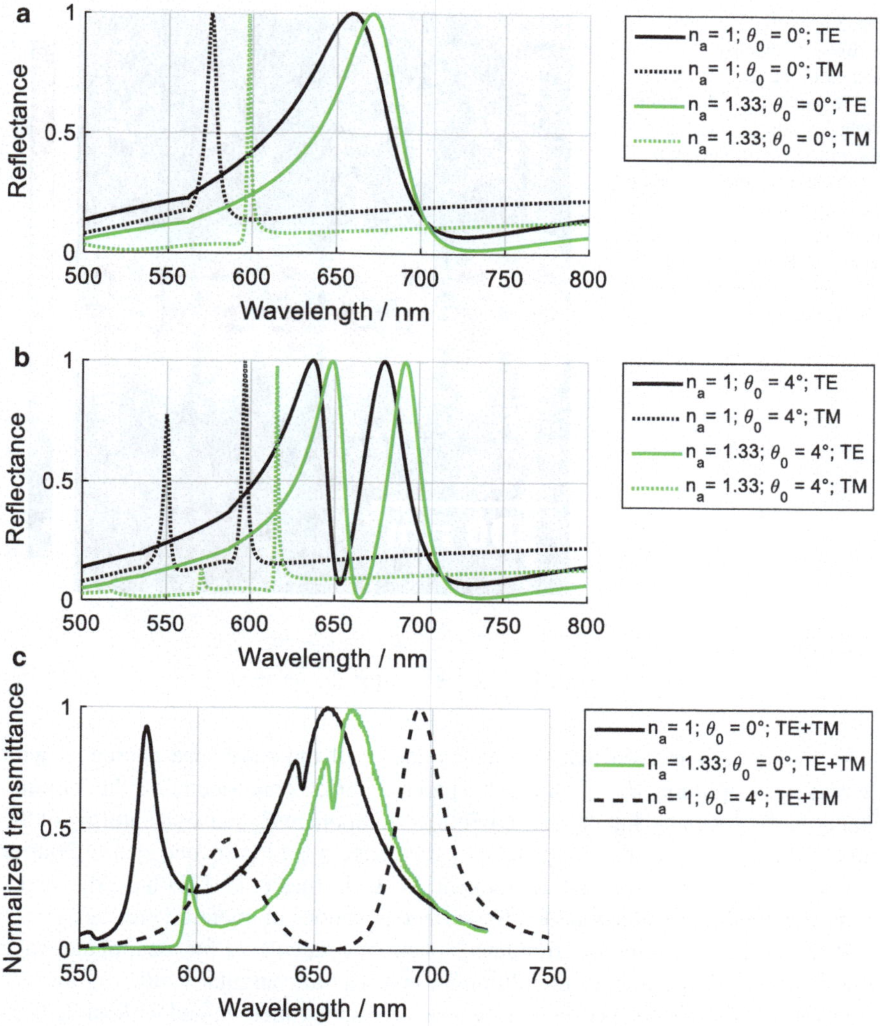

Fig. 5.7 **a** Simulated reflectance with wavelength for the grating in Fig. 5.4b for $\theta_0 = 0°$ (incidence angle in air) with air and water in the upper hemisphere. **b** Simulated reflectance for $\theta_0 = 4°$. **c** Experimental, normalized transmittance measured with crossed polarization filters as described in [21]

to a low quality factor, while a weak grating (with small refractive index contrast or small groove depth) shows a high quality factor. The design of the resonance quality factor and the bandgap width have been discussed in detail in the context of resonant waveguide grating filters [24, 28–30]. As examples, Fig. 5.6 shows the resonance spectra at normal incidence for gratings composed of two Fourier components with the following surface profile:

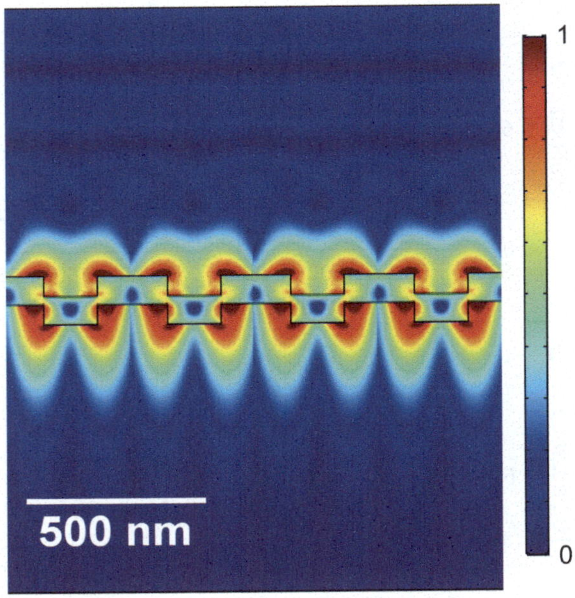

Fig. 5.8 Simulated and normalized electric near-field intensity for the TM resonance at 598 nm in water for normal incidence upon the structure given in Fig. 5.4b. For comparison of the dimensions of biological quantities of interest: protein layer ~1–20 nm, bacteria ~1–5 μm, higher cells ~5–20 μm [22]

$$s(x) = d_1 \sin\left(\frac{2\pi}{\Lambda}x\right) + d_2 \sin\left(2\frac{2\pi}{\Lambda}x + \phi_2\right) \tag{5.2}$$

In Fig. 5.6a it is seen that the quality factor of the resonance decreases with increasing amplitude d_1 of the first spatial Fourier component of the grating. Figure 5.6b illustrates that the second Fourier component d_2 causes a splitting of the mode. This splitting is also influenced by the phase ϕ_2 of the second spatial Fourier component relative to the first component. W. L. Barnes et al. [31] have discussed the corresponding results obtained for surface plasmons on metallic gratings.

It should be noted that the simulated spectra are calculated for ideal plane-wave illumination. In experiments, the illumination has a finite angular width. Figure 5.7c shows as an example an experimentally measured spectrum obtained with white light illumination limited in angle by two apertures. The spectrum shows a characteristic splitting of the TE resonance, which we often observe experimentally and attribute to an angular averaging of the bandstructure in Fig. 5.5b. In setting up the experiment, it needs to be considered that even a numerical aperture of NA $= 0.02$ corresponds to an opening angle of $\pm 1.15°$—and therefore has a considerable influence on the spectrum. In addition, the effect of the illumination incoherence needs to be considered [32].

Two-dimensional PCSs such as the one depicted schematically in Fig. 5.4c exhibit similar guided-mode resonances in the reflection and transmission spectrum of photonic crystal slabs. S. Fan and J. D. Joannopoulos [25] link the Fano-like line-shape of the resonances to the interference between directly reflected light and light scattered from the waveguide. S. G. Tikhodeev et al. [26] give a lucid description of the origin of the different resonances observed—and investigate the resonance quality factors of the different modes.

Next, let us consider the application of photonic crystal slabs for sensing. A change in the refractive index in the upper hemisphere of the photonic crystal slab causes a change in resonance wavelength, resonance angle, and quality factor. Figure 5.7 gives examples of the spectral changes for a change of the analyte in the upper hemisphere. The spectral or the angular response may be monitored, in order to measure changes in the refractive index. The experimental results are obtained with crossed polarization filters as described in [21]. Alternatively, the integrated intensity across a specific spectral or angular range may be tracked.

One commonly used measure of sensitivity is the amount that the guided-mode resonance shifts in wavelength, measured in nanometres, for a given change in the refractive index of the medium, specified therefore in units of nm/RIU, where RIU is the recognised abbreviation for 'refractive index units'. This sensitivity is also termed the bulk sensitivity and values for the resonances in Fig. 5.7 are given in Table 5.1.

Specifying sensitivity in nm/RIU may well be convenient from the characterisation viewpoint—and it may also be appropriate, e.g., where the bio-medical measurand is delivered in solution and no attempt is made to deposit the measurand on the surface of the sensor device. Typical bio-applications include the sensing of protein adsorption or cell activity at the surface of the photonic crystal slab. These events cause refractive index changes in a limited volume above the waveguide surface. Figure 5.8 shows an example of the electric near-field distribution and gives the dimensions of the objects of interest. Typical evanescent field penetration depths are on the order of 100 nm. For comparison dimensions, protein layers have a thickness of ~1–20 nm, bacteria ~1–5 μm, and higher cells ~5–20 μm [22].

The resonance wavelength change $\Delta\lambda$ or frequency change $\Delta\omega$, respectively, depend on the overlap of the electric field with the refractive index change, as described by J. Yang et al. [33]:

Table 5.1 Bulk sensitivity calculated from the spectra in Fig. 5.7

Mode	λ_1/nm at $n = 1$	λ_2/nm at $n = 1.33$	Sensitivity in nm/RIU $S_{\text{bulk}} = \frac{\Delta\lambda}{\Delta n} = \frac{\lambda_2 - \lambda_1}{1.33 - 1}$
Simulation, TE, 0°	659	671	36.4
Simulation, TM, 0°	576	598	66.7
Simulation, TE1, 4°	637	649	36.4
Simulation, TE2, 4°	679	692	39.4
Simulation, TM1, 4°	550	572	66.7
Simulation, TM2, 4°	596	616	60.6
Measurement, TE, 0°	655	666	33.3
Measurement, TM, 0°	578	597	57.6

$$\Delta\omega = -\omega \iiint_{V_p} \Delta\boldsymbol{\varepsilon}(\boldsymbol{r}, \omega)\tilde{\boldsymbol{E}}_{\mathrm{app}}(\boldsymbol{r}) \cdot \tilde{\boldsymbol{E}}(\boldsymbol{r})\mathrm{d}^3\boldsymbol{r} \quad (5.3)$$

Here, V_{p} is the perturbed volume, $\Delta\boldsymbol{\varepsilon}$ is the change in dielectric constant due to the perturbation, $\tilde{\boldsymbol{E}}$ is the unperturbed electric field vector of the quasi-normalized mode and $\tilde{\boldsymbol{E}}_{\mathrm{app}}$ is the approximate perturbed normalized electric field vector. In consequence, for high sensitivity, a strong electric field in the volume is required to change the refractive index. Comparing the electric field penetration into the analyte with the typical dimensions of biological quantities, it is seen that the structure in Fig. 5.8 is particularly suitable for the detection of thin protein layers at the surface of the PCS. For the detection of whole bacteria or even larger cells, a PCS with an electric field extending further into the analyte region should be designed. For this purpose, the use of low-index substrates has been suggested [34]. For the detection of thin protein layers at the surface, a small penetration depth is actually advantageous as it renders the sensor insensitive to non-specific refractive index variations in the sample volume and thereby reduces background noise. I. D. Block et al. [35] discuss the design of the PCS to match the particular sensing task. In order to characterize the sensor sensitivity for bio-applications, a sensitivity figure of merit that is matched to the task is more meaningful than the bulk sensitivity. In particular, a 'surface sensitivity' has been defined that describes the influence of a refractive index change at the surface on the spectral or angular resonance characteristics [22, 35, 36]. Table 5.2 gives the calculated surface sensitivity for the detection of a 20-nm thick bio-layer of refractive index $n_{\mathrm{bio}} = 1.38$.

It should be noted that sensitivity values scale with wavelength. Therefore, it is necessary to consider the resonance wavelength in judging how good is a particular sensitivity value. While the sensitivity is determined by the nanostructure, the detection limit depends on the employed read-out instrumentation that determines the noise and the resolution [37]. In the following sections, different realizations of sensor systems are discussed.

Table 5.2 Simulated surface sensitivity for the structure in Fig. 5.4b for a 20-nm thick biolayer of refractive index $n_{\mathrm{bio}} = 1.38$

Mode	λ_1/nm at $n = 1$	λ_2/nm at $n = 1.38$	Sensitivity in nm/RIU $S_{\mathrm{surf}} = \frac{\Delta\lambda}{\Delta n} = \frac{\lambda_2 - \lambda_1}{1.38 - 1}$
Simulation, TE, 0°	659.2	663.5	11.3
Simulation, TM, 0°	575.5	580.3	12.6

5.2.2 *Photonic Crystal Slab Biosensors*

As a first sensing system, we consider a sensor for refractive index measurements of liquids, as realized by M. Huang et al. [38]. As depicted in Fig. 5.9, the PCS is formed as a free-standing membrane with a two-dimensional hole array. Additionally, the PCS is combined with a fluidics system to allow for flow across or through the membrane. This sensor system reaches a record-high sensitivity of 510 nm/RIU, at a resonance wavelength around 850 nm, and a line width of ~10 nm in solution. This high sensitivity may be explained by the large field overlap with the analyte. In addition, the authors argue that the convective flow through the nanohole openings leads to improved analyte delivery to the surface, as compared with laminar flow across the surface. J. O. Grepstad et al. [39] employ a PCS membrane with sub-micron thickness and through holes to detect nanoparticles smaller than the diffraction limit of the optical microscope used. They demonstrate the detection and location of single particles with radius down to 75 nm in air.

Commercial systems used for high-throughput drug discovery and cellular assays based on PCS are already available on the market. Figure 5.10 shows, as an example, a 384-well microtiter plate with a one-dimensional PCS at the bottom of each well (as supplied by Epic® system, Corning Incorporated, Corning, NY, USA). The corresponding readout instrument employs spectral monitoring of the reflected light. N. Orgovan et al. [40] determine for this type of system a detection limit of 2.2×10^{-6} for the refractive index change, by employing serial dilution of glycerol solution. They obtain a surface sensitivity of 0.078 ng/cm^2 to adsorbed mass with layer-by-layer deposition of polyelectrolyte films. In cellular assays the mass redistribution in the cell within the penetration depth of the quasi-guided PCS mode is monitored (Y. Fang [41]). Upon stimulation, e.g., by bradykinin for A431 cells, a so called dynamic mass redistribution (DMR) occurs in the cell. This mass redistribution is accompanied by a refractive index change in the volume sampled by the quasi-guided PCS mode. Y. Fang [41] gives an overview of different types of cellular assay used for drug discovery.

In designing an instrument for cellular assays the scattering of light by the cells must be considered. The amount of scattering depends on the cell type and, in particular, on the surface density of the cells. Spectral and angular measurements of the guided-mode resonance have the advantage of being intensity independent. Y. Nazirizadeh et al. [42] have demonstrated that an intensity-based measurement may be utilized if resonance near cut-off is employed. Figure 5.11 depicts a cell adhesion and spreading experiment, together with the experimentally measured wavelength shift, resonance width, and resonance intensity, for two different modes—one near cut-off and one far away from cut-off. Both modes show a monotonic wavelength shift after addition of the cells, due to the attachment of cells to the surface. The resonance width of the non-cut-off mode shows an increase and a subsequent decrease, which can be attributed to the increased light scattering as cells attach to the surface. As the cell confluency at the surface passes a critical point, scattering decreases again, at high cell densities. This behaviour is reflected in a decrease in the resonance intensity

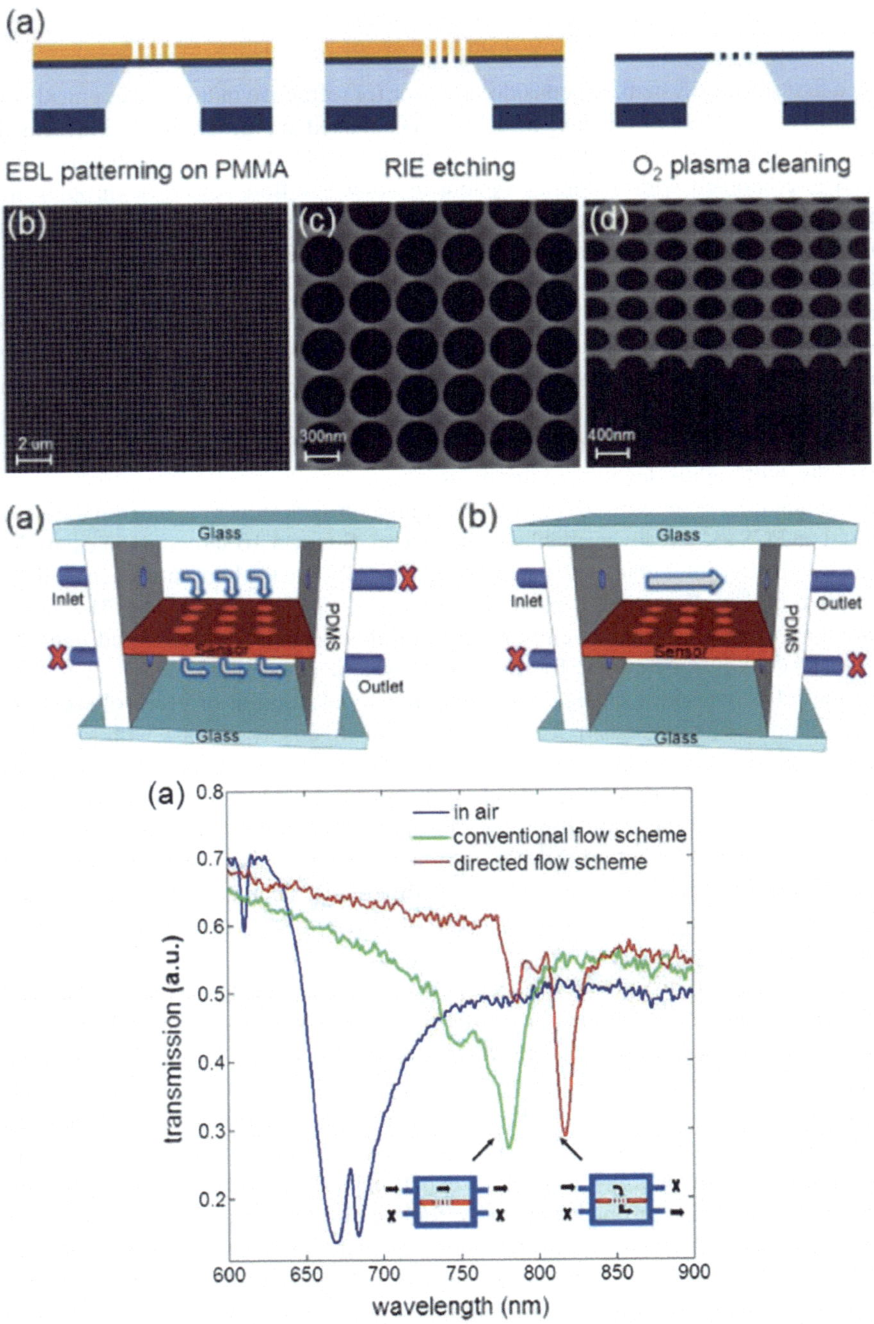

Fig. 5.9 PCS formed by free-standing membrane with nano-hole array and combined with fluidics. High sensitivity of 510 nm/RIU at 850 nm achieved. Reprinted with permission from [38]. Copyright (2009) Optical Society of America

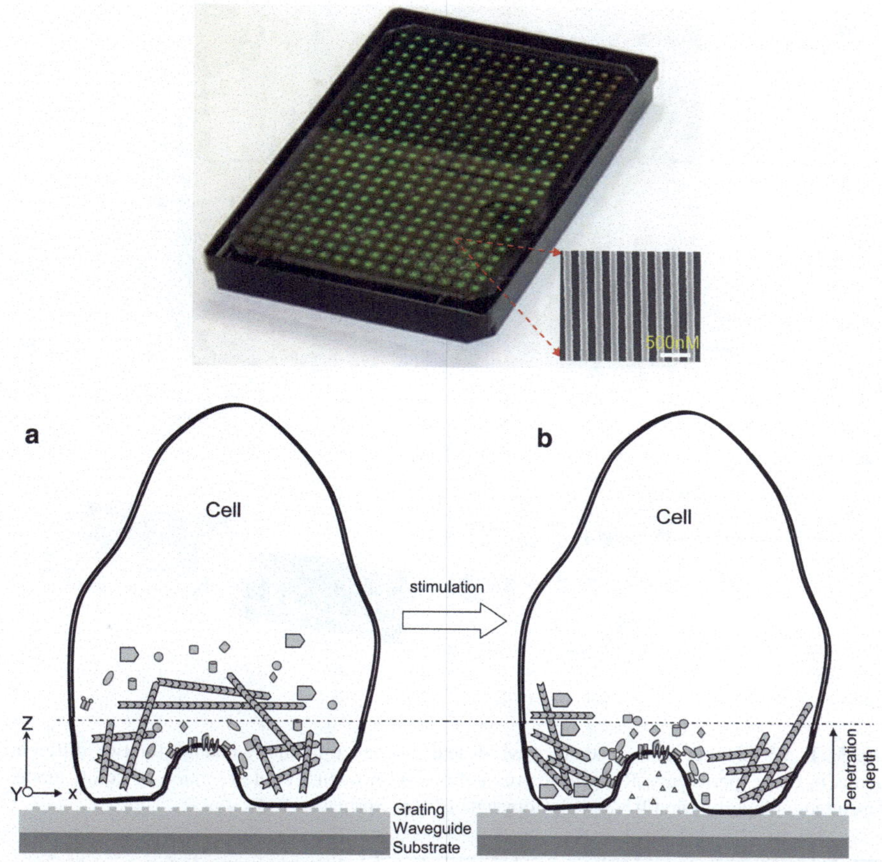

Fig. 5.10 **a** 384-well biosensor microplate (Epic® system, Corning Inc.). **b** Schematic of mass redistribution in cell upon stimulation. Reprinted with permission from [41]. Copyright (2006) Mary Ann Liebert, Inc.

of the non-cut-off mode, up to a minimum value and a subsequent intensity increase as scattering decreases for high cell confluency. The cut-off mode on the other hand shows an overall monotonic behaviour due to the additional effect of passing the cutoff wavelength. Comparing the resonance intensity change of the cut-off mode to the wavelength shift of the non-cutoff mode shows that the same information may be retrieved. Based on this result the most compact microtiter plate readout system to date was designed by employing simply a light-emitting diode and a photodiode for an intensity-based read-out of each of the wells [42].

Targeting a compact instrument design for mobile applications, D. Gallegos et al. [43] have constructed the smartphone adapter shown in Fig. 5.12. It uses the integrated camera of the smartphone to realize a spectrometer and carry out spectral

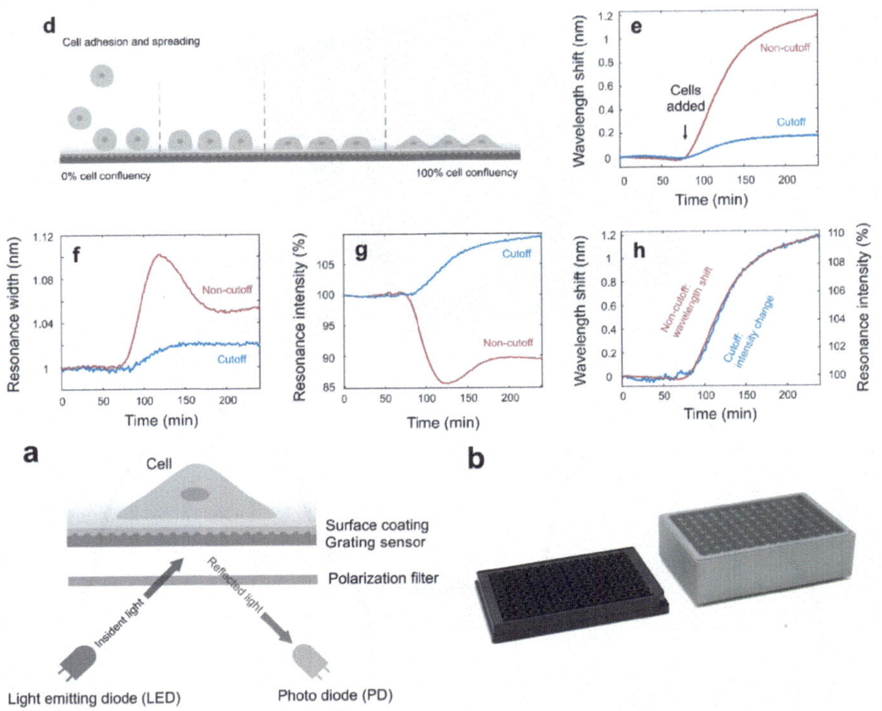

Fig. 5.11 Wavelength shift, resonance width and resonance intensity change for two different modes in cell experiments. Readout-instrument based on intensity interrogation of cutoff mode. Reprinted with permission from [42]. Copyright (2016) Springer

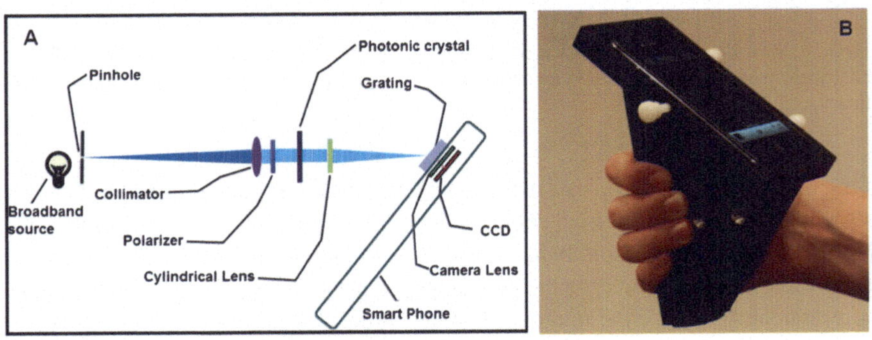

Fig. 5.12 Compact readout system using a spectrometer camera for retrieving spectral information. Reprinted with permission from [43]. Copyright (2013) The Royal Society of Chemistry

measurements on PCSs. Using custom software they obtained a wavelength accuracy for the resonance peak of 0.009 nm. The system is utilized for detection of an immobilized protein monolayer and for antibody binding experiments.

An intensity-based imaging readout system was demonstrated by S. Jahns et al. [13]. As depicted in Fig. 5.13 a functionalized PCS is imaged onto a camera. Protein binding to specifically-functionalized sites on the surface causes an intensity change. The surrounding unfunctionalized surface area is used for background correction. Specific detection of three different types of protein was demonstrated. On a 1-cm^2 surface area potentially $20 \times 20 = 400$ functionalized spots could be positioned with approximate spot diameters of 350 μm. This work has demonstrated that a point-of-care system for parallel detection of a panel of biomarkers is within reach. Figure 5.1 depicts how the system may be combined with a microfluidic chip requiring only a single drop of blood.

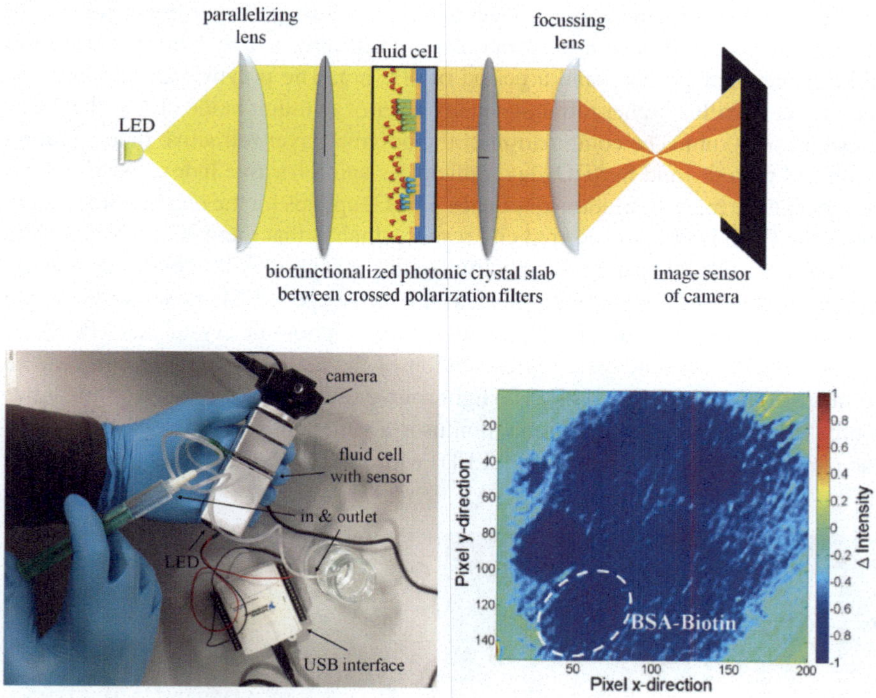

Fig. 5.13 Compact system employing a camera for intensity-based readout. False-colour image on the lower right shows an example of a difference intensity image after protein binding to specifically-functionalized circular positions. Reprinted with permission from [13]. Copyright (2015) Optical Society of America

5.2.3 Photonic Crystal Enhanced Microscopy (PCEM)

Microscopes of various kinds are in widespread use for a great variety of specific imaging tasks. The conventional optical microscope, in particular, is a 'natural' tool for use in biomedical applications—and versions of the conventional microscope that include detection and image formation by means of fluorescence or multiple-photon processes have been of vital importance in the major advances that have been made in recent years. In photonic crystal enhanced microscopy (PCEM), biological objects of various kinds may be studied through the use of more-or-less conventional microscope optics, but with illumination through a specially-prepared, periodically-nanostructured substrate. The imposition of a structured surface on the substrate on to which the biological objects are placed can give major enhancements in the qualities of the imaging process and the amount of information that can be extracted.

Figure 5.14 has two parts. Figure 5.14a is a schematic diagram of the grating structure—and shows the polymer layer that has been shaped by using nano-imprint lithography (NIL) to have a moderate depth (120 nm), a 1-to-1 mark-space ratio and a rectangular profile—and a period of 400 nm. The polymer grating has also been covered with a higher refractive index layer of titanium oxide (TiO_2) that has a thickness of 60 nm. The combination of the polymer layer refractive index, grating depth and period—and the TiO_2 layer thickness and refractive index, together with the fact that the refractive index of water ($n = 1.33$) applies for the region immediately above the PCS, lead to the desired end result, which is that there is nearly complete reflection of a selected part of the spectrum of the light used to illuminate the imaging region. Figure 5.14b is a scanning-electron microscope (SEM) image of an actual fabricated grating structure, i.e. a particular form of photonic crystal slab (PCS).

Figure 5.15 is a schematic representation of the actual instrument, taken (with permission) from reference [44]. The light source used in this work is a 'white light' source. More precisely: the illumination uses a sufficiently broadband, low coherence, source—a light emitting diode (LED) with a central peak emission wavelength

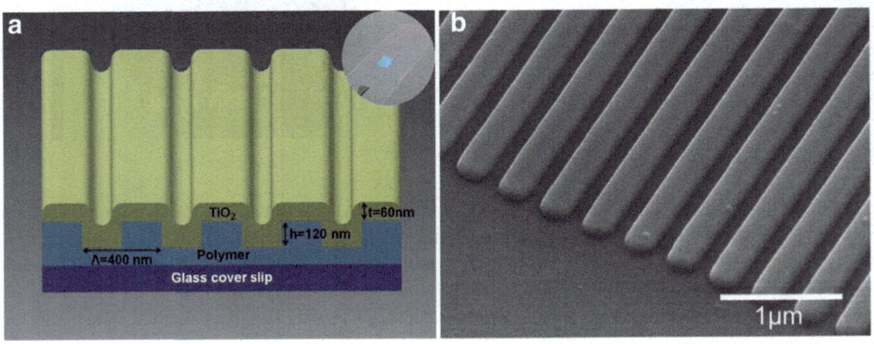

Fig. 5.14 Schematic diagram and scanning-electron microscope (SEM) image of PCS substrate for photonic crystal enhanced microscopy (PCEM). Reprinted with permission from [44]. Copyright (2013) The Royal Society of Chemistry

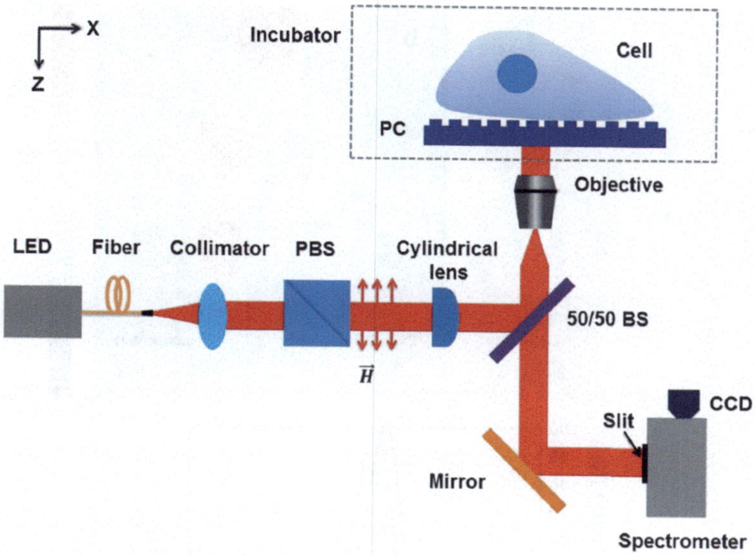

Fig. 5.15 Instrument for live cell imaging. Reprinted with permission from [44]. Copyright (2013) The Royal Society of Chemistry

of ~625 nm and a spectral width of ~50 nm. An important element in the instrument is the polarising beam-splitter (PBS). Because the diffraction properties of the grating/PCS structure are intrinsically polarisation dependent—and the polarisation of the light emitted by the LED is uncontrolled (i.e. random), the PBS selects that part of the light which, at a given moment, interacts resonantly with the surface.

Figure 5.15 also shows, schematically, a 'typical' object for investigation with the PCEM. Specifically, the object of interest is a cell with a characteristic diameter of several tens of micrometres (relating e.g. to the circle within which the entire cell can be located). The cell is a *murine dental cell*. An important aspect of cell imaging is that the refractive index is close to that of water, but the density of the different components has a direct correlation with the local refractive index. Information on the refractive index of cellular components has been reviewed by P. Y. Liu et al. [45].

The schematic diagram of Fig. 5.15 indicates clearly enough that the image derived from the cell adhering to the structure surface is projected through a spectrometer onto a CCD camera. But it is also important to note that a cylindrical lens is incorporated into the optical system and thereby transforms the beam that is illuminating the structured surface into a line focus. In consequence, the CCD behind the spectrometer captures the spectrum of a line on the sample surface simultaneously. Scanning is required to obtain a 2D image of the surface.

Figure 5.16 compares a bright-field image with an image obtained by the peak-wavelength measurement employing a PCS substrate. A much better contrast is

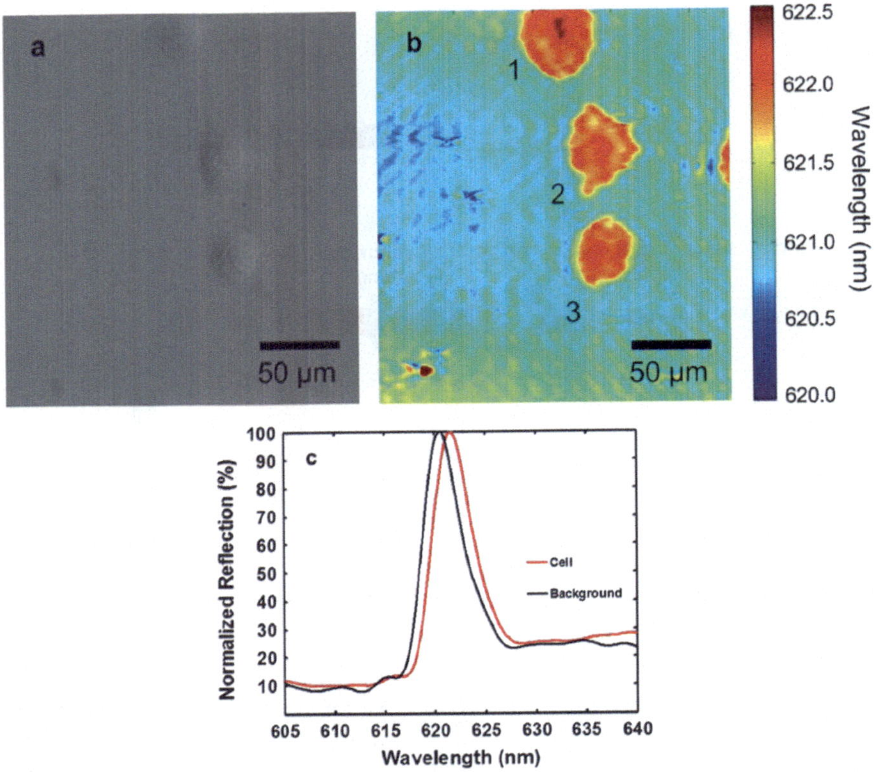

Fig. 5.16 **a** Bright-field image and **b** Peak-wavelength image of Panc-1 cells attached to a photonic crystal surface. **c** Example of spectral response. Reprinted with permission from [44]. Copyright (2013) The Royal Society of Chemistry

observed for the PCEM. As can be seen in the example spectrum, a typical wavelength shift of around 1 nm is measured for positions with cells when compared to surrounding PCS sites without cells.

An important consideration in employing a PCS for imaging applications is the resolution. Due to the coupling of light into waveguide modes, propagation occurs along the waveguide—and the resolution is thereby reduced. The propagation length is proportional to the quality factor of the quasi-guided mode and is therefore inversely proportional to the resonance linewidth [46]. An experimental demonstration of this effect is shown in Fig. 5.17. Two different photonic crystal slabs are considered, with linewidths of 17 nm and 4.5 nm, respectively. A USAF 1951 resolution standard pattern is etched into each of the photonic crystal slabs. Imaging is performed by evaluation of the resonance angle. It is clearly visible that the left image, which corresponds with a broader spectral linewidth, exhibits a higher spatial resolution, i.e., smaller features may be resolved. Experimentally, a spatial resolution of 3.9 μm is obtained for the broader spectral linewidth and 11 μm for the narrow one.

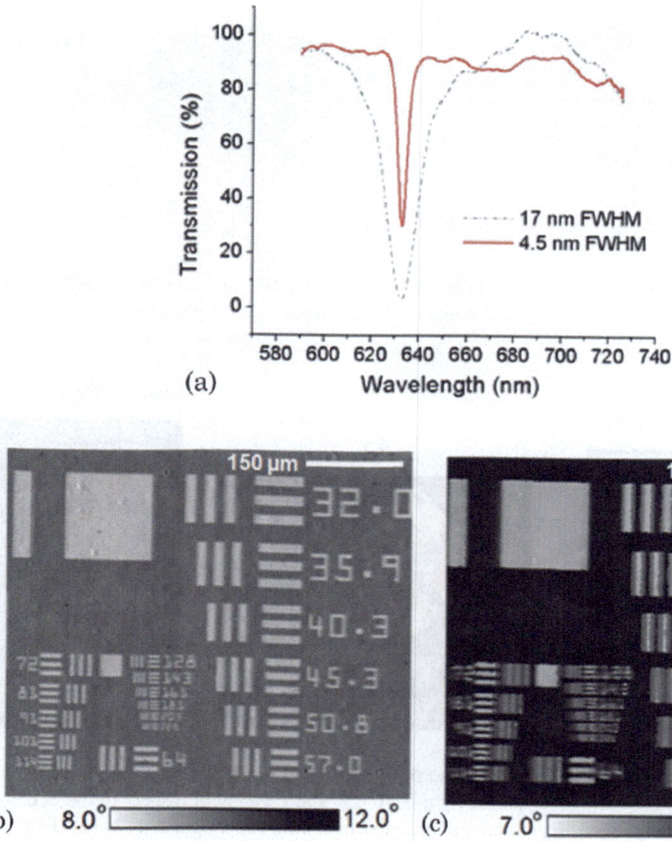

Fig. 5.17 **a** Resonance spectra for two different photonic crystal slabs. Images of USAF 1951 resolution standard patterns etched into the photonic crystal slabs taken by evaluation of the resonance angle for **b** the PCS with the broad and **c** with the narrow spectral linewidth. Reprinted with permission from [46]. Copyright (2009) Optical Society of America

G. J Triggs et al. [47] compared the resolution achieved with one-dimensional photonic crystal slabs to the case of a two-dimensional photonic crystal slab. As depicted in Fig. 5.18, the single grating vector G_x only allows for coupling into one propagation direction in the waveguide. On the other hand, two-dimensional PCSs offer more than one grating vector -and coupling is possible into multiple directions. A photoresist pattern consisting of lines with different pitch and orthogonal orientation is deposited on the photonic crystal slab, as shown in Fig. 5.19. Experimentally, a better resolution is obtained in the direction of the incident electric field vector. This result may be explained by the fact that the wave-vector of the incident beam is perpendicular to the electric field vector and therefore causes modal propagation in the direction perpendicular to the electric field vector. The coupling into the quasi-guided mode degrades the resolution. In the direction of the electric field vector, no propagation occurs and the resolution is determined by the optical

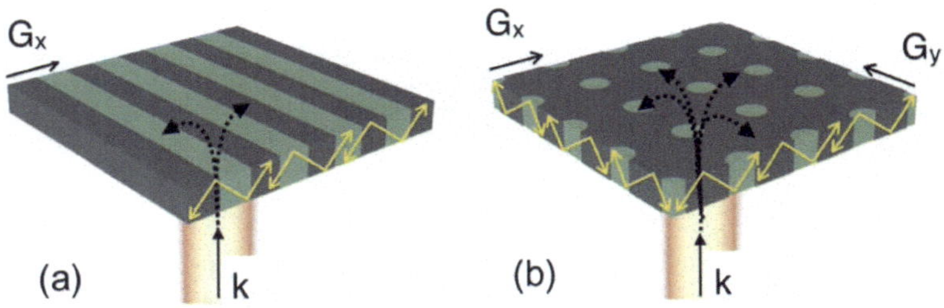

Fig. 5.18 Coupling of incident light into a photonic crystal slab for **a** a one-dimensional and **b** a two-dimensional nanostructure. Reprinted with permission from [47]. Copyright (2015) IEEE

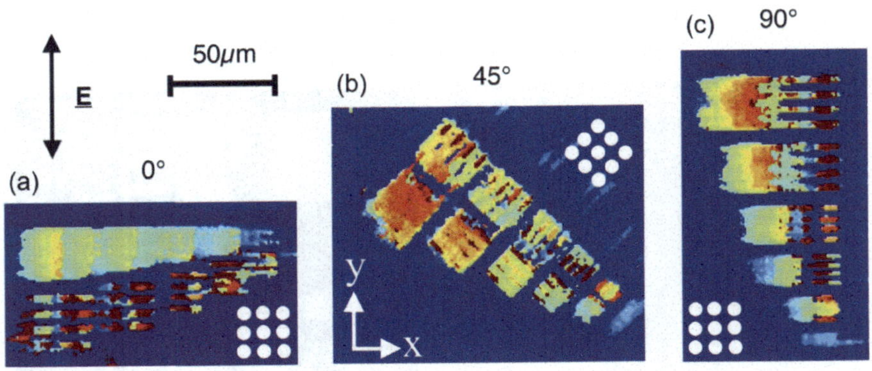

Fig. 5.19 Images obtained by evaluation of the peak-wavelength position for a two-dimensional photonic crystal slab. The sample is rotated by **a** 0°, **b** 45° and **c** 90° with respect to the depicted electric field vector orientation. Reprinted with permission from [47]. Copyright (2015) IEEE

imaging system. For a two-dimensional PCS mode, propagation may be excited in multiple directions—thereby degrading the spatial resolution, as compared with a one-dimensional grating structure.

Y. Nazirizadeh et al. [21] suggest that, in the case of sufficient refractive index contrast, PCEM may be conducted as a simple colour-based measurement. Crossed polarization filters are used to suppress background light that does not interact with the quasi-guided modes. Without the use of a spectrometer or an angle-scanning mechanism, a significantly increased contrast is obtained for objects at the surface. Additionally, for objects with known refractive index, a three-dimensional image may be reconstructed with PCEM [48].

In the sensing of biological molecules—and organic molecules more generally—characteristic resonant features directly related to the vibration of particular chemical bonds may be observed by scanning the wavelength of the light incident on the molecule and observing changes in the reflection, transmission or absorption spectrum of the light. By designing a PCEM at infrared wavelengths, these molecular resonances may be evaluated for further information. Finally, photonic crystal slabs have been successfully employed for enhanced fluorescence microscopy [49].

5.3 PhC Micro-cavities and Slow Light

By definition, slow light takes longer to transport optical energy or information over a given distance than normal 'fast' light. In terms of velocity, slow light is identified with a reduced group velocity—and, for normalized comparison purposes, one appropriate measure is the (phase) velocity of light in the medium from which the waveguide core is constructed. Slow light PhC structures, e.g. in the form of waveguide channels through block regions of 2D photonic crystal, have been demonstrated to exhibit experimental group velocities that are several tens of times smaller than the group velocity of guided light in the waveguide core material.

Slow light behaviour characteristically occurs in situations where the dispersion curve for optical propagation, defined in terms of ω versus β, has a strongly reduced slope. Here, β refers to the wavevector in the propagation direction.

Slow light occurs characteristically in the near zone-boundary regions of the PhC band-structure. In terms of the characteristic Bloch modes of the periodic structure of the photonic crystal, the group velocity decreases progressively as the zone-boundaries are approached from either side. In the limiting case, at the exact zone-boundary, the group velocity goes to zero.

Slow-light can lead, for instance, to enhanced non-linearity—and to enhanced sensitivity in sensing applications such as biomedical sensing. For applications of slow light in which broadband operation is desirable, e.g. the delay controlled processing of digital data, there are also dispersion control issues that can be addressed by local flattening of the dispersion curve, via designed local variation of the PhC lattice on either side of a channel waveguide in a PhC lattice. Such dispersion control issues are arguably of interest and relevance where it is desirable or essential to obtain a controlled amount of propagation delay for signals, e.g. pulses of light, that are spectrally broadband.

1D PhC structures in the form of more-or-less strong gratings created in slab waveguides can exhibit band-structure—and spectral regions where dispersion is strong—and propagation is slow.

Slow light propagation in PhC structures and waveguide channels through 2D PhC regions can also be considered as a particular situation in terms of the Bloch modes of the structure—in which the principle constituent modes in the forward and backward directions are strongly coupled through the intrinsic scattering processes of the PhC structure—and have comparable magnitudes. If the constituent forward and backward propagating modes are equal in magnitude, the (net) group velocity goes to zero—and the effective mass (or refractive index) goes to infinity.

In, for example, a slab waveguide that has a 1D grating/PhC structure imposed on it, there will be a finite band of frequencies—the stop-band—over which strong reflection occurs within the PhC lattice—or, equivalently, there is a strong reduction in transmission.

While the enhancement of sensitivity that is a characteristic feature of slow light propagation is potentially useful, the resonant microcavity approach is arguably superior to the slow light approach, because of the intrinsically compact path-folding

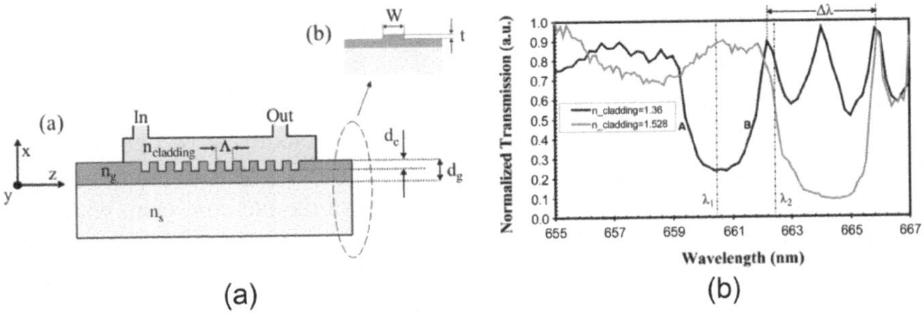

Fig. 5.20 **a** 1D grating PhC waveguide structure schematic showing fluid delivery cell. **b** Measured transmission stop-bands for two different upper cladding fluid refractive indices. Reprinted with permission from [51]. Copyright (2005) IEEE

involved in the resonant operation of a microcavity. Furthermore, in example situations described in the literature, e.g. by J. García-Rupérez et al. [50], the finite length slow light waveguide structure may become part of a Fabry–Perot cavity structure, because of partial reflection at each end of the slow light waveguide—i.e. a somewhat microcavity is generated. With a deliberately formed resonant microcavity, the greatly enhanced sensitivity of the structure is typically associated with strong spatial localisation that implies that only a very small amount of correctly located bio-material is required to shift the resonance frequency of the microcavity by a detectable amount.

W. C. L. Hopman et al. [51] demonstrated a compact *refractometric* sensor by launching light in the waveguide mode of a one-dimensionally periodic PhC—in the form of a shallow, ~76 μm long, grating etched into a ridge waveguide formed by depositing a high-index (2.01 μm) film of silicon nitride on a fairly thick (9 μm) silica ($n = 1.46$) buffer layer, on a silicon substrate. The structure is shown schematically in Fig. 5.20a. The basic sensing effect demonstrated, as shown in Fig. 5.20b, was a whole band shift (with $\Delta\lambda = 4$ nm) obtained by changing the modal effective refractive index of the waveguide via changing the refractive index of a liquid upper cladding—between the $n = 1.36$ of pure ethanol and $n = 1.528$ of a mixture of ethanol and benzyl-alcohol. This structure was evaluated as a temperature sensor, with a readily observable shift in the stop-band being obtained by changing the temperature of the fluid cladding between 20 and 142 °C. Much the same structure could be exploited as a compact biosensor, with changes in the waveguide modal index resulting from the presence of a bio-material in solution, delivered microfluidically—and the possibility of detecting bio-material deposited, e.g., via molecular binding processes.

Another version of a waveguide PhC sensor is described in the paper by N. Skivesen et al. [52]. In this paper, the sensor device used a waveguide channel through 2D hole-based PhC regions, with a core layer of silicon on a silica ($n = 1.44$) buffer layer—on a single crystal silicon substrate, i.e. close to a standard commercial SOI wafer section. The configuration is depicted in Fig. 5.21a. The basic refractometric action of the sensor structure was clearly and quantitatively demonstrated by observing the long wavelength transmission spectral edge position as the refractive

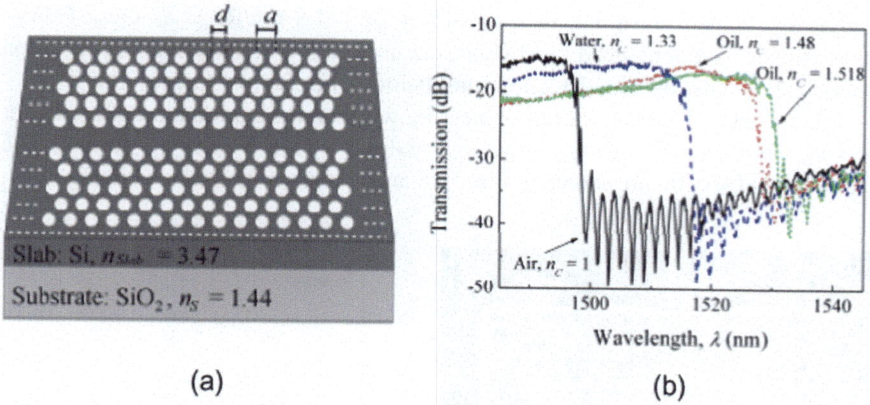

Fig. 5.21 **a** Schematic of PhC channel waveguide realised in SOI. **b** Pass-band edge wavelength shifts—with upper cladding refractive index as parameter. Reprinted with permission from [52]. Copyright (2007) Optical Society of America

index of the ambient medium was increased from that of air ($n = 1$) through that of water ($n = 1.33$) to the refractive indices of two different immersion oils ($n = 1.48$ and $n = 1.518$). Biosensing action was evaluated by investigating the effects of depositing bovine serum albumin (BSA), on to the sensor device, from solution in purified water. An edge wavelength transition of 0.2 nm was readily detected after deposition from a solution concentration of 10 μg/ml. The deposited layer thickness was estimated to be 2.6 nm for this situation.

Figure 5.22 shows the device presented by J. García-Rupérez et al. [50]. By using the band edge Fabry–Perot fringes seen in Fig. 5.22b a higher interaction with the

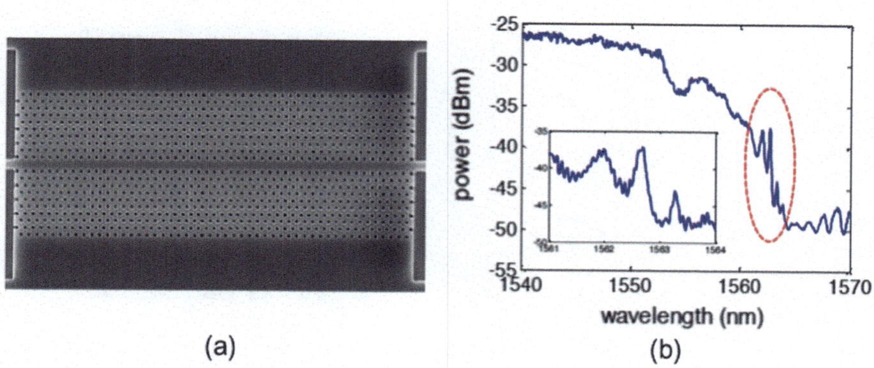

Fig. 5.22 **a** SEM image of the fabricated SOI planar photonic crystal waveguide. **b** Spectrum with Fabry–Perot fringes visible at the band edge. Reprinted with permission from [50]. Copyright (2010) Optical Society of America

analyte is obtained, resulting in a sensitivity of 174.8 nm/RIU. As these Fabry–Perot fringes have a reduced bandwidth, the uncertainty in detecting their position is reduced. A mass density detection limit below 2.1 pg/mm^2 is achieved.

E. Chow et al. [53] carried out pioneering work on the exploitation of biosensor devices based on microcavity structures realised within a 2D PhC format. The geometry and spectra are shown in Fig. 5.23 and results are reproduced in Fig. 5.24.

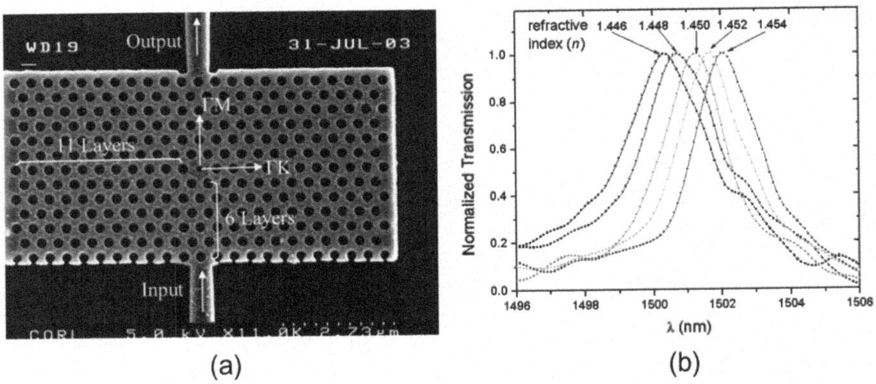

Fig. 5.23 **a** Scanning electron microscope view of a photonic crystal microcavity integrated with two ridge waveguides. The regular hole diameter is $d = 0.58\,a$, the defect hole diameter is $d' = 0.4\,a$, and $a = 440$ nm is the lattice constant of the photonic crystal. **b** Scanning electron microscope view of a photonic crystal microcavity integrated with two ridge waveguides. The regular hole diameter is $d = 0.58\,a$, the defect hole diameter is $d' = 0.4\,a$, and $a = 440$ nm is the lattice constant of the photonic crystal. Reprinted with permission from [53]. Copyright (2004) Optical Society of America

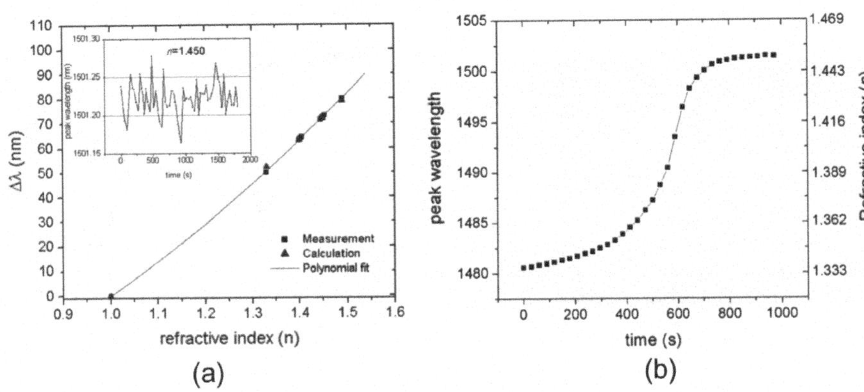

Fig. 5.24 **a** Resonant wavelength and the corresponding refractive index as a function of time during the evaporative process of water in a glycerol–water mixture. **b** Resonant wavelength shift Dl plotted as a function of ambient index n. The inset shows the fluctuation in the resonant wavelength measured at $n = 1.450$ over a time interval of 30 min. Reprinted with permission from [53]. Copyright (2004) Optical Society of America

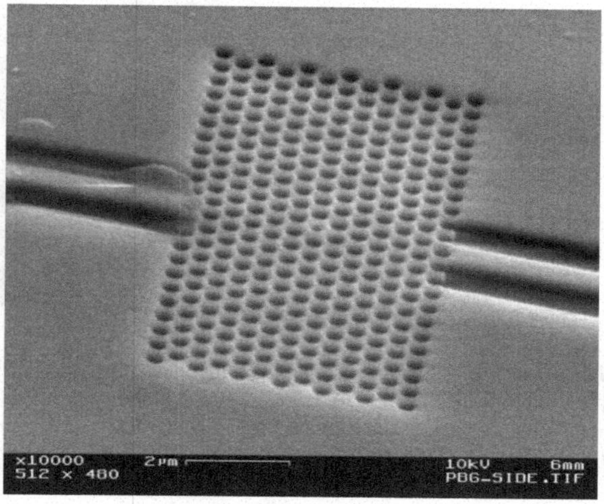

Fig. 5.25 Micrograph of the resonant microcavity created within a 2D hexagonal symmetry PhC region and fabricated in an SOI wafer section—together with input and output ridge waveguides. Reprinted with permission from [54]. Copyright (2007) Optical Society of America

M. R. Lee and P. M. Fauchet [54] have also described the operation of a sensor based on a microcavity that was realised within a 2D hole PhC region, as shown in the micrograph of Fig. 5.25. Although the microcavity design was non-optimum, resonance Q-factor values as large as 2000 were estimated. To demonstrate the potential capabilities of this device as a biosensor, measurements were carried out using small diameter (370 nm) latex spheres—to determine the shift in the resonance peak produced when the latex sphere lodges in the central hole of the microcavity region.

As already described in the introduction to this chapter—and shown in the SEM image of Fig. 5.2—arrays of 1D PhC microcavities have been used to demonstrate their possible application in a prototypical biosensing role [14]. As the figure shows, each microcavity was coupled to a feeder/output photonic wire waveguide. With the application of a measure of design, resonance Q-factor values of around 2000 were achieved, with the microcavity structure array immersed in fluid that had a refractive index close to that of water, i.e. $n = 1.33$.

This Q-factor value was sufficiently large to achieve useful sensitivity, e.g. detection of a refractive index change as small as 7×10^{-5} in microfluidically delivered test solutions, and readily resolve five different geometrically controlled resonance frequencies/wavelengths. (Fig. 5 of the reference suggests that many more than five resonances, e.g. fifty, could be achieved with much the same cavity design). Numerical simulation was used to calculate 10 attogram sensitivity for single-strand DNA binding in a target zone on an individual microcavity, as identified by a shift in the cavity resonance wavelength by 10^{-2} nm.

Y.-F. Chen et al. [55] developed the 1D PhC cavity significantly further. The optical trapping action of the resonant mode concentrates the deposition of the protein molecules at the centre of the microcavity. The trapping action can be described as the operation of a 'molecular nanotweezer'. Also shown in Fig. 5.26 are a top-view micrograph of the microcavity, with a small hole at the centre of the spacer

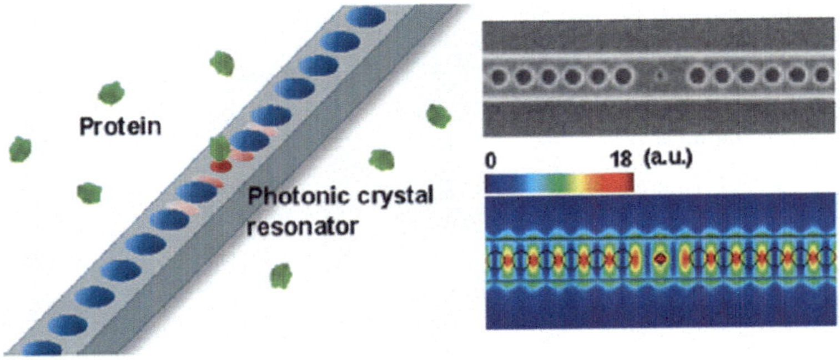

Fig. 5.26 Schematic representation of silicon nitride based 1d PhC microcavity resonator showing protein molecules in solution being attracted to the peak optical field strength region of 1.06 μm wavelength light launched into the microcavity. Reprinted with permission from [55]. Copyright (2012) American Chemical Society

section acting as an optical concentrator, together with a representation of the resonant modal field distribution. With the use of appropriately prepared surfaces on the microcavity structure, specific binding can be used to ensure that the local change in refractive index, after removal of the trapping light, is identifiable with the presence of a specific protein in the solution, via a shift in the microcavity resonance frequency/wavelength. With such cavities arranged in an array—and the application of distinctly different preparation processes to each microcavity in the array—the presence of a selected range of different proteins becomes possible. At the same time, the resonance frequencies of the individual microcavities can be chosen, leading to coding of the protein identification process via the resonance frequency/wavelength and the frequency/wavelength shift of the resonance produced by the sensing action. In the actual experiments carried out, Wilson disease molecules with a molecular weight of ~46 kDa were used. Resonance Q-factor values of 3000–5000 were clearly sufficiently high for successful demonstration of the capabilities of this approach.

S. Zlatanovic et al. [56] use a simple microcavity structure with a relatively low Q factor. In a label-free approach binding of anti-biotin to biotinylated-bovine serum albumin (b-BSA) is detected. The total sensor area is 50 μm^2, while the effective cavity area is only 0.27 μm^2. The resonance shift with concentration is shown in Fig. 5.27. The detection limit of 20 nM corresponds to 4.5 fg deposited on an area of approximately 50 μm^2. For an effective modal area of 0.272 μm^2, this corresponds to 80 molecules and a mass of 21 attogram of the anti-biotin.

As Fig. 5.27 shows, the response of this photonic crystal sensor indicated a clearly detectable resonance wavelength shift (0.5 nm) at a solution concentration of 3×10^{-9} M for the binding of anti-biotin to biotinylated–BSA that was deposited and immobilised on the carefully prepared surface of the PhC microcavity. From the slope of the main part of the graph, the value of the affinity constant was estimated to be 6.9×10^7 M^{-1}, which is considered to be a moderate value, e.g. c.f. the value of 10^{15} M^{-1} that has been cited for the streptavidin-biotin pair.

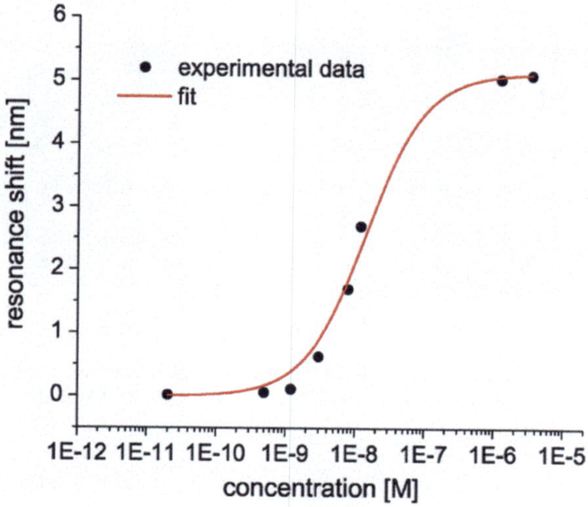

Fig. 5.27 Concentration response curve for anti-biotin measured with photonic crystal microcavity sensor. Reprinted with permission from [56]. Copyright (2009) Elsevier B.V.

W.-C. Lai et al. [57] have realised high-Q (values up to 26,700) microcavity structures in membrane PhC waveguides as shown in Fig. 5.28. Because of the fragility of an unsupported membrane, the buffer silica region from the SOI wafer section was retained. The significant optical absorption of water at the operating wavelength (~1.55 μm) represents an intrinsic barrier to the achievement of higher

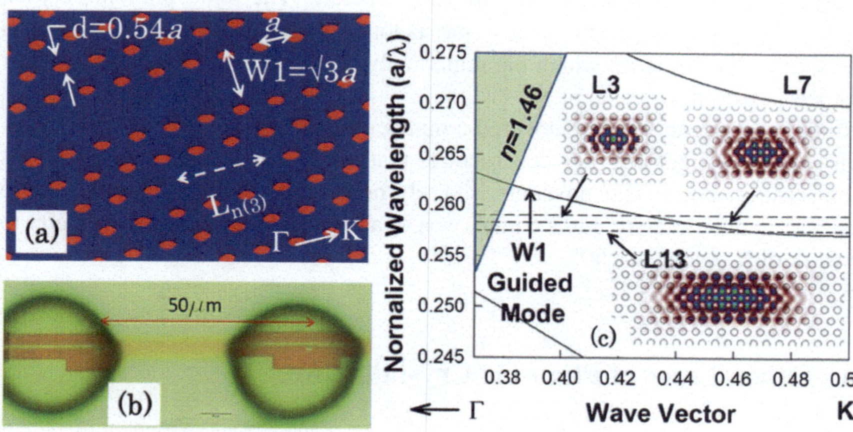

Fig. 5.28 **a** Schematic diagram of feeder W1 channel waveguide through 2D silicon supported nano-membrane PhC regions, coupled with $L_{n(3)}$ microcavity centered on the third row of the PhC region away from the waveguide channel region, **b** inkjet printed bio-molecule regions on PhC waveguide/microcavities, and **c** normalised frequency (inverse wavelength!) versus wave-vector (dispersion) for waveguide and microcavity modes, with resonant modal distributions. Reprinted with permission from [57]. Copyright (2012) Optical Society of America

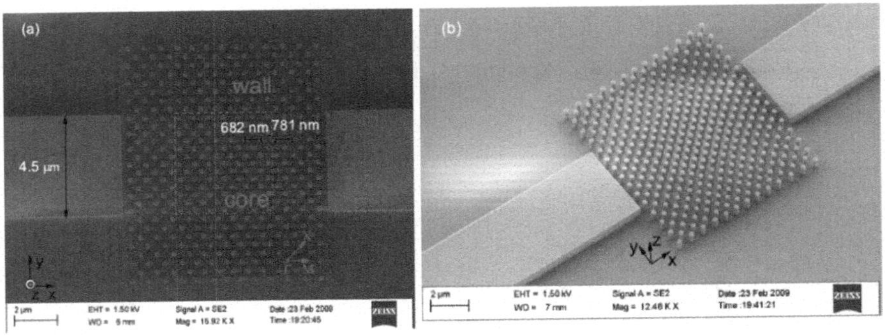

Fig. 5.29 **a** Top view of square lattice pillar PhC microcavity. **b** Slanting angle view of the same microcavity. Reprinted with permission from [58]. Copyright (2010) Optical Society of America

resonance Q-factor values—and also means that the potential advantages of using an unsupported membrane approach are not significant.

The photonic crystal microcavity structure realised by T. Xu et al. [58] is of interest because it is based on high aspect-ratio *pillars* in a square lattice, with a lattice period of 483 nm as shown in Fig. 5.29. The microcavity is formed by the combination of the square core area and a surrounding "wall" area that has a larger lattice period along one axis. Note that the main lattice axes are at a 45° angle with respect to the defining x–y axes of the core and wall domains. The pillar photonic crystal structure is particularly suitable for immersion in a fluid delivery and sensing environment, with a large effective surface area. Because the photonic crystal structure uses finite height pillars it is inherently somewhat leaky—but nevertheless resonance Q-factors as large as 27,000 were achieved with this structure. The pillar photonic crystal structure has a generic resemblance to the nanorod metamaterial described in A. V. Kabashin et al. [59], with which the remarkably high sensitivity of 30,000 nm per RIU was achieved.

In the text above, we have covered a number of interesting examples of microcavity and slow-light PhC device structures that have been used in demonstrations of biosensing action. This coverage is surely not exhaustive—other examples are presented in the literature. For a comparison of achieved detection limits we refer to the work of Y. Zou et al. [60].

5.4 Other Types of Photonic Crystal Sensor

The previous two sections have discussed sensors based on one- or two-dimensional photonic crystal structures implemented in a supported waveguide layer on a substrate or in a light-guiding membrane. In this section, we provide an overview of other approaches to photonic crystal biosensors. As described in Sects. 5.1 and 5.2 of this chapter, a one-dimensional (1D) photonic crystal region may be realized as a diffraction grating structure. An alternative 1D arrangement is a periodic layer

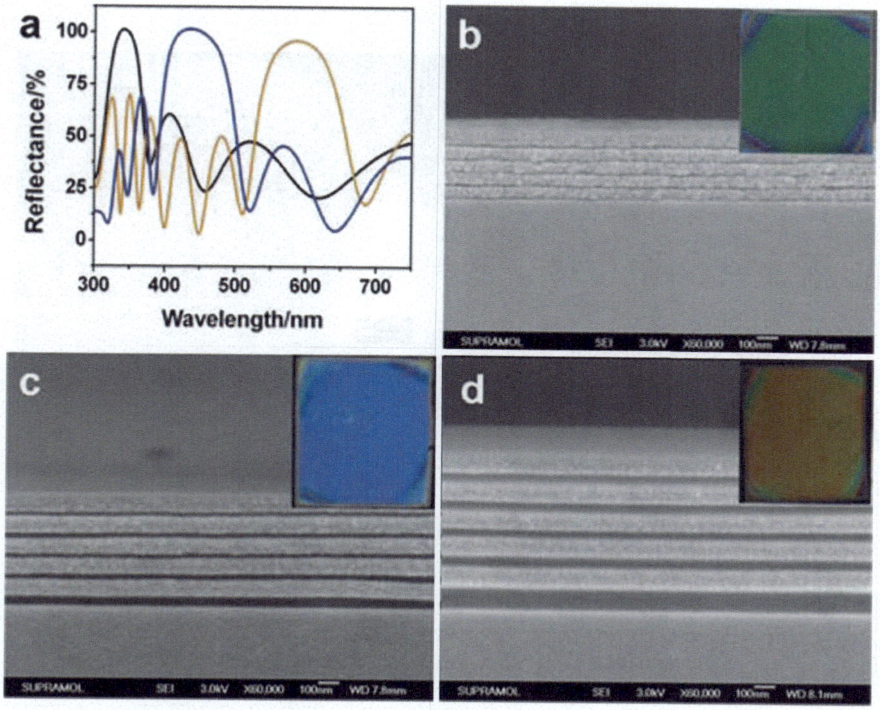

Fig. 5.30 1D photonic crystal—commonly called Bragg stack: **a** Reflection spectra for three different periods. **b** Cross-sectional SEM images for polymer and titania layers, respectively, of **b** 9 nm and 70 nm, **c** 20 nm and 80 nm, and **d** 57 nm and 89 nm. Inset photographs show sample size of 20 mm by 20 mm. Reprinted with permission from [62]. Copyright (2011) The Royal Society of Chemistry

sequence of two or more materials deposited on a substrate, although suspended membrane formats are also possible. Such structures are well known in the form of optical interference filters—and are commonly called Bragg mirrors [61]. Figure 5.30 shows a Bragg stack of different periods fabricated by the spin-coating of organic and titania nanoparticle sol layers, as well as example spectra for the three period case [62]. Adding a sensing cavity into the periodic sequence results in a Fabry–Perot resonator (called a Gires-Tournois resonator, if one mirror has a 100% reflectivity and the device is operated in reflection).

Photonic crystal structures have been integrated with fibers for several different types of sensor configuration. Here, distributed sensing, as well as lab-on-a-fiber applications, have been investigated. Two different configurations and combinations of these have been proposed. The periodic arrangement of refractive indices may be either in the longitudinal direction of the fiber or in the transverse cross section. The first configuration is known as a fiber Bragg grating (FBG)—and Fig. 5.31 (left) depicts an example structure of a microfiber Bragg grating (MFBG) [63]. Examples of the transverse periodic structuring of the fiber are the solid-core and hollow-core

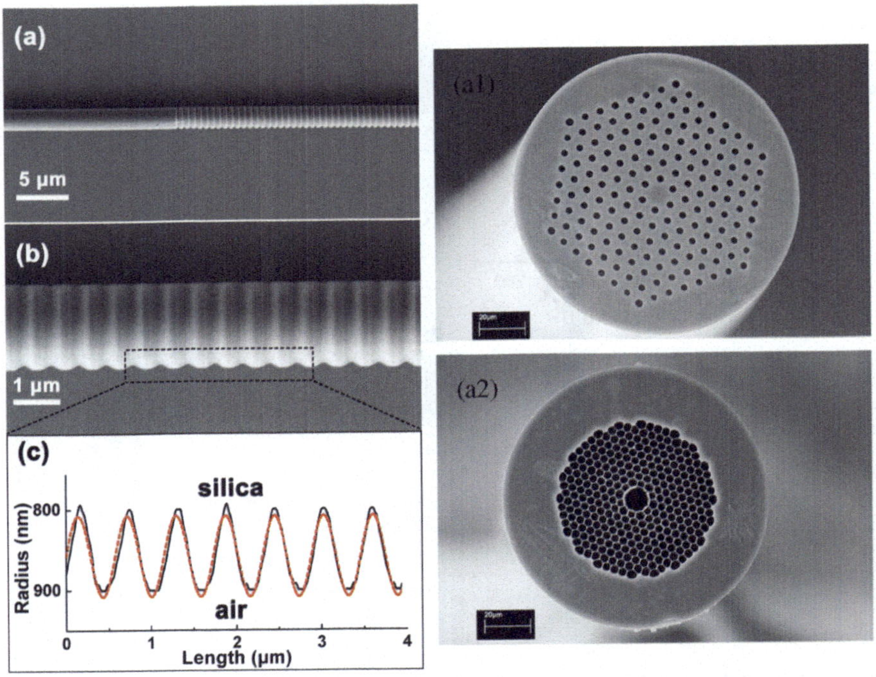

Fig. 5.31 (Left) Microfiber Bragg grating. Reprinted with permission from [63]. Copyright (2011) Optical Society of America. (Right) Solid-core (top) and hollow-core (bottom) photonic crystal fiber. Reprinted with permission from [64]. Copyright (2007) IEEE

photonic band gap fibers shown in Fig. 5.31 (right) [64]. Light guiding is provided by the photonic crystal band gap material around the hollow core. Sensitizing the air holes allows for the binding of biomaterial at specific positions. A combination of longitudinal and transverse periodic structuring has, for example, been demonstrated in [63]. For details on photonic crystal fiber sensors the reader is referred to Chap. 6 of this book. The remainder of this section is devoted to the class of three-dimensional (3D) photonic crystal structures used for sensing—and closely related 2D structures.

5.4.1 3D Photonic-Crystal Biosensors

Three-dimensional (3D) photonic crystals exhibit a periodic arrangement in the refractive index in all three spatial dimensions. Several different methods have been demonstrated for the fabrication of such 3D photonic crystals. Figure 5.3a shows an SEM image of a 3D photonic crystal fabricated using direct laser writing. The laser light illuminates a thick region of photoresist-type material at a wavelength that is greater than the single-photon polymerization threshold wavelength. For a tightly-focused laser beam the multi-photon polymerization threshold is reached locally in the focal region [19]. An alternative to direct laser writing is laser interference lithography. Polymer-based opal-type 3D photonic crystals have been fabricated using four-beam laser exposure [65]. Both fabrication methods are referred to as forms of top-down fabrication. A continuous layer of material is structured to obtain the 3D photonic crystal. A quite distinct fabrication process is the self-assembly of colloidal material into a regular lattice arrangement. Such a process is a form of bottom-up fabrication—where colloidal spheres are assembled into a large structure. Figure 5.3b shows an example of such a photonic crystal structure.

Silica colloids may be fabricated in solution by the Stöber process [66, 67]. Layer-by-layer deposition of hexagonally close-packed (in 2D) monolayers of silica colloids leads typically to the formation of 3D structures with face-centered cubic (fcc) symmetry. Fcc is favoured over hcp (in 3D) because of the forces involved in crystal formation, which are distance-dependent. This truth holds, even though the forces involved are many orders of magnitude smaller than those involved in the bonding of atoms in crystalline solid materials. Polymer-based opal-type 3D photonic crystals can also be readily fabricated. The processes for their realisation can resemble those used in forming silica-based opals, but with the polymer (near-)spheres typically formed in a separate process from the multi-layer deposition process involved in 3D photonic crystal formation.

Silica- or polymer-based opal-type photonic crystal structures typically provide a 3D periodic structure with the relatively low index contrast of approximately 1.5:1, assuming that the background refractive index is close to that of air ($n = 1$). Higher index contrast ratios are desirable in many situations. Therefore, inverted structures have been realised in e.g. *silicon*, taking advantage of the specifically different chemistries of *silica* and silicon [68]. The differential wet-etch chemistry exploited here is much the same as that used routinely in the silicon electronics industry. Silica is soluble in hydro-fluoric acid (HF), while silicon is soluble in sufficiently alkaline liquid—e.g. potassium hydroxide (KOH). Inversion of silica-based opal into polymeric inverse photonic crystal structures is also possible via selective chemistry. The HF that dissolves silica does not attack polymeric materials. Figure 5.3c shows an example of an inverted opal structure obtained by sol-gel formation of the inverted silica opal structure and subsequent removal of the polystyrene template with ethyl acetate [15].

Porosity is an important aspect when considering possible applications of 3D photonic crystal materials for sensing, whether explicitly chemical or biochemical in nature. The porosity scale may be labelled as nano- or micro- or even macro-porosity. Unfortunately, the terminology used is problematic—because 'nano-porous' structures based on inverted opal typically have pore sizes in the size range of hundreds of nanometres, i.e. they are really (sub-)micro-porous. In contrast, for instance, the pore sizes involved in what is now commonly called 'porous silicon' are much smaller. C. Pacholski [69] has reviewed photonic crystal sensors based on porous silicon. The structures realised may be multi-layers with alternating high-low-high-low... configurations (1D photonic crystals)—or 2D or 3D photonic crystals.

As for other formats, 3D photonic crystal structures can form the basis for bio-sensing and bio-medical sensing through a number of different approaches. These approaches include fluorescent molecular labelling of target species in immuno-reactions, label-free specific binding reactions—and modification of light emission from, e.g., deposited quantum dots. Colour change in 3D photonic crystal structures is one mechanism that can be exploited for bio-sensing, most obviously in intrinsically porous materials such as synthetic silica opal or inverse photonic crystal structures that are realised via inversion of synthetic opal. Colour changes can be produced, e.g., by *changes in the refractive index* of fluid material that permeates the pores of the 3D photonic crystal structure—or, alternatively, by *deposition of biomaterial* on the interior surfaces of the 3D photonic crystal—or by *swelling* (dilatation) *or contraction* of the photonic crystal lattice.

In 3D photonic-crystal hydrogel sensors the photonic crystal is embedded in the hydrogel [70, 71]. Fabrication chemistry needs to be carefully adjusted to allow for electrostatic self-assembly and to prevent disordering of the lattice by polymerization and functionalization of the hydrogel with recognition units [71]. Figure 5.32a shows the measurement principle based on the reversible contraction and swelling of the hydrogel causing a corresponding change in the lattice constant of the colloidal 3D photonic crystal. The readout is performed by colour evaluation in a reflection measurement. Such a sensing principle has been proposed for glucose sensing in tear fluid by integration of the hydrogel sensor into a contact lens [70]. An alternative to integration of the photonic crystal into the hydrogel is attaching the photonic crystal to surface of hydrogel. Under suitable conditions, hexagonally close-packed 2D photonic crystal lattices of nano-/micro-spheres can be produced that closely mimic the individual (111) planes of the fcc-symmetry 3D photonic crystal lattice. Figure 5.32b shows a schematic of such a 2D photonic crystal at the surface of a hydrogel [71].

Next, we shall consider one concrete example of a photonic hydrogel sensor for protein binding. As depicted in Fig. 5.33, the hydrogel is functionalized with biotin [18]. Upon the binding of the avidin, the hydrogel shrinks. This change in the period of the colloidal array is detected by monitoring the Debye diffraction ring. The sensor allows for a visual detection of 0.1 mg/ml avidin. The sensitivity is particularly high for low avidin concentrations. A picoMole (pM) to milliMole (mM) detection range has been given for the different types of photonic hydrogel sensor that have been cited in the review by A. K. Yetisen et al. [16]. A femtomole (fM) limit of detection was demonstrated by J. P. Walker and S. A. Asher [72]. The high sensitivity was achieved by using an enzyme—and the sensor is irreversible. An issue for photonic hydrogel sensors is their nonspecific response to temperature, pH and ionic concentration.

Figure 5.34 depicts a 3D photonic-crystal sensor that utilizes the superprism effect as a different means of detecting refractive index changes [73]. The light propagation angle in the photonic crystal changes as a function of the refractive index. In a theoretical and computational study, the exit position of the beam as a function of the refractive index is calculated by computing the normal to the dispersion surface to obtain the group velocity.

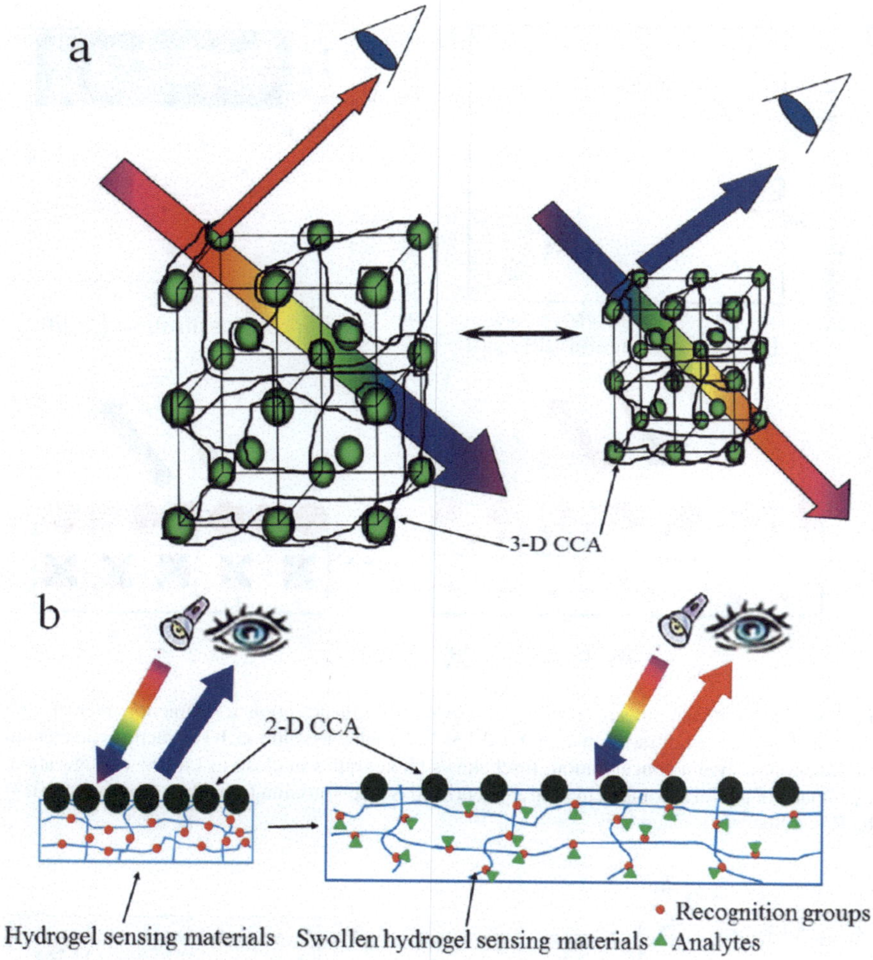

Fig. 5.32 **a** 3D photonic hydrogel sensor based on a crystalline colloidal array (CCA). A lattice change in the CCA is detected as a colour change. **b** 2-D photonic hydrogel sensor based on the same principle. Reprinted with permission from [71]. Copyright (2015) American Chemical Society

These last two sensing systems proposed by J.-T. Zhang et al. [18] and T. Prasad et al. [73] are inspiring because they offer alternative options for evaluating sensor signals besides the use of spectral measurements. J.-T. Zhang et al. [18] evaluate diffraction intensities and T. Prasad et al. [73] consider a spatial displacement. This demonstrates that system approaches beyond using a spectrometric evaluation are possible and may be advantageous for compact systems.

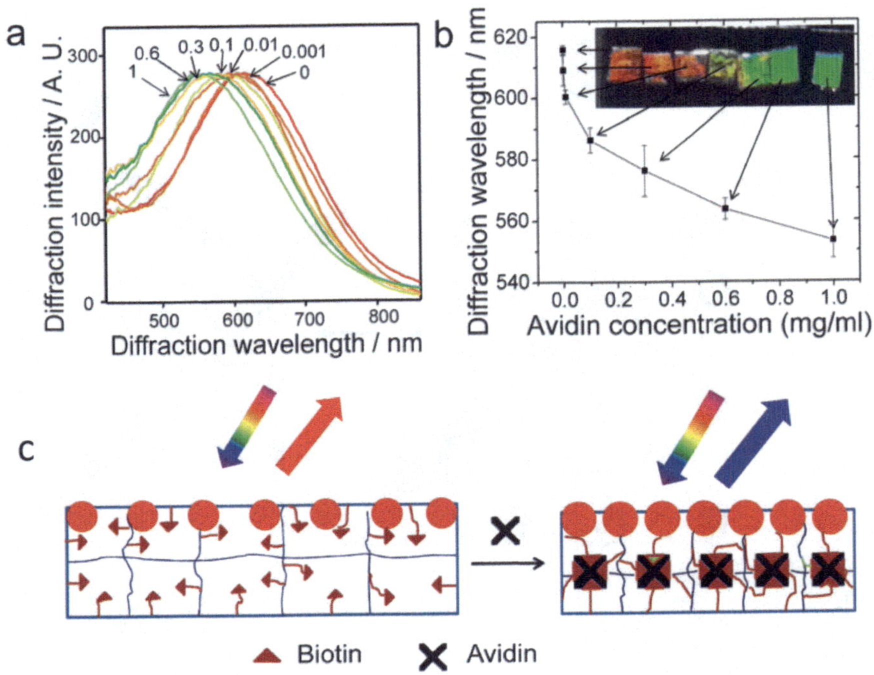

Fig. 5.33 2D biotin-functionalized hydrogel sensor for avidin detection. **a** Diffraction intensity as a function of avidin concentration in mg/ml in 0.1 M NaCl aqueous solution. **b** Diffraction wavelength as a function of avidin concentration. Inset shows photographs in close-to Littrow configuration. **c** Schematic of avidin binding in hydrogel. Reprinted with permission from [18]. Copyright (2013) The Royal Society of Chemistry

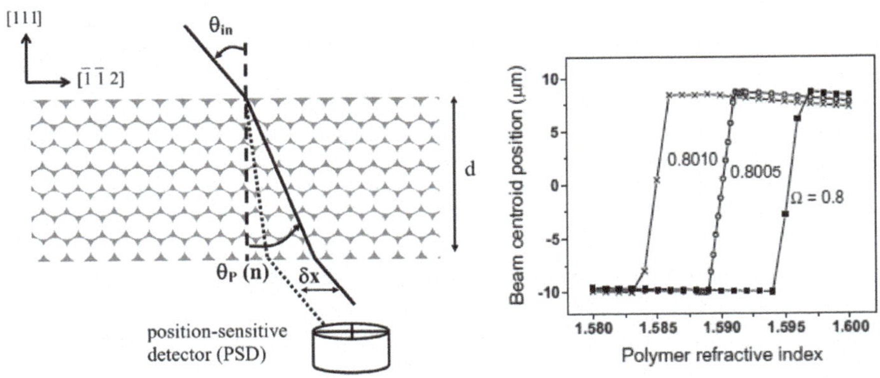

Fig. 5.34 3D photonic-crystal sensor employing superprism effect. The graph shows the calculated change of the exit beam position as a function of the refractive index. Reprinted with permission from [73]. Copyright (2006) Elsevier B.V.

5.5 Concluding Statement

This chapter has addressed the important topic of biomedical sensing based on photonic crystal (PhC) structures—attempting to provide a wide-ranging review, albeit not an exhaustive one. Our hope is that the interested reader will consult the literature that we have cited, in a critical manner—and be motivated to carry out their own further study and help to develop this field towards maturity. As other chapters in this book have shown, together with other work in the literature, photonic crystal structures are certainly not the only scientific and technological approach of interest for biomedical sensing applications. Despite the intrinsic attractions of photonic techniques, alternative approaches such as ones based on electrochemistry are available and strongly competitive [74]. Within the optical domain, surface plasmon resonance (SPR) sensing has an established position in label-free bio-medical sensing. In the appropriate (Kretschmann) configuration, a thin additional layer of bio-material modifies the exact angle (for a specified wavelength) at which light couples strongly to the surface plasmon mode of a thin shiny metallic layer deposited on a glass prism. The most typical metal used is gold, because of its high plasma resonance frequency and the narrowness of this resonance, allied to its chemical inertia. The SPR situation is termed a resonance because it is sharply defined in terms of the coupling angle of the light for plasmon propagation. The sharpness of the resonance—i.e. the narrowness of the range of angles over which coupling occurs for a specified optical wavelength—is primarily determined by the degree of collimation of the incident beam. In turn, the degree of collimation increases as the area of the sample that is illuminated by the coherent light beam used increases. In consequence, high resolution and sensitivity are synonymous with illumination of a large sample area—and therefore substantially inimical to miniaturisation.

Within the domain of optics or photonics based sensing techniques, other approaches based on plasmonic resonator structures [75], optoplasmonics [76, 77] meta-surfaces [78–80], stripe-waveguide ring resonators and stripe waveguide-based Mach-Zehnder interferometers [81–83] all offer attractive possibilities. These device structures may well be organised in arrays that have several thousand elements in them. Arrays of plasmonic devices (i.e. arrays of photonic 'atoms') supported by flat (or even gently curved) surfaces can be regarded as two-dimensional metamaterials or meta-surfaces. Arrays of such resonant 'atoms' are typically organised in regular patterns, but this is not an intrinsically essential requirement. In passing, we remark that periodic arrays of light scattering 'atoms' unavoidably have the potential to behave as photonic crystal structures and therefore to exhibit, for example, photonic bandgap behaviour. The conditions for exhibition of a photonic bandgap are determined by the wavelength used, in relation to the Bragg scattering angular dependence. Photonic-crystal slabs [44, 47, 48] such as those used in photonic-crystal enhanced microscopy (PCEM), are also a particular form of structured surface.

Measurement techniques based on some of these structures include 'surface enhanced Raman scattering' (SERS) [84, 85] and 'surface enhanced infra-red absorption' (SEIRA) spectroscopy [86–88]. SEIRA allows the spectroscopic identification of particular bio-molecules or classes of bio-molecules, via the specific chemical bond resonances that typically occur at mid-infrared wavelengths in organic materials. PCEM typically generates spectrally encoded images e.g. colour encoding of the depth of penetration of light into a cell—or local variations in the refractive index of the cell. Various spectroscopic and spectrometric approaches have been used in biomedical sensing and imaging—and have been demonstrated in the laboratory. In sufficiently well characterised situations, spectral scanning requirements may be either reduced or even eliminated. Organisation of bio-sensing devices in arrays or as structured surfaces implies the need for a multiply repeated scanning approach to the sensing process, with each element of an array being allocated a specific processing time during each field scan. In the structured surface (e.g. PCEM) approach, an unscanned 2D image may be simultaneously scanned spectrally—or one spatial dimension may be scanned and the second scanning dimension be encoded spectrally. In the present era of optical microscopy, electronic scanning, together with digital storage and processing, has become a standard approach—and approaches such as hyperspectral imaging are now routine.

Recently a review paper has appeared [89] that is concerned with resonant waveguide grating (RWG) structures—also known as guided-mode resonant (GMR) grating structures. Apart from the use of different nomenclature, these structures are essentially the same as one-dimensionally periodic photonic crystal slabs. This paper covers a wide range of theoretical/computational techniques, as well as addressing fabrication methods and a wide range of possible applications. These applications include fluorescence enhancement in bio-sensing. Another issue that we have not attempted to cover exhaustively is that of fluorescence labelling, as an alternative to reliance on, for example, specific binding in conjunction with the detection of small changes in refractive index that are identified and measured using resonant cavities, waveguide grating resonances or nulling interferometric approaches. In this chapter, the exploitation of refractive-index change (generalised, if need be, to mean change in the real part and/or the imaginary part of the complex refractive index) has played a central role. The most obvious alternative to the refractive-index based approach is the use of fluorescent labelling in the bio-analyte of interest. In particular, fluorescent labelling has been exploited in the 'classical' competition immunosensing reaction between antibodies and antigens, where the presence of specific antigens in a sample of analyte suppresses the fluorescence from identical antigens that have been fluorescently labelled and added into the analyte. P. C. Mathias et al. [90] have demonstrated that, by comparison with the situation for a plain glass substrate (i.e a microscope-slide), there is a remarkable (60-fold) enhancement of the fluorescence from the fluorophore cyanine-5, when it is spotted onto a streptavidin conjugated photonic crystal substrate. C. Wang et al. [91] have demonstrated detection of thrombin in the pM range with fluorescence enhancement in 3D PhC.

Only a few publications present a direct comparison with a standard technique or other techniques. Two publications providing a certain degree of comparability are the rotavirus detection performed by M. F. Pineda et al. [92] and B. Maeng et al. [93]. Both also perform a comparison with standard enzyme-linked immunosorbent assay (ELISA) tests. M. F. Pineda et al. [92] detect the rotavirus with a 1D grating PhC. They observe peak-wavelength shifts in the 10 pm-range. Even though B. Maeng et al. [93] report a much higher sensitivity, their limit of detection (LOD) appears rather similar to the one by M. F. Pineda et al. [92]. This demonstrates that, besides sensitivity, the system noise is another important factor for detecting low concentrations. In order to quantify the fundamental limit of detection, the fundamental noise sources need to be analyzed in addition to the system sensitivity. S. Kroker et al. [94] investigate photonic crystal designs that are optimized for low noise behavior.

In due course, i.e. on a ten or twenty year timescale, new and decisive winners in biomedical sensor device technology will surely have emerged. Success will result from the resolution of many issues, not least of which will be the ability to generate and protect the intellectual property aspects of a particular type of photonics based biomedical sensor device. Whatever the scientific merits or intellectually interesting content of a particular biomedical optical sensing concept or sensor device, the vital practical issues of reliability, reproducibility, manufacturability and cost will need to be addressed as central aspects of product development. Such issues should only be considered as providing a starting point in the quest for a successful, widely exploitable and useful approach to biomedical sensing. The qualities of the initial samples of bio-analyte and their pre-filtering, purification and preparation (e.g. the dilution of blood samples in water) for assessment will be of vital importance. Eventually various systems issues, including human factors and the techniques used for signal and data processing, will be of predominant importance for full implementation.

We conclude that photonic crystal structures in various forms offer many interesting possibilities for application in bio-medical sensing. Time, aided by a considerable measure of human endeavour, will tell if this promise will actually be fulfilled.

References

1. E. Yablonovitch, Inhibited spontaneous emission in solid-state physics and electronics. Phys. Rev. Lett. **58**, 2059–2062 (1987)
2. S. John, Strong localization of photons in certain disordered dielectric superlattices. Phys. Rev. Lett. **58**, 2486–2489 (1987)
3. J. Cooper, A. Glidle, R. De La Rue, Photonic crystal structures in molecular biosensing. Opt. Photonics News 26–31 (2010)
4. B. Troia, A. Paolicelli, F. De Leonardis, V.M.N. Passaro, Photonic crystals for optical sensing : a review. Adv. Photonic Cryst. 241–295 (2013)
5. M.C. Estevez, M. Alvarez, L.M. Lechuga, Integrated optical devices for lab-on-a-chip biosensing applications. Laser Photonics Rev. **6**, 463–487 (2012)
6. V.M.N. Passaro, B. Troia, M. La Notte, F. De Leonardis, P. Bari, Chapter 5: Chemical sensors based on photonic structures, in *Advances in Chemical Sensors*, ed. by W. Wang (InTech, 2012). ISBN: 978-953-307-792-5
7. D. Threm, Y. Nazirizadeh, M. Gerken, Photonic crystal biosensors towards on-chip integration. J. Biophotonics **5**, 601–616 (2012)
8. X. Fan, I.M. White, S.I. Shopova, H. Zhu, J.D. Suter, Y. Sun, Sensitive optical biosensors for unlabeled targets: a review. Anal. Chim. Acta **620**, 8–26 (2008)

9. R.V. Nair, R. Vijaya, Photonic crystal sensors: an overview. Prog. Quantum Electron. **34**, 89–134 (2010)
10. K. Chen, T. Duy Dao, T. Nagao, Tunable nanoantennas for surface enhanced infrared absorption spectroscopy by colloidal lithography and post-fabrication etching. Sci. Rep. **7**, 1–8 (2017)
11. W. Wu, A. Katsnelson, O.G. Memis, H. Mohseni, A deep sub-wavelength process for the formation of highly uniform arrays of nanoholes and nanopillars. Nanotechnology **18** (2007)
12. A.V. Whitney, B.D. Myers, R.P. Van Duyne, Sub-100 nm triangular nanopores fabricated with the reactive ion etching variant of nanosphere lithography and angle-resolved nanosphere lithography. Nano Lett. **4**, 1507–1511 (2004)
13. S. Jahns, M. Bräu, B.-O. Meyer, T. Karrock, S.B. Gutekunst, L. Blohm, C. Selhuber-Unkel, R. Buhmann, Y. Nazirizadeh, M. Gerken, Handheld imaging photonic crystal biosensor for multiplexed, label-free protein detection. Biomed. Opt. Express **6**, 3724 (2015)
14. S. Mandal, D. Erickson, Nanoscale optofluidic sensor arrays. Opt. Express **16**, 1623 (2008)
15. Y. Nishijima, K. Ueno, S. Juodkazis, V. Mizeikis, H. Misawa, T. Tanimura, K. Maeda, Inverse silica opal photonic crystals for optical sensing applications. Opt. Express **15**, 12979 (2007)
16. A.K. Yetisen, H. Butt, L.R. Volpatti, I. Pavlichenko, M. Humar, S.J.J. Kwok, H. Koo, K.S. Kim, I. Naydenova, A. Khademhosseini, S.K. Hahn, S.H. Yun, Photonic hydrogel sensors. Biotechnol. Adv. **34**, 250–271 (2016)
17. J.P. Couturier, M. Sütterlin, A. Laschewsky, C. Hettrich, E. Wischerhoff, Responsive inverse opal hydrogels for the sensing of macromolecules. Angew. Chemie Int. Ed. **54**, 6641–6644 (2015)
18. J.-T. Zhang, X. Chao, X. Liu, S.A. Asher, Two-dimensional array Debye ring diffraction protein recognition sensing. Chem. Commun. **49**, 6337 (2013)
19. M. Deubel, G. Von Freymann, M. Wegener, S. Pereira, K. Busch, C.M. Soukoulis, Direct laser writing of three-dimensional photonic-crystal templates for telecommunications. Nat. Mater. **3**, 444–447 (2004)
20. D. Rosenblatt, A. Sharon, A.A. Friesem, Resonant grating waveguide structures. IEEE J. Quantum Electron. **33**, 2038–2059 (1997)
21. Y. Nazirizadeh, T. Becker, J. Reverey, C. Selhuber-Unkel, D.H. Rapoport, U. Lemmer, M. Gerken, Photonic crystal slabs for surface contrast enhancement in microscopy of transparent objects. Opt. Express **20**, 14451 (2012)
22. R. Horvath, L.R. Lindvold, N.B. Larsen, Reverse-symmetry waveguides: theory and fabrication. Appl. Phys. B Lasers Opt. **74**, 383–393 (2002)
23. P.J. Bock, P. Cheben, J.H. Schmid, J. Lapointe, A. Delâge, S. Janz, G.C. Aers, D.-X. Xu, A. Densmore, T.J. Hall, Subwavelength grating periodic structures in silicon-on-insulator: a new type of microphotonic waveguide. Opt. Express **18**, 20251–20262 (2010)
24. D.L. Brundrett, E.N. Glytsis, T.K. Gaylord, J.M. Bendickson, Effects of modulation strength in guided-mode resonant subwavelength gratings at normal incidence. J. Opt. Soc. Am. A **17**, 1221 (2000)
25. S. Fan, J.D. Joannopoulos, Analysis of guided resonances in photonic crystal slabs. Phys. Rev. B Condens. Matter Mater. Phys. **65**, 1–8 (2002)
26. S.G. Tikhodeev, A.L. Yablonskii, E.A. Muljarov, N.A. Gippius, T. Ishihara, Quasiguided modes and optical properties of photonic crystal slabs. Phys. Rev. B Condens. Matter Mater. Phys. **66**, 451021–4510217 (2002)
27. G.A. Turnbull, P. Andrew, M.J. Jory, W.L. Barnes, I.D.W. Samuel, Relationship between photonic band structure and emission characteristics of a polymer distributed feedback laser. Phys. Rev. B **64**, 125122 (2001)
28. S.S. Wang, R. Magnusson, Theory and applications of guided-mode resonance filters. Appl. Opt. **32**, 2606 (1993)
29. S.M. Norton, G.M. Morris, T. Erdogan, Experimental investigation of resonant-grating filter lineshapes in comparison with theoretical models. J. Opt. Soc. Am. A **15**, 464 (1998)
30. D.K. Jacob, S.C. Dunn, M.G. Moharam, Normally incident resonant grating reflection filters for efficient narrow-band spectral filtering of finite beams. J. Opt. Soc. Am. A Opt. Image Sci. Vis. **18**, 2109–2120 (2001)

31. W.L. Barnes, T.W. Preist, S.C. Kitson, J.R. Sambles, Physical origin of photonic energy gaps in the propagation of surface plasmas on gratings. Phys. Rev. B: Condens. Matter **54**, 6227–6244 (1996)
32. T. Frank, O. Buchnev, T. Cookson, M. Kaczmarek, P. Lagoudakis, V.A. Fedotov, Discriminating between coherent and incoherent light with planar metamaterials. Nano Lett. **19**, 6869–6875 (2019)
33. J. Yang, H. Giessen, P. Lalanne, Simple analytical expression for the peak-frequency shifts of plasmonic resonances for sensing. Nano Lett. **15**, 3439–3444 (2015)
34. R. Horvath, H.C. Pedersen, N. Skivesen, D. Selmeczi, N.B. Larsen, Monitoring of living cell attachment and spreading using reverse symmetry waveguide sensing. Appl. Phys. Lett. **86**, 1–3 (2005)
35. I.D. Block, N. Ganesh, M. Lu, B.T. Cunningham, A sensitivity model for predicting photonic crystal biosensor performance. IEEE Sens. J. **8**, 274–280 (2008)
36. Y. Nazirizadeh, F. von Oertzen, K. Plewa, N. Barié, P.-J. Jakobs, M. Guttmann, H. Leiste, M. Gerken, Sensitivity optimization of injection-molded photonic crystal slabs for biosensing applications. Opt. Mater. Express **3**, 556 (2013)
37. I.M. White, X. Fan, On the performance quantification of resonant refractive index sensors. Opt. Express **16**, 1020 (2008)
38. M. Huang, A.A. Yanik, T.-Y. Chang, H. Altug, Sub-wavelength nanofluidics in photonic crystal sensors. Opt. Express **17**, 24224 (2009)
39. J.O. Grepstad, P. Kaspar, O. Solgaard, I.-R. Johansen, A.S. Sudbø, Photonic-crystal membranes for optical detection of single nano-particles, designed for biosensor application. Opt. Express **20**, 7954 (2012)
40. N. Orgovan, B. Kovacs, E. Farkas, B. Szabó, N. Zaytseva, Y. Fang, R. Horvath, Bulk and surface sensitivity of a resonant waveguide grating imager. Appl. Phys. Lett. **104** (2014)
41. Y. Fang, Label-free cell-based assays with optical biosensors in drug discovery. Assay Drug Dev. Technol. **4**, 583–595 (2006)
42. Y. Nazirizadeh, V. Behrends, A. Prósz, N. Orgovan, R. Horvath, A.M. Ferrie, Y. Fang, C. Selhuber-Unkel, M. Gerken, Intensity interrogation near cutoff resonance for label-free cellular profiling. Sci. Rep. **6**, 24685 (2016)
43. D. Gallegos, K.D. Long, H. Yu, P.P. Clark, Y. Lin, S. George, P. Nath, B.T. Cunningham, Label-free biodetection using a smartphone. Lab Chip **13**, 2124 (2013)
44. W. Chen, K.D. Long, M. Lu, V. Chaudhery, H. Yu, J.S. Choi, J. Polans, Y. Zhuo, B.A.C. Harley, B.T. Cunningham, Photonic crystal enhanced microscopy for imaging of live cell adhesion. Analyst **138**, 5886 (2013)
45. P.Y. Liu, L.K. Chin, W. Ser, H.F. Chen, C.-M. Hsieh, C.-H. Lee, K.-B. Sung, T.C. Ayi, P.H. Yap, B. Liedberg, K. Wang, T. Bourouina, Y. Leprince-Wang, Cell refractive index for cell biology and disease diagnosis: past, present and future. Lab Chip **16**, 634–644 (2016)
46. I.D. Block, P.C. Mathias, S.I. Jones, L.O. Vodkin, B.T. Cunningham, Optimizing the spatial resolution of photonic crystal label-free imaging. Appl. Opt. **48**, 6567–6574 (2009)
47. G.J. Triggs, M. Fischer, D. Stellinga, M.G. Scullion, G.J.O. Evans, T.F. Krauss, Spatial resolution and refractive index contrast of resonant photonic crystal surfaces for biosensing. IEEE Photonics J. **7** (2015)
48. Y. Nazirizadeh, J. Reverey, U. Geyer, U. Lemmer, C. Selhuber-Unkel, M. Gerken, Material-based three-dimensional imaging with nanostructured surfaces. Appl. Phys. Lett. **102**, 011116 (2013)
49. W. Chen, K.D. Long, H. Yu, Y. Tan, J.S. Choi, B.A. Harley, B.T. Cunningham, Enhanced live cell imaging via photonic crystal enhanced fluorescence microscopy. Analyst **139**, 5954–5963 (2014)
50. J. García-Rupérez, V. Toccafondo, M.J. Bañuls, J.G. Castelló, A. Griol, S. Peransi-Llopis, Á. Maquieira, Label-free antibody detection using band edge fringes in SOI planar photonic crystal waveguides in the slow-light regime. Opt. Express **18**, 24276–24286 (2010)
51. W.C.L. Hopman, P. Pottier, D. Yudistira, J. Van Lith, P.V. Lambeck, R.M. De La Rue, A. Driessen, H.J.W.M. Hoekstra, R.M. De Ridder, Quasi-one-dimensional photonic crystal as a

compact building-block for refractometric optical sensors. IEEE J. Sel. Top. Quantum Electron. **11**, 11–15 (2005)
52. N. Skivesen, A. Têtu, M. Kristensen, J. Kjems, L.H. Frandsen, P.I. Borel, Photonic-crystal waveguide biosensor. Opt. Express **15**, 3169 (2007)
53. E. Chow, A. Grot, L.W. Mirkarimi, M. Sigalas, G. Girolami, Ultracompact biochemical sensor built with two-dimensional photonic crystal microcavity. Opt. Lett. **29**, 1093–1095 (2004)
54. M.R. Lee, P.M. Fauchet, Two-dimensional silicon photonic crystal based biosensing platform for protein detection. Opt. Express **15**, 4530 (2007)
55. Y.-F. Chen, X. Serey, R. Sarkar, P. Chen, D. Erickson, Controlled photonic manipulation of proteins and other nanomaterials. Nano Lett. **12**, 1633–1637 (2012)
56. S. Zlatanovic, L.W. Mirkarimi, M.M. Sigalas, M.A. Bynum, E. Chow, K.M. Robotti, G.W. Burr, S. Esener, A. Grot, Photonic crystal microcavity sensor for ultracompact monitoring of reaction kinetics and protein concentration. Sens. Actuators B Chem. **141**, 13–19 (2009)
57. W.-C. Lai, S. Chakravarty, Y. Zou, R.T. Chen, Silicon nano-membrane based photonic crystal microcavities for high sensitivity bio-sensing. Opt. Lett. **37**, 1208 (2012)
58. T. Xu, N. Zhu, M.Y.-C. Xu, L. Wosinski, J.S. Aitchison, H.E. Ruda, Pillar-array based optical sensor. Opt. Express **18**, 5420–5425 (2010)
59. A.V. Kabashin, P. Evans, S. Pastkovsky, W. Hendren, G.A. Wurtz, R. Atkinson, R. Pollard, V.A. Podolskiy, A.V. Zayats, Plasmonic nanorod metamaterials for biosensing. Nat. Mater. **8**, 867–871 (2009)
60. Y. Zou, S. Chakravarty, D.N. Kwong, W.-C. Lai, X. Xu, X. Lin, A. Hosseini, R.T. Chen, Cavity-waveguide coupling engineered high sensitivity silicon photonic crystal microcavity biosensors with high yield. IEEE J. Sel. Top. Quantum Electron. **20**, 171–180 (2014)
61. H.A. Macleod, *Thin-Film Optical Filters* (Institute of Physics Publishing, 2001)
62. Z. Wang, J. Zhang, J. Li, J. Xie, Y. Li, S. Liang, Z. Tian, C. Li, Z. Wang, T. Wang, H. Zhang, B. Yang, Colorful detection of organic solvents based on responsive organic/inorganic hybrid one-dimensional photonic crystals. J. Mater. Chem. **21**, 1264–1270 (2011)
63. Y. Liu, C. Meng, A.P. Zhang, Y. Xiao, H. Yu, L. Tong, Compact microfiber Bragg gratings with high-index contrast. Opt. Lett. **36**, 3115 (2011)
64. L. Xiao, M.S. Demokan, W. Jin, Y. Wang, C.L. Zhao, Fusion splicing photonic crystal fibers and conventional single-mode fibers: microhole collapse effect. J. Lightwave Technol. **25**, 3563–3574 (2007)
65. M. Campbell, D.N. Sharp, M.T. Harrison, R.G. Denning, A.J. Turberfield, Fabrication of photonic crystals for the visible spectrum by holographic lithography. Nature **404**, 53–56 (2000)
66. E.B.W. Stöber, A. Fink, Controlled growth of monodisperse silica spheres in the micron size range. Jounal Colloid Interface Sci. **26**, 62–69 (1968)
67. N. Plumeré, A. Ruff, B. Speiser, V. Feldmann, H.A. Mayer, Stöber silica particles as basis for redox modifications: particle shape, size, polydispersity, and porosity. J. Colloid Interface Sci. **368**, 208–219 (2012)
68. A. Blanco, E. Chomski, S. Grabtchak, M. Ibisate, S. John, S.W. Leonard, C. Lopez, F. Meseguer, H. Miguez, J.P. Mondia, G.A. Ozin, O. Toader, H.M. van Driel, Large-scale synthesis of a silicon photonic crystal with a complete three-dimensional bandgap **405**, 437–440 (2000)
69. C. Pacholski, Photonic crystal sensors based on porous silicon. Sensors **13**, 4694–4713 (2013)
70. V.L. Alexeev, S. Das, D.N. Finegold, S.A. Asher, Photonic crystal glucose-sensing material for noninvasive monitoring of glucose in tear fluid. Clin. Chem. **50**, 2353–2360 (2004)
71. Z. Cai, N.L. Smith, J.T. Zhang, S.A. Asher, Two-dimensional photonic crystal chemical and biomolecular sensors. Anal. Chem. **87**, 5013–5025 (2015)
72. J.P. Walker, S.A. Asher, Acetylcholinesterase-based organophosphate nerve agent sensing photonic crystal. Anal. Chem. **77**, 1596–1600 (2005)
73. T. Prasad, D.M. Mittleman, V.L. Colvin, A photonic crystal sensor based on the superprism effect. Opt. Mater. (Amst.) **29**, 56–59 (2006)
74. S. Laschi, M. Mascini, Planar electrochemical sensors for biomedical applications. Med. Eng. Phys. **28**, 934–943 (2006)

75. B. Lahiri, A.Z. Khokhar, R.M. De La Rue, S.G. McMeekin, N.P. Johnson, Asymmetric split ring resonators for optical sensing of organic materials. Opt. Express **17**, 1107 (2009)
76. F. De Angelis, M. Patrini, G. Das, I. Maksymov, M. Galli, L. Businaro, L.C. Andreani, E. Di Fabrizio, A hybrid plasmonic–photonic nanodevice for label-free detection of a few molecules. Nano Lett. **8**, 2321–2327 (2008)
77. J. Xavier, S. Vincent, F. Meder, F. Vollmer, Advances in optoplasmonic sensors—combining optical nano/microcavities and photonic crystals with plasmonic nanostructures and nanoparticles. Nanophotonics **7**, 1–38 (2018)
78. A.V. Kildishev, A. Boltasseva, V.M. Shalaev, Planar photonics with metasurfaces. Science (80-) **339**, 1232009–1232009 (2013)
79. K.E. Chong, I. Staude, A. James, J. Dominguez, S. Liu, S. Campione, G.S. Subramania, T.S. Luk, M. Decker, D.N. Neshev, I. Brener, Y.S. Kivshar, Polarization-independent silicon metadevices for efficient optical wavefront control. Nano Lett. **15**, 5369–5374 (2015)
80. F. Callewaert, V. Velev, P. Kumar, A.V. Sahakian, K. Aydin, Inverse-designed broadband all-dielectric electromagnetic metadevices. Sci. Rep. **8**, 1–8 (2018)
81. A. Yalçin, K.C. Popat, J.C. Aldridge, T.A. Desai, J. Hryniewicz, N. Chbouki, B.E. Little, O. King, V. Van, S. Chu, D. Gill, M. Anthes-Washburn, M.S. Ünlü, B.B. Goldberg, Optical sensing of biomolecules using microring resonators. IEEE J. Sel. Top. Quantum Electron. **12**, 148–154 (2006)
82. D.X. Xu, A. Densmore, A. Delâge, P. Waldron, R. McKinnon, S. Janz, J. Lapointe, G. Lopinski, T. Mischki, E. Post, P. Cheben, J.H. Schmid, Folded cavity SOI microring sensors for high sensitivity and real time measurement of biomolecular binding. Opt. Express **16**, 15137 (2008)
83. A. Densmore, M. Vachon, D.-X. Xu, S. Janz, R. Ma, Y.-H. Li, G. Lopinski, A. Delâge, J. Lapointe, C.C. Luebbert, Q.Y. Liu, P. Cheben, J.H. Schmid, Silicon photonic wire biosensor array for multiplexed real-time and label-free molecular detection. Opt. Lett. **34**, 3598 (2009)
84. H.-Y. Wu, C.J. Choi, B.T. Cunningham, Plasmonic nanogap-enhanced Raman scattering using a resonant nanodome array. Small **8**, 2878–2885 (2012)
85. S.H. Yazdi, I.M. White, A nanoporous optofluidic microsystem for highly sensitive and repeatable surface enhanced Raman spectroscopy detection. Biomicrofluidics **6**, 014105 (2012)
86. J. Kundu, F. Le, P. Nordlander, N.J. Halas, Surface enhanced infrared absorption (SEIRA) spectroscopy on nanoshell aggregate substrates. Chem. Phys. Lett. **452**, 115–119 (2008)
87. R. Bukasov, J.S. Shumaker-Parry, Silver nanocrescents with infrared plasmonic properties as tunable substrates for surface enhanced infrared absorption spectroscopy. Anal. Chem. **81**, 4531–4535 (2009)
88. A. Tittl, A. Leitis, M. Liu, F. Yesilkoy, D.-Y. Choi, D.N. Neshev, Y.S. Kivshar, H. Altug, Imaging-based molecular barcoding with pixelated dielectric metasurfaces. Science (80-) **360**, 1105–1109 (2018)
89. G. Quaranta, G. Basset, O.J.F. Martin, B. Gallinet, Recent advances in resonant waveguide gratings. Laser Photonics Rev. **12**, 1800017 (2018)
90. P.C. Mathias, H.-Y. Wu, B.T. Cunningham, Employing two distinct photonic crystal resonances to improve fluorescence enhancement. Appl. Phys. Lett. **95**, 021111 (2009)
91. C. Wang, C.Y. Lim, E. Choi, Y. Park, J. Park, Highly sensitive user friendly thrombin detection using emission light guidance from quantum dots-aptamer beacons in 3-dimensional photonic crystal. Sens. Actuators B Chem. **223**, 372–378 (2016)
92. M.F. Pineda, L.L.-Y. Chan, T. Kuhlenschmidt, C.J. Choi, M. Kuhlenschmidt, B.T. Cunningham, Rapid specific and label-free detection of porcine rotavirus using photonic crystal biosensors. IEEE Sens. J. **9**, 470–477 (2009)
93. B. Maeng, Y. Park, J. Park, Direct label-free detection of Rotavirus using a hydrogel based nanoporous photonic crystal. RSC Adv. **6**, 7384–7390 (2016)
94. S. Kroker, J. Dickmann, C.B. Rojas Hurtado, D. Heinert, R. Nawrodt, Y. Levin, S.P. Vyatchanin, Brownian thermal noise in functional optical surfaces. Phys. Rev. D **96**, 022002 (2017)

Chapter 6
Optical Fiber Sensors

Ming Ding and Gilberto Brambilla

Abstract Fiber optics represents a platform suitable for the monitoring of numerous physical properties. In biology and medicine, optical fibers have found a range of applications ranging from diagnosis to therapy such as cavitational and endoscopic laser surgery. This chapter reviews optical fibers and their wealth of applications in biomedical sensing: from cellular microenvironment to pH, gas, temperature, pressure, and blood flow.

6.1 Optical Fibers

6.1.1 Solid Core Fibers

Optical fibers are cylindrical waveguides that generally consist of an inner glass core with a refractive index n_1, surrounded by a glass or polymeric cladding of lower refractive index n_2—and are often covered by layers of a plastic coating (Fig. 6.1).

Optical fibers are frequently classified as single-mode fibers or multi-mode fibers according to the number of modes supported by the core. If the core refractive index profile is uniform, the fiber is called a "step-index fiber", while if it gradually decreases along the radial coordinate, the fiber is called a "graded-index fiber". In glass optical fibers, the core is typically made from a silicate glass doped with oxides of germanium, phosphorus and other elements in smaller amounts, while the cladding is made of pure silica or fluorosilicate glasses. Index guiding fibers have also been

M. Ding
School of Instrument Science and Opto-electronics Engineering, Beihang University, Beijing 100191, People's Republic of China
e-mail: mingding@buaa.edu.cn

G. Brambilla (✉)
Optoelectronics Research Centre, University of Southampton, Southampton SO17 1BJ, UK
e-mail: gb2@orc.soton.ac.uk

R. De La Rue et al. (eds.), *Biomedical Optical Sensors*,
Biological and Medical Physics, Biomedical Engineering,
https://doi.org/10.1007/978-3-030-48387-6_6

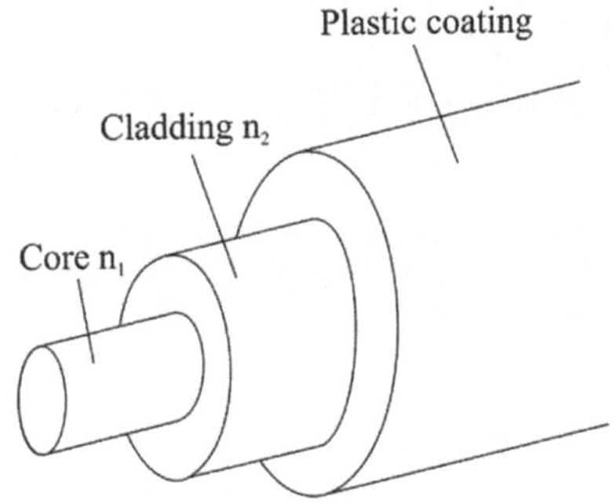

Fig. 6.1 Optical fiber geometry

fabricated from polymers, natural fibers (i.e. silk) and liquid materials, but their applications in the biomedical field have been somewhat restricted.

6.1.2 Microstructured Fibers

Glass and polymer fibers have also been manufactured from a single solid material. These fibers, often called microstructured fibers or photonic crystal fibers (PCFs), are structures where guiding is achieved by a regular spatial distribution of micrometre size air holes [120]. By tailoring the hole sizes and positions, microstructured fibers can vary the dispersion and confinement—and even confine light in a lower refractive index hollow core. Microstructured fibers are often classified into two categories: solid-core and hollow-core fibers.

In the solid-core microstructured fibers (Fig. 6.2a), waveguiding occurs through total internal reflection, in a similar manner to that of traditional optical fibers, with a central core that has a refractive index larger than the average refractive index of the cladding. However, there is no definite boundary between the solid-core region and the air-hole cladding region.

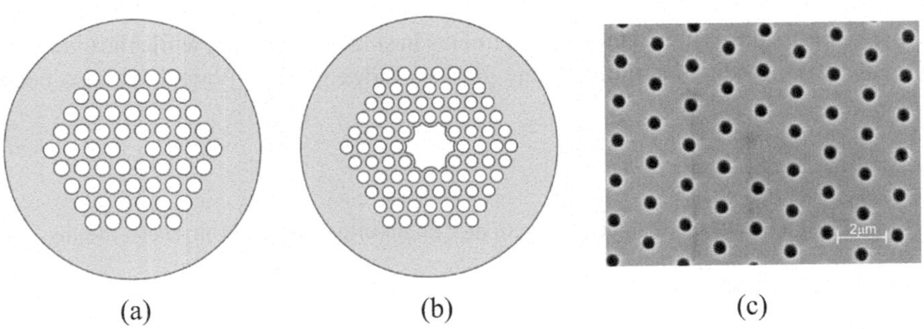

Fig. 6.2 Schematic diagram of **a** a solid-core PCF and **b** a hollow-core photonic crystal fiber (PCF); **c** SEM image of the PCF core region

In hollow-core microstructured fibers (Fig. 6.2b), the air-guided mode has an effective index that is lower than the effective index of the cladding. This type of fiber requires strict control of the periodic cladding structures, since guiding strongly depends on the fiber geometry, with the holes and the related pitch having micrometre dimensions (Fig. 6.2c). Because of the limited applications to biomedical sensing, only fibers having a core refractive index that is larger than the cladding refractive index will be considered in this chapter. These optical fibers are often called index guiding fibers.

6.1.3 Optical Micro/Nanofibers

Optical micro/nanofibers (MNFs) are optical fibers that have been tapered until their uniform waist region has a size comparable to the wavelength. MNFs are fabricated by heating conventional optical fibers to the softening temperature under a moderate degree of stretching [16]. This results in a micrometric optical fiber that is connected by biconical tapers [13, 134] to other fiberized components. MNFs have been manufactured in numerous different optical materials, including bismuthate [17] lead silicate [17], phosphate [146], tellurite [146], and chalcogenide glasses [92]—and a variety of polymers [47, 49, 54, 102, 159]. The extraordinary optical properties exhibited by MNFs include large evanescent fields, strong optical confinement, flexibility, configurability and robustness. Such desirable features have made MNFs an excellent platform for applications in sensing. Light propagation in MNFs is easily explained by refractive index guiding, where the core is effectively the optical fiber material and the cladding is represented by the surrounding medium, usually air or an aqueous solution.

6.1.4 Light Propagation

The propagation of light in index guiding fibers is usually explained either in terms of rays or of waves [106, 129]. In the ray theory, a light ray entering the fiber from a medium with a refractive index n_0 is reflected at the boundary between the fiber core and the fiber cladding (Fig. 6.3). If the incident angle is smaller than θ_{NA}, then the reflection angle at the core/cladding interface is greater than the critical angle θ_c, and the light is totally reflected; θ_{NA} is therefore called the maximum acceptance angle. If the medium outside the optical fiber is air with a refractive index value of $n_0 \approx 1$, the application of Snell's law at the air-fiber and at the core-cladding interfaces provides an expression for θ_{NA}:

$$\sin\theta_{NA} = \sqrt{n_1^2 - n_2^2} = n_1 n \sqrt{2\Delta} = \mathrm{NA} \tag{6.1}$$

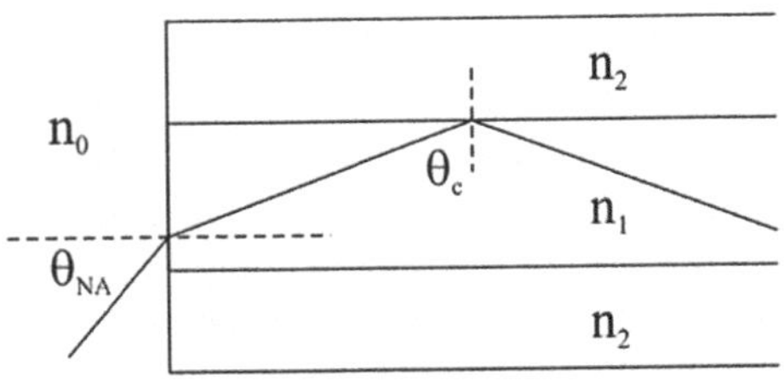

Fig. 6.3 Ray propagation by total internal reflection at the core/cladding interface in optical fibers. θ_c, θ_{NA}, n_1, n_2, and n_0 represent the critical angle at the core-cladding interface, the incident angle at the optical fiber end surface and the refractive indices of the core, the cladding and the air, respectively

where NA is the numerical aperture of the fiber and $\Delta = (n_1 - n_2)/n_1$ is the normalized refractive index difference.

In wave theory, the solution of Maxwell's equations provides orthonormal field distributions called "modes". Assuming a step index profile, the solution of Maxwell's equations in cylindrical coordinates leads to the propagation equation:

$$\frac{d^2\psi}{dr^2} + \frac{1}{r}\frac{d\psi}{dr} + \frac{1}{r^2}\frac{d\psi}{d\phi} + \left(k^2 - \beta^2\right) = 0 \tag{6.2}$$

where ψ is the wave function of the guided light, r is the radial coordinate, ϕ is the azimuthal coordinate, k is the wave vector in the medium and β is the modal propagation constant.

If the wave-function is assumed to be a propagating wave, then the solutions of (6.2) are expressed by linear combinations of Bessel functions. The boundary conditions require that the optical field has a finite value on the axis ($r = 0$) and tends to zero for very large distances from the fiber axis ($r \to \infty$), providing an easy mathematical formulation for the longitudinal field components:

$$\begin{aligned} &AJ_v(ur/a)e^{iv\phi} \quad \text{for}\, r < a \text{ (in the core)} \\ &BK_v(wr/a)e^{iv\phi} \quad \text{for}\, r > a \text{ (in the cladding)} \end{aligned} \tag{6.3}$$

where $J_v(ur/a)$ and $K_v(wr/a)$ are Bessel functions of the first and second kind, and $u^2 = \left(k_1^2 - \beta^2\right)a^2$, $k_1 = 2\pi n_1/\lambda_0$, $w^2 = \left(\beta^2 - k_2^2\right)a^2$, $k_1 = 2\pi n_2/\lambda_0$ and $w^2 + u^2 = V^2$.

The functions in (6.3) are called guided modes and are quantised. While J_v represents the fraction of the mode propagating inside the core, K_v describes the fraction of the mode which lies outside the core and is often called the "evanescent field". A specific geometry and refractive index profile can support only selected values of the propagation constant β. The parameter V is related only to the optical fiber geometry:

$$V = \frac{2\pi a}{\lambda_0}(NA)^2 \tag{6.4}$$

and determines the number of modes supported by an optical fiber waveguide: at $V < 2.405$, only a single-mode, designated as HE_{11}, is supported, while at $V > 2.405$ the number of modes increases with increasing V and asymptotically tends to $V^2/2$. It is important to note that the fibers that are commonly called ‘single-moded’ in reality support two modes with orthogonal polarizations—and the HE_{11} mode is therefore said to be degenerate. Moreover, the choice of angular position is arbitrary, as the mode is a continuously angularly degenerate mode—i.e. it has infinite degeneracy in the azimuthal coordinate.

In most practical fibers, the normalized index difference Δ is on the order of 1% and therefore the approximation $n_2/n_1 \approx 1$ is applicable, which is called the weak-guidance approximation [127, 128]: the modes are called linearly polarized [44] and designated as *LP* modes. The fundamental mode is designated as LP_{01}.

6.1.5 Optical Fiber Sensor Technologies

Fiber optics represents a platform that can be applied to the sensing of a number of physical properties, including displacement, temperature, pressure, rotation, sound, strain, magnetic field, electric field, radiation, flow, liquid level, chemical analysis, and vibration, to cite just a few examples.

Optical fiber sensors are commonly grouped into *distributed* or *point* sensors, according to whether their sensing can be performed along the length of the fiber or at a specific single position, usually at one extremity of the fiber. They can also be divided according to the output used to evaluate a measurand: phase or intensity [28, 43, 45, 46].

Intensity-based sensors rely on external physical perturbations, or mechanical transducers in close contact with the fiber, to produce a change in the transmitted intensity, usually through increased attenuation, bending, reflection, scattering, or fluorescence. Intensity-modulated sensors can use large core fibers (or fiber bundles) and cheap detection systems—and have therefore found widespread application in industry.

Phase-based sensors evaluate the change in the phase of the signal propagating in a sensing fiber, usually by comparison with a reference in an interferometer. Figure 6.4 shows a schematic diagram of a typical Mach-Zehnder interferometer: light from a laser source is split into a reference fiber unaffected by external perturbations and a sensing fiber that is exposed to the surrounding environment—and the light is then recombined by means of a fiber coupler, before reaching a detector. If the two components are exactly in phase upon recombination, they interfere constructively—while, if they are out of phase, destructive interference occurs. The two components are usually deliberately placed in quadrature, meaning that a small change in the environmental conditions results in the maximum relative change in intensity.

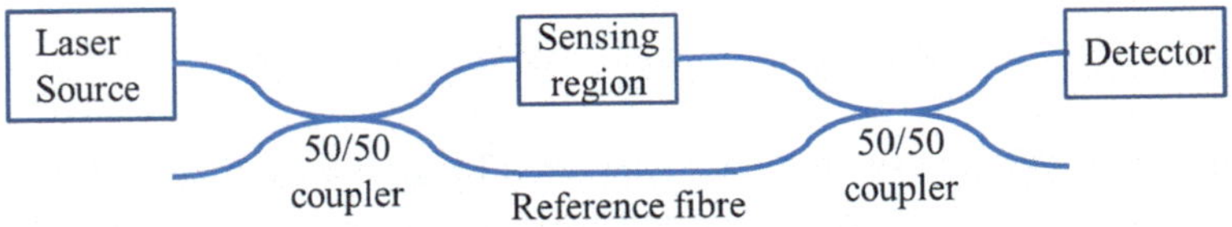

Fig. 6.4 Schematic of a sensor based on the Mach-Zehnder interferometer. Light is injected from a laser source into a 50/50 coupler, where it is split in two equal parts traveling through a reference fiber and a sensing fiber, respectively. In the sensing fiber the signal experiences a phase delay which, when recombined with the reference at the second 50/50 coupler, is measured by the detector in the form of an intensity change due to interference

In general, phase-based sensors have a higher accuracy and a much larger dynamic range than intensity-based sensors, since phase-differences can be measured with extreme precision, often to down to the tens of μrad level, within a very broad range. Yet, phase-modulated sensors often require a more expensive detection system—and therefore have application in the distributed sensing market, where the cost per unit of length is small.

Amongst the point sensors, a significant fraction of the industrial implementations has relied on fiber gratings, which are structures inscribed in the fiber core by periodically changing the refractive index. There are two types of gratings: Fiber Bragg Gratings (FBGs) and Long Period Gratings (LPGs). While the former has a period comparable to the wavelength of the light propagating in the core—and couples forward propagating core modes with backward propagating core modes, the latter is characterised by a period that is orders of magnitude larger than the wavelength and couples forward propagating core modes with cladding modes [37]. Because of the relatively strong photosensitivity effects exhibited by optical fibers when exposed to ultraviolet light, fiber grating fabrication has widely exploited ultraviolet lasers, often in conjunction with phase masks (for FBGs) or amplitude masks (for LPGs).

6.2 Sensors

Biological and medical sensing have benefited extensively from the use of optical fibers, because of their minimal cross section, negligible loss and extreme flexibility. Indeed, it is widely recognised that optical fibers found applications in medical imaging well before their application in telecommunications: in 1954 fiber bundles were successfully deployed in endoscopy [62]—providing easy access to otherwise inaccessible parts of the human body and thereby paving the way to beam delivery for endoscopic laser surgery.

This section will review the wealth of applications of optical fibers in biosensors that have arisen since their first deployment in endoscopy. This review will include biomolecule sensing, blood flow, gas, pressure, pH, and temperature.

6.2.1 Biomolecules

Optical fiber biosensors mostly use surface functionalization with biological molecules, such as antibodies or enzymes, that show high selectivity for a specific analyte. When the analyte binds to the functionalised surfaces, there is a change in an optical property (typically absorption, refractive index, luminescence, fluorescence emission or quenching) that can then be used for detection. Biomolecular sensing can be carried out in two distinct configurations, i.e. by exploiting the mode evanescent field along the direction of propagation—or in the so-called optrode configuration.

In the optrode configuration, the distant fiber end is functionalised with recognition biomolecules (Fig. 6.5) and the optical fiber simply delivers light to the sensing region at the fiber tip—and then collects it, after reflection, back to the detector. Functionalizing molecules are frequently encapsulated inside a polymer, a membrane or beads [11, 15, 41, 57, 85, 147] that often also perform the additional task of concentrating the analyte to be detected. Although the sensor selectivity in optrode biosensors can be extremely high, the sensitivity is limited by the size of the functionalised area and by the concentration of recognition molecules at the surface.

When the signal to be detected is considerably smaller than that used for excitation, collection can be performed using a second fiber, a CCD (charge coupled device) camera or an inverted fluorescence microscope. This method found large application in the simultaneous detection of multiple analytes using fiber bundles [2, 39, 57], where each single core transmits information about a single analyte. Microstructured fibers have been widely deployed in sensors exploiting fluorescence because the air holes provide easily functionalisable surfaces in close proximity to the core evanescent field. α-streptavidin and α-CRP (cancer-reactive protein) antibodies have been functionalised inside polymer microstructured fibers and fluorescence—and have provided a detection limit of 80 nM when analysing a sample volume of 27 μL [64].

A sensor configuration exploiting evanescent fields was first proposed in 1965 [60]—and later applied to immunoassays [80]. In this class of sensors, the evanescent field overlaps with recognition molecules bound to the fiber surface (Fig. 6.6). As the mode propagating in the core has an extremely small overlap with the cladding surface, part of the cladding is usually removed from the fiber until the guided mode has a significant overlap with the molecules in the solution. The evanescent

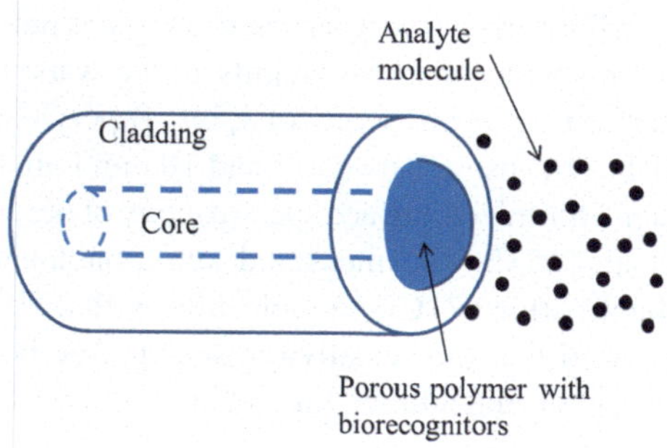

Fig. 6.5 The optrode configuration: a biological recognitor is immobilized at the distant end of an optical fiber, where it interacts with the analyte molecules in solution

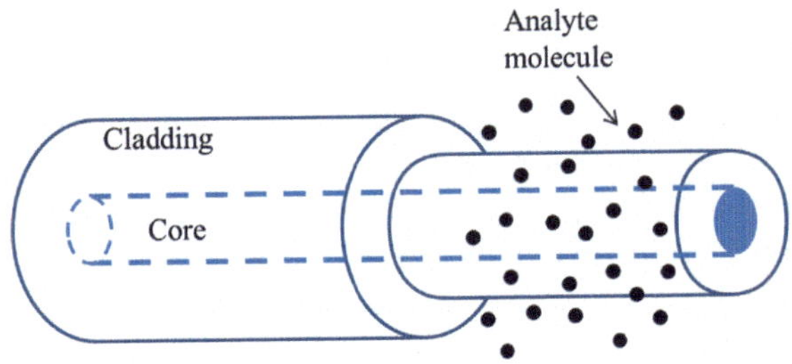

Fig. 6.6 Evanescent wave biosensors: part of the cladding is etched away and interaction between the mode propagating in the core and the analyte occurs at the side surface of the fiber where recognition molecules can be immobilized

field is extremely sensitive to the refractive index at the sensor surface—and any change that modifies the refractive index of the surface, such as binding of analyte to an immobilized recognition molecule, will result in the change of the propagation properties of the mode. The evanescent field generated within the medium or cladding decays exponentially with the distance from the core surface and its penetration depth, defined as the distance at which the field strength decays to $1/e$ of its value at the interface, is comparable to the wavelength of light. The detectable optical changes are therefore limited to a small region in close proximity to the fiber surface, with relatively little interference from the bulk solution.

Because of the large surface area available in their voids, microstructured fibers have found a prompt application in biomolecular sensing. Selective detection of α-streptavidin and Cy5-labelled α-CRP antibodies in series, at preselected positions, was demonstrated using a single piece of microstructured fiber fabricated from a cyclic olefin copolymer and serial fluorescence-based selective sensing [36]. The sensitivity and detection limit of sensors based on microstructured fibers are strongly dependent on the overlap between the mode and the analyte—and the fiber cross section geometry can be optimised for aqueous or gas detection [56]. Long period gratings have been used to improve the overlap in refractometers, providing a maximum sensitivity of 1500 nm/RIU at a refractive index of 1.33, and a detectable index change of $2 \cdot 10^{-5}$, two orders of magnitude larger than the sensitivity observed for a long-period grating in a standard optical fiber [117]. Long period gratings in microstructured fibers have also been used to determine the thicknesses of a monolayer of poly-L-lysine and double-stranded DNA, with a sensitivity of 1.4 nm/1 nm in terms of the shift in resonance wavelength in nanometre per nanometre thickness (1 nm) of the biomolecular layer [116].

MNFs have a very strong evanescent field and have therefore been used to detect biomolecules in close vicinity to the waist region (Fig. 6.7). Stiebeiner et al. [135] performed spectroscopy of 3,4,9,10-perylenetetracarboxylic dianhydride molecules (PTCDA) using between 1 and 10 mm long MNFs that had as little as 100 nm diameter. Optimized surface spectroscopy of organic molecules was performed by Warken et al. [153], who measured sub-monolayers of 3,4,9,10-perylene-tetracarboxylic dianhydride (PTCDA) molecules with a 500 nm diameter, 3 mm long, MNF—and showed that the sensitivity exceeds free-beam absorption spectroscopy by several orders of magnitude. Takiguchi et al. [140] used MNFs with a diameter of 410 nm

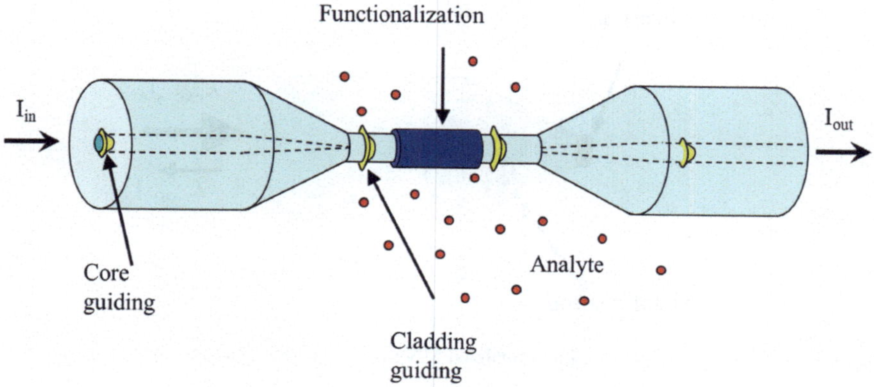

Fig. 6.7 Schematic diagram of a coated MNF for bio-chemical sensing

for saturated absorption spectroscopy of the acetylene (C_2H_2) ν1 + ν3 band transitions. Vishnoi et al. [149] demonstrated a spectrophotometer based on a tapered multimode fiber and used it to measure dye solutions with concentrations as low as 1 ppm in water. A 900 nm diameter MNF embedded in a 125 mm wide micro-channel with a detection length of 2.5 cm was used by Zhang et al. [163] to measure the absorbance of methylene blue (MB)—with a detection limit of 50 pM within the operational range of 0 to5 nM. The sensor was also functionalised to test bovine serum albumin (BSA) and achieved a detection limit of 10 fg.mL^{-1}.

A 3.69 μm-diameter, 7.1 mm-long MNF was used by Wiejata et al. [157] to detect a fluorescein solution at concentrations of 10–60 μM by exciting it at a wavelength of 460 nm and collecting the fluorescence emitted at 516 nm. Tian et al. [144] used a micro-channel chip designed for the label-free bio-testing of an IgG antibody-antigen pair to detect biomolecules, exploiting the modal interference that occurs along a non-adiabatic MNF. Wang et al. [152] modelled a bio-sensor using the effective phase shift in a fiber Bragg grating inscribed in a MNF. Proteins and DNA adsorbed on the surface were detected with a detection limit of 3.3 pg.mm^{-2}—by using the stop-band degeneracy and its application to refractive index sensing.

6.2.2 *Single Cell Environments*

Cellular micro-environments at the single cell level have been monitored by using MNF tips [141], since these allowed for a respective reduction in probe size, sample volume and response time of three, six and two orders of magnitudes with respect to the equivalent conventional optical sensor based on conventional 125 μm diameter optical fibers. In its easiest configuration, the sensor head is excited by light launched into its fiber pigtail and the resulting evanescent field at the MNF tip is used to excite target molecules bound to the antibody molecules (Fig. 6.8); the fluorescence

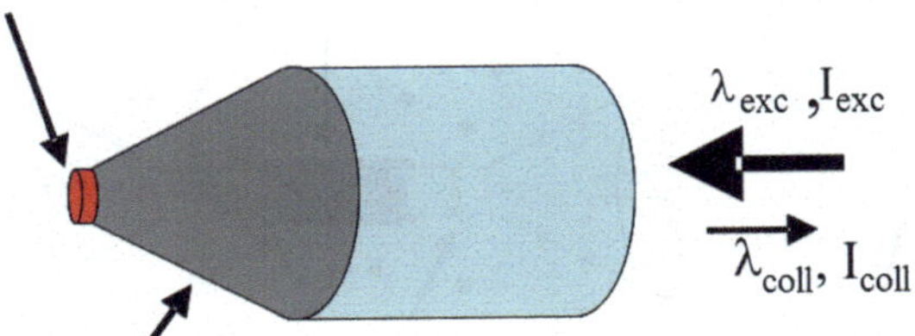

Fig. 6.8 Schematic diagram of a functionalized MNF tip for selective bio-chemical sensing

emission from the analyte molecules is then collected and analysed in the same fiber pigtail. MNF tips of 30–50 nm diameter [150] were covalently bound with antibodies that selectively target analyte molecules within the cellular microenvironment: this included benzo[α]pyrene [71], benzo[α]pyrene tetrol (BPT) [27] and caspase-9 [70]. A detection limit of ~0.64×10^{-11} M was reported for BPT.

DNA/RNA detection without the need for a dye-labelled target molecule or an intercalation reagent in the testing solution was demonstrated using fluorescence-based MNF biosensors [88]. Experimental results showed that the concentration and the mass detection limits for a sub-micron diameter MNF tip are 10 nM and 0.27 aM, respectively. MNF tips have also been used to detect telomerase in the nucleus of living MCF-7 cancer cells [165]. The encapsulation of dextran-linked fluorescence indicators in an organic hydrogel [111] enabled the co-immobilization of different indicators.

An intensity-based sensor in a compact micro-ball lens structure fabricated at the cleaved tip of a MNF coupler was proposed [55] for sensing the glucose concentration in deionized water. A sensitivity of 0.26 dB/% was achieved for a concentration change from 0 to 12 volume %, with the output intensity of the sensor decreasing linearly from -57.4 to -60.5 dBm.

6.2.3 Potential of Hydrogen (pH) Measurement

pH is often considered to be the single most important indicator in medicine, as it is strictly related to the correct operation of many organs of the human body. pH is frequently measured by a chromophore or by absorption-based indicators. In the first configurations demonstrated, optical fiber pH sensors used phenol red, covalently bound to polyacrylamide microspheres, as an indicator. The microspheres were inserted in a cellulose dialysis tube connected to two large-core plastic fibers, which were in turn inserted into a blood vessel through a needle. In vivo testing was carried out by detecting extracellular acidosis during regional ischemia in dog hearts [138], conjunctival pH [1], and pH gradients in canine myocardial ischemia [155]. The first fiberized sensor for the simultaneous monitoring of pH, oxygen and carbon dioxide concentrations relied on three optical fibers enclosed in a polymer tube [42]. The pH measurement was carried out using hydroxypyrene trisulphonic acid attached to a cellulose matrix at the fiber tip.

The measurement of pH level in the stomach and oesophagus is extremely important for monitoring of the human foregut. One of the first sensors demonstrated used two fluorophores, fluorescein and eosin, immobilized in amino-ethyl cellulose particles fixed on polyester foil [7, 103]. Another sensor used two absorbance dyes, meta-cresol purple and bromophenol blue, attached to polyacrylamide microspheres. Although the accuracy achieved, better than 0.1 pH units, satisfied clinical requirements, the response time, between 1 and 6 min depending on the pH step, was considered too long—so long that it would prevent detection of any rapid change of pH, like the detection of gastro-oesophageal reflux, where pH changes occur over time scales shorter than 1 min). A third sensor [8] used two chromophores, bromophenol blue (BPB) and thymol blue (TB), fixed at the end of plastic optical fibers, to form a very thin pH-sensitive layer with an accuracy of the order of 0.05 pH units. A pH sensor using an acid-base indicator covalently immobilized on the core of a 200/380 μm fiber demonstrated a detection limit of 0.05 pH units [9].

A multi-layered assembly of poly(allylamine hydrochloride) and poly(acrylic acid) on a 30 μm diameter, 10 mm length MNF was exploited as a pH sensor by detecting the wavelength shift associated with changes in the pH level [130]—and provided a sensitivity of 87.5 nm/pH unit between pH 4 and 6, a response time of 60 s and a detection limit of 0.05 pH units.

6.2.4 Oxygen and Carbon Dioxide

The continuous monitoring of oxygen (O_2) and carbon dioxide (CO_2) is very important in many biomedical fields, such as cardiovascular and cardiopulmonary systems, and they are therefore the two chemical parameters that have been most investigated for in vivo applications.

Carbon dioxide (CO_2) sensors rely on measurement of the pH of carbonate solutions, as these depend on the amount of CO_2 dissolved in them. All optrodes developed for measurement of blood CO_2 concentration exploit the same dye as utilized for pH measurement, fixed at the fiber end and covered by a membrane permeable to CO_2. In situ measurements use fluorophores, such as hydroxypyrene trisulphonic acid, dissolved in a bicarbonate buffer solution enclosed by a membrane permeable to CO_2. Many optical fiber CO_2 sensors have been described for the detection of CO_2 in blood or in human breath [42, 59, 97]. An optical fiber sensor for the continuous detection of gastric CO_2 has a measurement range of 0–14 kPa, with a resolution of less than 0.1 kPa and an accuracy of 0.27 kPa.

The measurement of oxygen concentration is often performed either spectroscopically, by targeting haemoglobin, or by using a fluorophore the efficiency of which is strongly affected by the presence of oxygen. The maximum oxygen concentration in blood was the first quantity to be measured optically—by exploiting the different absorption spectra of the haemoglobin and the oxyhaemoglobin in the near infrared [69]. Optical fiber sensors that are capable of being inserted in arteries and veins rely on the collection of reflected or absorbed light in the greatest absorption regions of haemoglobin and oxyhaemoglobin—and the oxygen saturation is calculated as the ratio of the absorptions at the two different wavelengths. The use of multiple wavelengths or of the whole spectrum was proposed [91, 96] to discriminate other haemoglobin derivates. Non-invasive sensors use the light transmitted through body extremities, such as earlobes, toes, or fingertips, but suffer from strong signals from adjacent tissues [122]. Blood oxygen saturation has also been measured in teeth [94] by measuring reflectance in the visible to near-infrared spectral range (typically at 660 and 850 nm). The use of spectrophotometer analysers [40] enabled the real-time measurement of intracapillary haemoglobin oxygenation and concentration, local oxygen uptake rate, local capillary blood flow, changes in subcellular particle sizes, and capillary wall permeability. The tissue was illuminated by a Xenon arc lamp via a bifurcated fiber bundle—and the back-scattered light provided in situ, real-time spectra of small volumes of tissue.

Haemoglobin is fully saturated at $\approx$100 Torr—and its use for the indirect measurement of oxygen concentration has limited application in the case of the respiration of gas mixtures with oxygen contents larger than 20%, as routinely used in anaesthesia. The first oxygen sensor involving optrodes used perylene-dibutyrate as a fluorophore fixed on amberlite resin beads at the end of two plastic optical fibers that were coated by a hydrophobic membrane permeable to oxygen [109, 133]. In vivo testing was carried out in dog eyes by measuring the fluorophore quenching due to arterial oxygen. A non-invasive sensor measured the oxygen flow at the skin surface to provide information on the oxygen flow inside the underlying tissue [61]: two optrodes based on a ruthenium complex were placed in contact with the skin—and used to measure, in real time, the difference in oxygen pressure across a membrane. An oxygen sensor using MNFs relied on the fluorescence quenching of tris-(1,10-phenanthroline)ruthenium(II) chloride in the presence of oxygen or dissolved oxygen [119] and provided a detection limit of 1×10^{-17} mol with response times shorter than 1 s, in volumes as small as 100 fL.

6.2.5 Temperature

Because of the immunity of fibers to electromagnetic interference, optical fiber thermometry has found application in the controlled heating of biological tissues—e.g. hyperthermia for cancer treatment, the thermal mapping during photoradiation therapy of malignant tumors and the thermodilution technique for blood flow measurement. A number of mechanisms have been proposed, including all-fiber devices [18], the use of transducers [33] or wavelength modulators [75, 108]. In one of its early demonstrations, a reflective fiber Fabry–Perot resonator [75] that was formed by coating the fiber end-face with a dielectric was used for continuous temperature monitoring in hyperthermia systems, covering the temperature range 25–45 °C, with a resolution better than 0.1 °C. Subsequently, a thermographic phosphor containing rare-earth atoms was bonded to the end of an optical fiber (~ 400 μm core diameter)—and the variation of its fluorescence with the temperature was utilized as a transduction sensing mechanism [156] in the −50 to 200 °C operational range, with an accuracy of 0.1 °C. Long period gratings [66] and fiber Bragg gratings (FBG) have also been exploited for thermometry with a sensitivity up to 175 pm/°C [115]. FBG thermometry has been used to perform distributed measurements in soft-tissue [121] by means of an array of ten gratings. The biocompability of the optical fiber sensor has been a major issue, with a number of coatings such as polyimide, polydimethylsiloxane (PDMS), ethylene-tetrafluoroethylene (or Tefzel®), and poly-tetrafluoroethylene (or Teflon) being investigated for biomedical applications [14, 118]

MNFs have also been used in temperature sensing, in the form of a simple modal interferometer [166], by coating the waist region with a material having a high thermo-optic coefficient. In this design, the interference pattern strongly depends on the refractive index, which in turn depends on the ambient temperature—with an operational range from −20 to 80 °C. The lyotropic liquid crystal decylammonium chloride/water/NH4CI (DACI-LLC) was combined with an MNF to produce a thermometer with an operating range from 40 to 80 °C [148]. Coating the MNF with a sol-gel thin-film that entrapped quantum dots provided variations in photoluminescence with temperature in the range from 30 to 70 °C, with a sensitivity of 1%/°C [162, 163]. An in-line Mach-Zehnder using two cascaded MNFs to excite and recombine the cladding mode with the core mode provided a sensitivity of 0.08 nm/°C over the temperature range from 20 to 60 °C [86, 90]. A 100 μm long intrinsic Fabry–Perot etalon fabricated using a femtosecond laser on the waist of a 30 μm diameter MNF [164] provided the means to measure temperature in the range from 0 to 100 °C, with a sensitivity of 12 pm/°C. Temperature and strain discrimination were achieved by exploiting multimode interference in a tapered core-less multimode fiber [3]: the temperature and strain resolutions were 1.6°C and 5.6 με, respectively. A tapered bend-resistant fiber, with different phase shifts between the inner and the outer cladding modes, provided a temperature sensitivity of −0.0253 rad/°C [89].

The combination of MNFs with Bragg gratings as end-reflectors showed a sensitivity of 9.7 pm/°C between 20 and 70 °C [65]. An MNF coupler tip thermometer was also demonstrated at temperatures up to 1283 °C—with a 2-dimensional resolution of <200 μm, a sensitivity of 11.96 pm/°C—and a detection limit of 0.836 °C for a wavelength resolution of 10 pm [31]. A 4.4 μm long micro-cavity fabricated within an MNF tip section smaller than 10 μm in diameter provided a reflective Fabry–Perot modal interferometer with a temperature sensitivity of 20 pm/°C between 19 and 520 °C, and a detection limit of 0.58 °C [77]. A similar temperature probe using a first-order, 36.6 μm long, FBG fabricated by focused ion-beam machining in a 6.5 μm diameter MNF demonstrated a similar sensitivity of 20 pm/°C and was operated at temperatures up to 500 °C [78]. A shorter device combining an MNF tip and a FBG exhibited a similar temperature sensitivity [38].

Tapered photonic crystal fiber coated with a layer of liquid crystal provided a sensitivity of 0.2 nm/°C within an operational range of 50–78 °C [114, 125]. Microfiber coil resonators embedded in Teflon with opposite thermo-optic coefficients were proposed and demonstrated as a temperature-insensitive device [20, 21] and showed a sensitivity of <6 pm/°C at room temperature.

6.2.6 Pressure

Although pressure sensors have found many applications in medicine [113, 118], most optical fiber pressure sensors have been designed for inclusion in catheters. The early sensors relied on monitoring the movement of mirrors mechanically coupled to pressure-sensitive membranes, which results in the modulation of light reflected and coupled into the output [82, 87, 95]. Applications included intra-muscular [26, 101], intra-cranial [107], cardiovascular [126] and skeletal muscle [124] pressure monitoring, airway pressure monitoring in young [131, 132] and in mechanically ventilated patients [29]—and even urological neurology and urodynamics. A typical fiber optic pressure sensor device is based on a pressure-balancing system: a pressure-sensitive membrane, located in a side hole of the catheter tip, is attached to a cantilever mirror. Under pressure, the membrane deflects the cantilever mirror with respect to the end of the fiber bundle and proportionally alters the intensity of the back-reflected light. Membranes have been positioned either at the distal end of the optical fiber [76, 100] or on the side [110, 142], in order to decrease the artefacts related to biological flow. Another typical fiber optic pressure sensor is based on a bifurcated fiber optic bundle with a pressure-sensitive membrane diaphragm at the end [51]. Although the sensor operated up to a pressure of 3000 mm of mercury (mmHg), the device response was highly nonlinear above 1000 mmHg. Physiological measurements could be performed in the range from 100 to 300 mmHg, with a typical linearity of 0.5% and a temperature shift of 0.1%/°C [50]. This micro-transducer was deployed in the measurement of pressure profiles in urethral conducts [58] and of intra-uterine pressure during labour [136, 137]—and provided results in agreement with those obtained using a catheter-tip bridge strain-gauge transducer [143]. The use of an external Fabry–Perot cavity bonded to the end of the fiber [25, 139] has made it possible to improve the pressure sensitivity, at the expense of the size. Diaphragms made of polymers [22, 23] or silica [34] have been added at the ends of optical fibers that were previously etched to form Fabry–Perot cavities, providing a more robust solution—with a dynamic range of 0–1200 kPa and resolutions of 10 Pa and 30 Pa, respectively.

Other optical fiber pressure transducers have relied on mirror bending rather than displacement [81]: in this design, the light is brought to the diaphragm surface by a cylindrical array of fibers and the reflected light from the diaphragm is then distributed over an array of collection fibers arranged concentrically around the illumination fibers. The pressure is calculated from the ratio of the amount of light received by an outer ring of collection fibers to that from an inner ring with more collection fibers.

More recent manometers have exploited Bragg gratings [5], which enables multiplexing [10] up to a series of seventy-two sensing elements—with a distance of 1 cm separating the adjacent sensing elements. Optical fiber manometry in catheters has shown extremely high sensitivity, enabling immediate applications in colonic diagnostics [4, 6, 32] and high-resolution mapping of the upper gastrointestinal tract [151].

Multiplexing has also enabled easy embedding of sensor fibers in pressure mats [112] with applications in pedobarography [52], which was previously carried out using critical reflection at the interface between plastic and glass [12]. Long period gratings in polymeric fibers [19] have been proposed for pressures up to 150 mbar and showed a sensitivity of 10.5 pm/mbar. Hollow-core microstructured fibers have been used for pressure sensing by selectively blocking cladding air holes by glue sealing and femtosecond laser machining [63]. By immersing the fiber end into a liquid, a compressible Fabry–Pérot cavity was formed in the fiber core of the PCF. A pressure sensitivity of 18.15 nm/kPa was demonstrated experimentally in the pressure range from 110 to 130 kPa.

6.2.7 Blood Flow

The velocity of blood flow in vessels is an important measurable quantity for medical applications, such as thrombosis and strokes, and it has been monitored using laser Doppler flowmetry [67, 79, 98], which relates the Doppler frequency shift of laser light scattered by moving blood to its flow velocity. The fiberized version of laser Doppler flowmetry allows for invasive measurements of blood flow [68]. Figure 6.9 shows the basic scheme for fiber-optic laser Doppler flowmetry: light from a laser is guided by an optical fibre onto the tissue or vascular network being studied, where it is scattered and partially absorbed by the tissue [72]. Light scattered by moving blood cells suffers a Doppler shift, which is analysed by the detector and is then processed to derive the blood flow rate through spectral analysis [99].

This technique has been widely applied in cardiology and has even been used to evaluate phasic blood flow in deeper intramyocardial layers [99]—having been

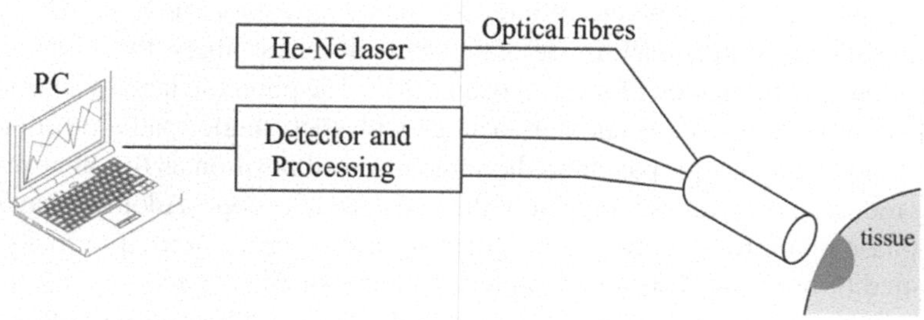

Fig. 6.9 Basic schematic of fiber-optic laser Doppler flowmetry

previously restricted to the epimyocardium because of methodological limitations. It has also been used to visualize and measure non-invasively the middle cardiac vein flow [53]. The use of feedback optical stabilization has been shown to improve significantly the Doppler signal in self-mixing intra-arterial measurements [30]. On the other hand, the target scattering properties have proved to be a major issue that hinders the sensor detection limit [160]. Self-mixing laser Doppler performed on extra-corporeal blood flow has shown that the Doppler shift is generated by erythrocyte cells [104, 105].

6.2.8 Endoscopy

Although many different types of sensor have been developed for applications in the biomedical realm, it is fair to state that the most successful application of optical fibers in healthcare is represented by the endoscope, which has been integrated with gas/liquid delivery lines and cutting tools for keyhole surgery and investigation of symptoms, mostly of the gastro-intestinal and respiratory tracts.

Even though in its original conception the endoscope was designed in the form of an array of small diameter optical fibers that transferred, in a coherent manner, a visual image over a distance of the order of 1 m [35], the most recent research has aimed to replace the use of numerous optical fiber cores (each representing a single pixel) with a single optical fiber supporting thousands of micrometric cores supported by thin struts [158] or with a single solid fiber supporting multiple modes [24, 48]. Multicore fibers made from suspended cores with air claddings have an extremely large numerical aperture—and each core can guide modes at visible wavelengths in a very small, sub-micron diameter—thus reducing the overall size of a single pixel. The use of polarization maintaining cores [73], semi-random arrangements [74] and proximal detection [154] in these multicore fibers allowed also for adaptive multiphoton endomicroscopy, which it has been suggested could allow for internal cell-scale examination, even during neurosurgery.

The use of one multimode fiber (MMF) allows the reduction of the endoscope cross-section by a factor of 20x, thus increasing the endoscope flexibility, while preserving a high resolution [93]. MMF endoscopy has quite suddenly become available because of the development of a device, named the photonic lantern [83, 84], that is capable of relating each mode of an array of optical fibers, to a single mode (or finite linear combination of modes) in an MMF. The photonic lantern has made it possible to transfer the information associated with each single spatial position to a different mode that can propagate in the same physical position as for other modes, thereby reducing the overall footprint of the average endoscope. Additional analysis on the phase relationship between the different modes has been used to focus light transmitted through an MMF to an unprecedentedly small size, to remove lenses at the endoscope ends and to reconstruct images behind the fibers using holographic techniques [24] (Fig. 6.10).

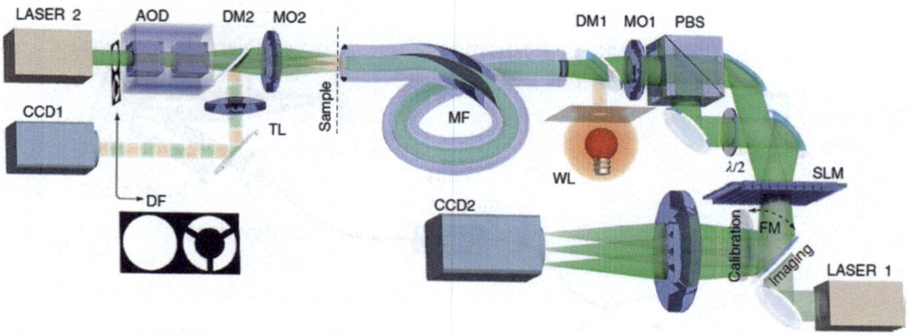

Fig. 6.10 Schematic of the microscopy based on multimode fibres (reproduced with permission from Cižmár and Dholakia [24])

Endoscopy has seen significant development in extending its operational range towards longer wavelengths [123], allowing for thermal imaging of internal organs and easier detection of tumours. Although the intrinsic fragility of chalcogenide glasses and their relatively poor compatibility with biological matter seems to be a major hurdle for the deployment of mid-IR endoscopes in the operating theatre, suitable packaging would solve these issues and allow for additional information on thermal imaging to be made readily available during surgery.

In order to visualize remotely and noninvasively the tissue microstructure and operation with micrometre resolution, optical coherence tomography (OCT) has been integrated within endoscopy, providing real time 3D imaging. Figure 6.11 shows a configuration where an external Michelson interferometer has been used to compensate the optical path difference between the reference and the signal coming from the sample in the endoscopic Fizeau sensing interferometer, which included a graded index lens at its end.

A more significant development in endoscopy could possibly come from the deployment of high power lasers in surgery procedures. In addition to the optical fiber bundles utilised to deliver imaging, current endoscopes include light delivery systems to illuminate the organs—and a channel to deliver medical instruments to manipulate and cut tissues. High power lasers have been used for a variety of material processing tasks and it is not inconceivable to think that in future they will replace the medical instrumentation used within endoscopes to cut and incise biological tissue. Indeed, carbon dioxide (CO_2) and Neodynium YAG lasers have both been advocated for reduction of bleeding during surgery and post-operative ecchymosis [145].

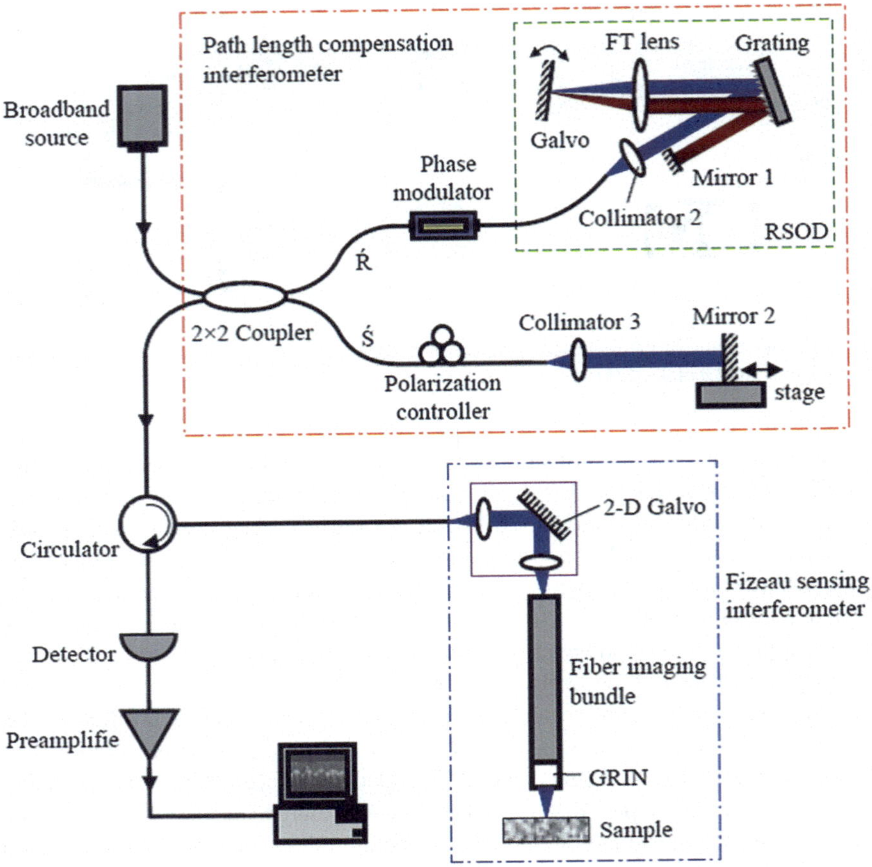

Fig. 6.11 Schematic of an endoscopic OCT with outside path compensator (reproduced with permission from Yang et al. [161])

References

1. E. Abraham, S.E. Fink, D.R. Markle, G. Pinholster, M. Tsang, Continuous monitoring of tissue pH with a fiberoptic conjunctival sensor. Ann. Emerg. Med. **14**, 840–844 (1985)
2. F.P. Anderson, W.G. Miller, Fiber optic immunochemical sensor for continuous, reversible measurement of phenytoin. Clin. Chem. **34**, 1417–1421 (1988)
3. R.M. André, C.R. Biazoli, S.O. Silva, M.B. Marques, C.M.B. Cordeiro, O. Frazão, Strain-temperature discrimination using multimode interference in tapered fiber. IEEE Photonics Technol. Lett. **25**, 155–158 (2013)
4. J.W. Arkwright, I.D. Underhill, S.A. Maunder, N. Blenman, M.M. Szczesniak, L. Wiklendt, I.J. Cook, D.Z. Lubowski, P.G. Dinning, Design of a high-sensor count fibre optic manometry catheter for in-vivo colonic diagnostics. Opt. Express **17**, 22423–22431 (2009)
5. J.W. Arkwright, N.G. Blenman, I.D. Underhill, S.A. Maunder, M.M. Szczesniak, P.G. Dinning, I.J. Cook, In-vivo demonstration of a high resolution optical fiber manometry catheter for diagnosis of gastrointestinal motility disorders. Opt. Express **17**, 4500–4508 (2009)

6. J.W. Arkwright, N.G. Blenman, I.D. Underhill, S.A. Maunder, N.J. Spencer, M. Costa, S.J. Brookes, M.M. Szczesniak, P.G. Dinning, A fibre optic catheter for simultaneous measurement of longitudinal and circumferential muscular activity in the gastrointestinal tract. J. Biophotonics **4**(4), 244–251 (2011)
7. F. Baldini, P. Bechi, S. Bracci, F. Cosi, F. Pucciani, In vivo optical-fibre pH sensor for gastro-oesophageal measurements. Sens. Actuators B Chem. **29**, 164–168 (1995)
8. F. Baldini, S. Bracci, F. Cosi, P. Bechi, F. Pucciani, Controlled-pore glasses embedded in plastic optical fibers for gastric pH sensing purposes. Appl. Spectrosc. **48**, 549–552 (1994)
9. F. Baldini, L. Ciaccheri, A. Falai, A.G. Mignani, J. Rayss, G. Sudolski, Thymol blue immobilized on tapered fibres as an optical transducer for pH sensing. Chem. Biochem. Environ. Fiber Sens. X **3540**, 28–33 (1999)
10. M. Becker, M. Rothhardt, K. Schröder, S. Voigt, J. Mehner, A. Teubner, T. Lüpke, C. Thieroff, M. Krüger, C. Chojetzki, Characterization of fiber Bragg grating-based sensor array for high resolution manometry. Proc. SPIE **8439** (2012)
11. E. Benito-Pane, M. Granda Valdes, B. Glahn-Martinez, M.C. Moreno Bondi, Fluorescence based fiber optic and planar waveguide biosensors. A review. Anal. Chim. Acta **943**, 17–40 (2016)
12. R.P. Betts, T. Duckworth, I.G. Austin, S.P. Crocker, S. Moore, Critical light reflection at a plastic/glass interface and its application to foot pressure measurements. J. Med. Eng. Technol. **4**(3), 136–142 (1980)
13. T.A. Birks, Y.W. Li, The shape of fiber tapers. J. Lightwave Technol. **10**, 432–438 (1992)
14. D.R. Biswas, Optical fiber coatings for biomedical applications. Opt. Eng. **31**(7), 1400 (1992)
15. S.M. Borisov, O.S. Wolfbeis, Optical biosensors. Chem. Rev. **108**(2), 423–461 (2008)
16. G. Brambilla, Optical fibre nanowires and microwires: a review. J. Opt. **12**, 043001 (2010)
17. G. Brambilla, E. Koizumi, X. Feng, D.J. Richardson, Compound-glass optical nanowires. Electron. Lett. **41**, 400–402 (2005)
18. M. Brenci, G. Conforti, R. Falciai, A.G. Mignani, A.M. Scheggi, All-fibre temperature sensor. Int. J. Opt. Sens. **1** (1986)
19. I.L. Bundalo, R. Lwin, L. Leon-Saval, A. Argyros, All-plastic fiber-based pressure sensor. Appl. Opt. **55**(4), 811–816 (2016)
20. Y. Chen, Y. Ming, W. Guo, F. Xu, Y.-Q. Lu, Temperature characteristics of microfiber coil resonators embedded in Teflon, in *Passive Components and Fiber-Based Devices VIII, Proceedings of SPIE-OSA-IEEE Asia Communications and Photonics* (2012), p. 830711-1-4
21. Y. Chen, F. Xu, Y.Q. Lu, Teflon-coated microfiber resonator with weak temperature dependence. Opt. Express **19**, 22923–22928 (2011)
22. E. Cibula, D. Donlagic, Miniature fiber-optic pressure sensor with a polymer diaphragm. Appl. Opt. **44**(14), 2736–2744 (2005)
23. E. Cibula, S. Pevec, B. Lenardič, É. Pinet, D. Donlagic, Miniature all-glass robust pressure sensor. Opt. Express **17**(7), 5098–5106 (2009)
24. T. Cižmár, K. Dholakia, Exploiting multimode waveguides for pure fibre-based imaging. Nat. Commun. **3**, 1027 (2012)
25. P.S. Cottler, W.R. Karpen, D.A. Morrow, K.R. Kaufman, Performance characteristics of a new generation pressure microsensor for physiologic applications. Ann. Biomed. Eng. **37**(8), 1638–1645 (2009)
26. A.G. Crenshaw, J.R. Styf, S.J. Mubarak, A.R. Hargens, A new "transducer-tipped" fiber optic catheter for measuring intramuscular pressures. J. Orthop. Res. **8**(3), 464–468 (1990)
27. B.M. Cullum, G.D. Griffin, G.H. Miller, T. Vo-Dinh, Intracellular measurements in mammary carcinoma cells using fiber-optic nanosensors. Anal. Biochem. **277**, 25–32 (2000)
28. C.M. Davis, E.F. Carome, M.H. Weik, S. Ezekiel, R.E. Einzig, *Fiber Optic Sensor Technology Handbook* (Dynamic Systems, Reston, Virginia, 1982)
29. R.A. De Blasi, G. Conti, M. Antonelli, M. Bufi, A. Gasparetto, A fibre optics system for the evaluation of airway pressure in mechanically ventilated patients. Intensive Care Med. **18**(7), 405–409 (1992)

30. F.F.M. de Mul, L. Scalise, A.L. Petoukhova, M. van Herwijnen, P. Moes, W. Steenbergen, Glass-fiber self-mixing intra-arterial laser Doppler velocimetry: signal stability and feedback analysis. Appl. Opt. **41**(4), 658–667 (2002)
31. M. Ding, P.F. Wang, G. Brambilla, A microfiber coupler tip thermometer. Opt. Express **20**, 5402–5408 (2012)
32. P.G. Dinning, L. Wiklendt, I. Gibbins, V. Patton, P. Bampton, D.Z. Lubowski, I.J. Cook, J.W. Arkwright, Low-resolution colonic manometry leads to a gross misinterpretation of the frequency and polarity of propagating sequences: Initial results from fiber-optic high-resolution manometry studies. Neurogastroenterol. Motil. **25**, 640–649 (2013)
33. A.W. Domanski, T.R. Wolinski, W. Borys, Fiber-optic liquid crystalline high-sensitivity temperature sensor. SPIE Proc. Fiber Opt. Laser Sensor VIII **1169**, 573–581 (1990)
34. D. Donlagic, E. Cibula, All-fiber high-sensitivity pressure sensor with SiO_2 diaphragm. Opt. Lett. **30**(16), 2071–2073 (2005)
35. J.M. Edmonson, History of the instruments for gastrointestinal endoscopy. Gastrointest. Endosc. **37**, S27–S56 (1991)
36. G. Emiliyanov, P.E. Hoiby, L.H. Pedersen, O. Bang, Selective serial multi-antibody biosensing with TOPAS microstructured polymer optical fibers. Sensors **13**(3), 3242–3251 (2013)
37. T. Erdogan, Fiber grating spectra. J. Lightwave Technol. **15**(8), 1277–1294 (1997)
38. J. Feng, M. Ding, J.L. Kou, F. Xu, Y.Q. Lu, An optical fiber tip micrograting thermometer. IEEE Photonics J. **3**, 810–814 (2011)
39. J.A. Ferguson, T.C. Boles, C.P. Adams, D.R. Walt, A fiber-optic DNA biosensor microarray for the analysis of gene expression. Nat. Biotechnol. **14**, 1681–1684 (1996)
40. K.H. Frank, M. Kessler, K. Appelbaum, W. Dummler, The Erlangen micro-lightguide spectrophotometer EMPHO-I. Phys. Med. Biol. **34**, 1883–1900 (1989)
41. S.M. Gautier, L.J. Blum, P.R. Coulet, Multi-function fibre-optic sensor for bioluminescent flow determination of ATP or NADH. Anal. Chim. Acta **235**, 243–253 (1990)
42. J.L. Gehrich, D.W. Lubbers, N. Opitz, D.R. Hansmann, W.W. Miller, J.K. Tusa, M. Yafuso, Optical fluorescence and its application to an intravascular blood-gas monitoring-system. IEEE Trans. Biomed. Eng. **33**, 117–132 (1986)
43. T.G. Giallorenzi, J.A. Bucaro, A. Dandridge, G.H. Sigel, J.H. Cole, S.C. Rashleigh, R.G. Priest, Optical fiber sensor technology. IEEE J. Quantum Electron. **18**, 626–665 (1982)
44. D. Gloge, Weakly guiding fibers. Appl. Opt. **10**, 2252–2258 (1971)
45. K.T.V. Grattan, B.T. Meggitt, *Optical Fiber Sensor Technology: Fundamentals* (Kluwer Academic Publishers, Nertherlands, 2000)
46. K.T.V. Grattan, T. Sun, Fiber optic sensor technology: an overview. Sens. Actuators Phys. **82**, 40–61 (2000)
47. F.X. Gu, L. Zhang, X.F. Yin, L.M. Tong, Polymer single-nanowire optical sensors. Nano Lett. **8**, 2757–2761 (2008)
48. R.Y. Gu, R.N. Mahalati, J.M. Kahn, Design of flexible multi-mode fiber endoscope. Opt. Express **23**(21), 26905–26918 (2015)
49. M.L. Guo, J.C. Shi, B.J. Li, Polymer-based micro/nanowire structures for three-dimensional photonic integrations. Opt. Lett. **33**, 2104–2106 (2008)
50. T. Hansen, A fiberoptic micro-tip pressure transducer for medical applications. Sens. Actuators **4**, 545–554 (1983)
51. T. Hansen, A. Munkhaugen, Fiber-optic sensors for medical and electrotechnical applications using bifurcated fiber bundles. IEEE J. Quantum Electron. **17**, 2530 (1981)
52. J.Z. Hao, K.M. Tan, S.C. Tjin, C.Y. Liaw, P.R. Chaudhuri, X. Guo, C. Lu, Design of a foot-pressure monitoring transducer for diabetic patients based on FBG sensors, in *The 16th Annual Meeting of the IEEE Lasers and Electro-Optics Society, 2003. LEOS 2003* (2003)
53. K. Harada, M. Tamura, M. Toyono, Noninvasive visualization and measurement of middle cardiac vein flow by transthoracic Doppler echocardiography. Pediatr. Cardiol. **27**(6), 679–684 (2006)
54. S.A. Harfenist, S.D. Cambron, E.W. Nelson, S.M. Berry, A.W. Isham, M.M. Crain, K.M. Walsh, R.S. Keynton, R.W. Cohn, Direct drawing of suspended filamentary micro- and nanostructures from liquid polymers. Nano Lett. **4**, 1931–1937 (2004)

55. S.W. Harun, A.A. Jasim, H.A. Rahman, M.Z. Muhammad, H. Ahmad, Micro-ball lensed fiber-based glucose sensor. IEEE Photonics J. **13**, 348–350 (2013)
56. S. Haxha, A. Teyeb, F.A. Malek, E.K. Akowuah, I. Dayoub, Design of environmental biosensor based on photonic crystal fiber with bends using finite element method. Opt. Photonics J. **5**(3), 69–78 (2015)
57. B.G. Healey, L. Li, D.R. Walt, Multianalyte biosensors on optical imaging bundles. Biosens. Bioelectron. **12**, 521–529 (1997)
58. P. Hilton, C.J. Mayne, Urethral pressure measurement: a comparison of profiles obtained by conventional and fibre-optic microtransducers. Neurol. Urodyn. **8**(5), 481–489 (1989)
59. T. Hirschfeld, F. Miller, S. Thomas, H. Miller, F. Milanovich, R.W. Gaver, Laser-fiber-optic optrode for real-time invivo blood carbon-dioxide level monitoring. J. Lightwave Technol. **5**, 1027–1033 (1987)
60. T. Hirshfeld, Total reflection fluorescence. Can. J. Spectrosc. **10**, 128 (1965)
61. G.A. Holst, T. Koster, E. Voges, D.W. Lubbers, Flox—an oxygen-flux-measuring system using a phase-modulation method to evaluate the oxygen-dependent fluorescence lifetime. Sens. Actuators B Chem. **29**, 231–239 (1995)
62. H.H. Hopkins, N.S. Kapany, A flexible fiberscope using static scanning. Nature **173**, 39–41 (1954)
63. M. Hou, Y. Wang, S. Lui, J. Guo, Z. Li, P. Lu, Sensitivity-enhanced pressure sensor with hollow-core photonic crystal fiber. J. Lightwave Technol. **32**(23), 4035–4039 (2014)
64. J.B. Jensen, P.E. Hoiby, G. Emiliyanov, O. Bang, L.H. Pedersen, A. Bjarklev, Selective detection of antibodies in microstructured polymer optical fibers. Opt. Express **13**(15), 5883–5889 (2005)
65. C.K. Ji, C.L. Zhao, J. Kang, X.Y. Dong, S.Z. Jin, Multiplex and simultaneous measurement of displacement and temperature using tapered fiber and fiber Bragg grating. Rev. Sci. Instrum. **83**, 053109 (2012)
66. J. Jung, H. Nam, J.H. Lee, N. Park, B. Lee, Simultaneous measurement of strain and temperature by use of a single-fiber Bragg grating and an erbium-doped fiber amplifier. Appl. Opt. **38**, 2749–2751 (1999)
67. F. Kajiya, O. Hiramatsu, K. Mito, Y. Ogasawara, K. Tsujioka, A study of coronary circulation by laser doppler velocimetry. Med. Prog. Technol. **12**, 77–85 (1987)
68. F. Kajiya, O. Hiramatsu, K. Mito, Y. Ogasawara, K. Mito, K. Tsujioka, An optical-fiber laser doppler-velocimeter and its application to measurements of coronary blood-flow velocities, in *Regulation of Coronary Blood Flow* (Springer Japan KK, 1991), pp. 11–23
69. N.S. Kapany, N. Silbertrust, Fiber optics spectrophotometer for in-vivo oximetry. Nature **204**, 138–142 (1964)
70. P.M. Kasili, J.M. Song, T. Vo-Dinh, Optical sensor for the detection of caspase-9 activity in a single cell. J. Am. Chem. Soc. **126**, 2799–2806 (2004)
71. R.M. Kasili, B.M. Cullum, G.D. Griffin, T. Vo-Dinh, Nanosensor for in vivo measurement of the carcinogen benzo[a]pyrene in a single cell. J. Nanosci. Nanotechnol. **2**, 653–658 (2002)
72. D. Kilpatrick, F. Kajiya, Y. Ogasawara, Fibre optic laser Doppler measurement of intravascular velocity. Australas. Phys. Eng. Sci. Med. **11**, 5–14 (1988)
73. Y. Kim, S. Warren, J. Knight, M. Neil, C. Paterson, J. Stone, C. Dunsby, P. French, Adaptive multiphoton endomicroscope incorporating a polarization-maintaining multicore optical fiber. IEEE J. Sel. Top. Quantum Electron. **22**(3), 6800708 (2016)
74. Y. Kim, S. Warren, F. Favero, J. Stone, J. Clegg, M. Neil, C. Paterson, J. Knight, P. French, C. Dunsby, Semi-random multicore fibre design for adaptive multiphoton endoscopy. Opt. Lett. **26**(3), 3661–3673 (2018)
75. R. Kist, S. Drope, H. Wolfelschneider, Fiber-Fabry-Perot (FFP) thermometer for medical applications. Proc. Soc. Photo Opt. Instrum. Eng. **514**, 165–170 (1984)
76. K. Kobayashi, H. Okuyama, T. Kato, T. Yasuda, Fiberoptic catheter-tip micromanometer. Iyodenshi To Seitai Kogaku **15**(7), 465–472 (1977)
77. J.L. Kou, J. Feng, L. Ye, F. Xu, Y.Q. Lu, Miniaturized fiber taper reflective interferometer for high temperature measurement. Opt. Express **18**, 14245–14250 (2010)

78. J.L. Kou, S.J. Qiu, F. Xu, Y.Q. Lu, Demonstration of a compact temperature sensor based on first-order Bragg grating in a tapered fiber probe. Opt. Express **19**, 18452–18457 (2011)
79. V.I. Krasovskii, I.N. Feofanov, P.I. Ivashkin, M.A. Kazaryan, A fiber-optic Doppler blood flow-velocity sensor. St. Petersburg Polytech. Univ. J. Phys. Math. **3**(1), 35–38 (2017)
80. M.N. Kronick, W.A. Little, A new immunoassay based on fluorescence excitation by internal reflection spectroscopy. J. Immunol. Methods **1975**, 235–240 (1975)
81. C.M. Lawson, V.J. Tekippe, Fiber-optic diaphragm-curvature pressure transducer. Opt. Lett. **8**, 286–288 (1983)
82. A. Lekholm, L.H. Lindström, Optoelectronic transducer for intravascular measurements of pressure variations. Med. Biol. Eng. Comput. **7**(3), 333–335 (1969)
83. S.G. Leon-Saval, T.A. Birks, J. Bland-Hawthorn, M. Englund, Multimode fiber devices with single-mode performance. Opt. Lett. **30**(19), 2545–2547 (2005)
84. S.G. Leon-Saval, N.K. Fontaine, R. Amezcua-Correa, Photonic lantern as mode multiplexer for multimode optical communications. Opt. Fiber Technol. **35**, 46–55 (2017)
85. A. Leung, P. Mohana Shankar, R. Mutharasan, A review of fiber-optics biosensors. Sens. Actuators B: Chem. **125**(2), 688–703 (2007)
86. Q. Li, S.S. Wang, Y.T. Chen, M. Yan, L.M. Tong, M. Qiu, Experimental demonstration of plasmon propagation, coupling, and splitting in silver nanowire at 1550-nm wavelength. IEEE J. Sel. Top. Quantum Electron. **17**, 1107–1111 (2011)
87. L.H. Lindström, Miniaturized pressure transducer intended for intravascular use. IEEE Trans. Bio-Med. Eng. **BME-17**(3), 207–219 (1970)
88. X.J. Liu, W. Farmerie, S. Schuster, W.H. Tan, Molecular beacons for DNA biosensors with micrometer to submicrometer dimensions. Anal. Biochem. **283**, 56–63 (2000)
89. P. Lu, J. Harris, Y. Xu, Y. Lu, L. Chen, X. Bao, Simultaneous refractive index and temperature measurements using a tapered bend-resistant fiber interferometer. Opt. Lett. **37**, 4567–4569 (2012)
90. P. Lu, L.Q. Men, K. Sooley, Q.Y. Chen, Tapered fiber Mach-Zehnder interferometer for simultaneous measurement of refractive index and temperature. Appl. Phys. Lett. **94**, 131110 (2009)
91. D.W. Lubbers, Chemical in vivo monitoring by optical sensors in medicine. Sens. Actuators B Chem. **11**, 253–262 (1993)
92. E.C. Magi, L.B. Fu, H.C. Nguyen, M.R.E. Lamont, D.I. Yeom, B.J. Eggleton, Enhanced Kerr nonlinearity in sub-wavelength diameter As_2Se_3 chalcogenide fiber tapers. Opt. Express **15**, 10324–10329 (2007)
93. R.N. Mahalati, R.Y. Gu, J.M. Kahn, Resolution limits for imaging through multi-mode fiber. Opt. Express **21**(2), 1656–1668 (2013)
94. M. Makiniemi, H. Kopola, K. Oikarinen, E. Herrala, A novel fibre optic dental pulp vitalometer, in *Proceedings of Medical Sensors II and Fiber Optic Sensors*, vol. 2331 (1995), pp. 140–148
95. H. Matsumoto, M. Saegusa, K. Saito, K. Mizoi, Development of a fiber optic catheter tip pressure transducer. J. Med. Eng. Technol. **2**, 239–242 (1978)
96. M.J. Milano, K.Y. Kim, Diode array spectrometer for the simultaneous determination of hemoglobin in whole blood. Anal. Chem. **49**, 555–561 (1977)
97. A. Mills, A. Lepre, L. Wild, Breath-by-breath measurement of carbon dioxide using a plastic film optical sensor. Sens. Actuators B Chem. **39**, 419–425 (1997)
98. K. Mito, Y. Ogasawara, O. Hiramatsu, K. Tsujioka, F. Kajiya, A laser Doppler catheter for monitoring both phase and mean coronary vein flow. Heart Vessels **6**(1), 1–8 (1990)
99. K. Mito, Y. Ogasawara, O. Hiramatsu, Y. Wada, K. Tsujioka, F. Kajiya, Evaluation of blood flow velocity waveforms in intramyocardial artery and vein by laser Doppler velocimeter with an optical fiber, in *Microcirculation in Circulatory Disorders*, ed. by H. Manabe, B.W. Zweifach, K. Messmer (Springer, Tokyo, 1988), pp. 525–528
100. S. Morikawa, Fiberoptic catheter-tip pressure transducer. Iyodenshi To Seitai Kogaku **10**(1), 36–39 (1972)

101. M. Nakhostine, J.R. Styf, S. van Leufen, A.R. Hargens, D.H. Gershuni, Intramuscular pressure varies with depth: the tibialis anterior muscle studied in 12 volunteers. Acta Orthop. Scand. **64**(3), 377–381 (1993)
102. A.S. Nain, J.C. Wong, C. Amon, M. Sitti, Drawing suspended polymer micro-/nanofibers using glass micropipettes. Appl. Phys. Lett. **89**, 183105–183107 (2006)
103. E.J. Netto, J.I. Peterson, M. Mcshane, V. Hampshire, A fiber-optic broad-range pH sensor system for gastric measurements. Sens. Actuators B Chem. **29**, 157–163 (1995)
104. M. Norgia, A. Pesatori, L. Rovati, Self-mixing laser Doppler: a model for extracorporeal blood flow measurement, in *2010 IEEE Instrumentation and Measurement Technology Conference (I2MTC)*, Austin, TX, pp. 304–307, 3–6 May 2010
105. M. Norgia, A. Pesatori, L. Rovati, Self-mixing laser Doppler spectra of extracorporeal blood flow: a theoretical and experimental study. IEEE Sens. J. **12**, 552–557 (2012)
106. K. Okamoto, *Fundamentals of Optical Waveguides* (Academic Press Ed., 2000)
107. R.C. Ostrup, T.G. Luerssen, L.F. Marshall, M.H. Zornow, Continuous monitoring of intracranial pressure with a miniaturized fiberoptic device. J. Neurosurg. **67**(2), 206–209 (1987)
108. C. Ovren, M. Adolfsson, B. Hok, Fiberoptic systems for temperature and vibration measurements in industrial applications. Opt. Lasers Eng. **5** (1984)
109. J.I. Peterson, R.V. Fitzgerald, D.K. Buckhold, Fiber-optic probe for in vivo measurement of oxygen partial pressure. Anal. Chem. **56**, 62–67 (1984)
110. E. Pinet, A. Pham, S. Rioux, Miniature fiber optic pressure sensor for medical applications: an opportunity for intra-aortic balloon pumping (IABP) therapy, in *Proceedings SPIE Volume 5855, 17th International Conference on Optical Fibre Sensors* (2005). https://doi.org/10.1117/12.623806
111. M. Plaschke, M. Geyer, J. Reichert, H.J. Ache, Submicron fiber-optic sensors for calcium-ions and pH with internal calibration. Chem. Biochem. Environ. Fiber Sens. IX **3105**, 31–37 (1997)
112. N. Pleros, G.T. Kanellos, G. Papaioannou, Optical fiber sensors in orthopedic biomechanics and rehabilitation, in *9th International Conference on Information Technology and Applications in Biomedicine, 2009. ITAB 2009* (2010). https://doi.org/10.1109/itab.2009.5394386
113. S. Poeggel, D. Tosi, D. Duraibabu, G. Leen, D. McGrath, E. Lewis, Optical fibre pressure sensors in medical applications. Sensors **15**, 17115–17148 (2015)
114. G. Rajan, S. Mathews, G. Farrell, Y. Semenova, A liquid crystal coated tapered photonic crystal fiber interferometer. J. Opt. **13**, 015403 (2011)
115. Y.J. Rao, B. Hurle, D.J. Webb, D.A. Jackson, L. Zhang, I. Bennion, In-situ temperature monitoring in NMR machines with a prototype in-fiber Bragg grating sensor system, in *12th Optical Fiber Sensors (OFS-12)*, Williamsburg (1997)
116. L. Rindorf, J.B. Jensen, M. Dufva, L.H. Pedersen, P.E. Høiby, O. Bang, Photonic crystal fiber long-period gratings for biochemical sensing. Opt. Express **14**, 8224–8231 (2006)
117. L. Rindorf, O. Bang, Highly sensitive refractometer with a photonic-crystal-fiber long-period grating. Opt. Lett. **33**, 563–565 (2008)
118. P. Roriz, O. Frazao, A.B. Lobo-Ribeiro, J.L. Santos, J.A. Simoes, Review of fiber-optic pressure sensors for biomedical and biomechanical applications. J. Biomed. Opt. **18**(5), 050903 (2013)
119. Z. Rosenzweig, R. Kopelman, Development of a submicrometer optical-fiber oxygen sensor. Anal. Chem. **67**, 2650–2654 (1995)
120. P. Russell, Photonic crystal fibers. Science **299**, 358–362 (2003)
121. E. Samset, T. Mala, R. Ellingsen, I. Gladhaug, O. Søreide, E. Fosse, Temperature measurement in soft tissue using a distributed fibre Bragg-grating sensor system. Minim. Invasive Ther. Allied Technol. **10**(2), 89–93 (2009)
122. C. Scoggin, L. Nett, T.L. Petty, Clinical evaluation of a new ear oximeter. Heart Lung **6**, 121–126 (1977)

123. A.B. Seddon, A prospective for new mid-infrared medical endoscopy using chalcogenide glasses. Int. J. Appl. Glass Sci. **2**(3), 177–191 (2011)
124. O.M. Sejersted, A.R. Hargens, K.R. Kardel, P. Blom, O. Jensen, L. Hermansen, Intramuscular fluid pressure during isometric contraction of human skeletal muscle. J. Appl. Physiol. **56**(2), 287–295 (1984)
125. Y. Semenova, L. Bo, S. Mathews, P. Wang, Q. Wu, G. Farrell, Spectral tuning of a microfiber coupler with a liquid crystal overlay, in *OFS 22nd International Conference on Optical Fiber Sensors* (2012), p. 842184-1-4
126. R. Singh, A.J. Ranieri Jr., H.R. Vest Jr., D.L. Bowers, J.F. Dammann Jr., Simultaneous determinations of cardiac output by thermal dilution, fiberoptic and dye-dilution methods. Am. J. Cardiol. **25**(5), 579–587 (1970)
127. A.W. Snyder, Asymptotic expression for eigenfunctions and eigenvalues of dielectric optical waveguides. IEEE Trans. Microw. Theory Tech. **MTT-17**, 1130–1138 (1969)
128. A.W. Snyder, R.M. De La Rue, Asymptotic solution of eigenvalue equations for surface waveguide structures. IEEE Trans. MTT **MTT-18**, 650–651 (1970)
129. A.W. Snyder, J.D. Love, *Optical Waveguide Theory* (Springer ed., 1983)
130. A.B. Socorro, I. Del Villar, J.M. Corres, F.J. Arregui, I.R. Matias, Tapered single-mode optical fiber pH sensor based on lossy mode resonances generated by a polymeric thin-film. IEEE Sens. J. **12**, 2598–2603 (2012)
131. S. Sondergaard, S. Karason, A. Hanson, K. Nilsson, S. Hojer, S. Lundin, O. Stenqvist, Direct measurement of intratracheal pressure in pediatric respiratory monitoring. Pediatr. Res. **51**(3), 339–345 (2002)
132. S. Sondergaard, S. Karason, S. Lundin, O. Stenqvist, Fibre-optic measurement of tracheal pressure in paediatric endotracheal tubes. Eur. J. Anaesthesiol. **18**, 24–25 (2001)
133. E. Stefansson, J.I. Peterson, Y.H. Wang, Intraocular oxygen-tension measured with a fiber-optic sensor in normal and diabetic dogs. Am. J. Physiol. **256**, H1127–H1133 (1989)
134. A. Stiebeiner, R. Garcia-Fernandez, A. Rauschenbeutel, Design and optimization of broadband tapered optical fibers with a nanofiber waist. Opt. Express **18**, 22677–22685 (2010)
135. A. Stiebeiner, O. Rehband, R. Garcia-Fernandez, A. Rauschenbeutel, Ultra-sensitive fluorescence spectroscopy of isolated surface-adsorbed molecules using an optical nanofiber. Opt. Express **17**, 21704–21711 (2009)
136. L. Svennningsen, J. Øystein, Application of fiberoptics to the clinical measurement of intrauterine pressure in labor journal. Acta Obstet. Gynecol. Scand. **65**(6), 551–555 (1985)
137. L. Svennningsen, J. Øystein, M.S. Dodgson, A fibreoptic pressure transducer for intrauterine monitoring, in *Fetal Physiological Measurements* (1986), pp. 15–21
138. G.A. Tait, R.B. Young, G.J. Wilson, D.J. Steward, D.C. Macgregor, Myocardial pH during regional ischemia—evaluation of a fiber-optic photometric probe. Am. J. Physiol. **243**, H027–H1031 (1982)
139. S. Takeuchi, H. Tohara, H. Kudo, K. Otsuka, H. Saito, H. Uematsu, K. Mitsubayashi, An optic pharyngeal manometric sensor for deglutition analysis. Biomed. Microdevice **9**(6), 893–899 (2007)
140. M. Takiguchi, Y. Yoshikawa, T. Yamamoto, K. Nakayama, T. Kuga, Saturated absorption spectroscopy of acetylene molecules with an optical nanofiber. Opt. Lett. **36**, 1254–1256 (2011)
141. W.H. Tan, Z.Y. Shi, S. Smith, D. Birnbaum, R. Kopelman, Submicrometer intracellular chemical optical fiber sensors. Science **258**, 778–781 (1992)
142. J.B. Taylor, B. Lown, M. Polanyi, In vivo monitoring with a fiber optic catheter. J. Am. Med. Assoc. **221**(7), 667–673 (1972)
143. K.F. Tham, S. Arulkumaran, S. Chua, C. Anandakumar, P. Singh, S.S. Ratnam, A comparison between fibreoptic and catheter-tip bridge strain gauge transducers for measurement of intrauterine pressure in labour. J. Obstet. Gynaecol. **17**(1), 83–87 (1991)
144. Y. Tian, W.H. Wang, N. Wu, X.T. Zou, X.W. Wang, Tapered optical fiber sensor for label-free detection of biomolecules. Sensors **11**, 3780–3790 (2011)

145. A.K. Tomer, A. Singh, A. Gupta, Lasers in endoscopic surgery. Int. J. Prev. Clin. Dent. Res. **3**(1), 55–57 (2015)
146. L.M. Tong, L.L. Hu, J.J. Zhang, J.R. Qiu, Q. Yang, J.Y. Lou, Y.H. Shen, J.L. He, Z.Z. Ye, Photonic nanowires directly drawn from bulk glasses. Opt. Express **14**, 82–87 (2006)
147. B.J. Tromberg, M.J. Sepaniak, T. Vo-Dinh, G.D. Griffin, Fiber-optic chemical sensors for competitive binding fluoroimmunoassay. Anal. Chem. **59**, 1226–1230 (1987)
148. C. Veilleux, R.J. Black, J. Lapierre, L.W. Reeves, Nematic liquid-crystal clad tapered optical fiber with temperature sensing properties. J. Appl. Phys. **67**, 6648–6653 (1990)
149. G. Vishnoi, T.C. Goel, P.K.C. Pillai, Spectrophotometric studies of chemical species using tapered core multimode optical fiber. Sens. Actuators B Chem. **45**, 43–48 (1997)
150. T. Vo-Dinh, Nanobiosensors: probing the sanctuary of individual living cells. J. Cell. Biochem. 154–161 (2002)
151. S. Voigt, M. Rothhardt, M. Becker, T. Lupke, C. Thieroff, A. Teubner, J. Mehner, Homogeneous catheter for esophagus high-resolution manometry using fiber Bragg gratings. Proc. SPIE **7559**, 75590B (2010)
152. G.G. Wang, P.P. Shum, H.P. Ho, X. Yu, D.J.J. Hu, Y. Cui, L.M. Tong, C.L. Lin, Modeling and analysis of localized biosensing and index sensing by introducing effective phase shift in microfiber Bragg grating (mu FBG). Opt. Express **19**, 8930–8938 (2011)
153. F. Warken, E. Vetsch, D. Meschede, M. Sokolowski, A. Rauschenbeutel, Ultra-sensitive surface absorption spectroscopy using sub-wavelength diameter optical fibers. Opt. Express **15**, 11952–11958 (2007)
154. S.C. Warren, Y. Kim, J.M. Stone, C. Mitchell, J.C. Knight, M.A.A. Neil, C. Paterson, P.M.W. French, C. Dunsby, Adaptive multiphoton endomicroscopy through a dynamically deformed multicore optical fiber using proximal detection. Opt. Express **24**(19), 21474–21484 (2016)
155. R.M. Watson, D.R. Markle, Y.M. Ro, S.R. Goldstein, D.A. Mcguire, J.I. Peterson, R.E. Patterson, Transmural pH gradient in canine myocardial ischemia. Am. J. Physiol. **246**, H232–H238 (1984)
156. K.A. Wickersheim, M.H. Sun, Fiberoptic thermometry and its applications. J. Microw. Power Electromagn. Energy **22**, 85–94 (1987)
157. P.J. Wiejata, P.M. Shankar, R. Mutharasan, Fluorescent sensing using biconical tapers. Sens. Actuators B Chem. **96**, 315–320 (2003)
158. H.A.C. Wood, K. Harrington, T.A. Birks, J.C. Knight, J.M. Stone, High-resolution air-clad imaging fibers. Opt. Lett. **43**(21), 5311–5314 (2018)
159. X.B. Xing, Y.Q. Wang, B.J. Li, Nanofiber drawing and nanodevice assembly in poly(trimethylene terephthalate). Opt. Express **16**, 10815–10822 (2008)
160. B. Yang, W. Zhang, J. Zhou, H. Gui, L. Lu, B. Yu, Influence of external objects scattering property on self-mixing signal inside fiber laser. Opt. Int. J. Light Electron Opt. **125**(9), 2160–2163 (2014)
161. Y. Yang, Z. Ding, J. Meng, L. Wu, Z. He, T. Wu, M. Chen, Common path endoscopic optical coherence tomography with outside path length compensation. Proc. SPIE **6826**, 68261S (2007)
162. F.J. Zhang, B. Wang, F.F. Pang, T.Y. Wang, A luminescent temperature sensor based on a tapered optical fiber coated with quantum dots. Opt. Sens. Biophotonics II **7990**, 312–313 (2011)
163. L. Zhang, P. Wang, Y. Xiao, H.K. Yu, L.M. Tong, Ultra-sensitive microfibre absorption detection in a microfluidic chip. Lab Chip **11**, 3720–3724 (2011)
164. X. Zhang, J. Li, Y. Li, W. Wang, F. Pang, Y. Liu, T. Wang, Sensing properties of intrinsic Fabry-Perot interferometers in fiber tapers, in *OFS 22nd International Conference on Optical Fiber Sensors* (2012), p. 842189-1-4
165. X.T. Zheng, C.M. Li, Single living cell detection of telomerase over-expression for cancer detection by an optical fiber nanobiosensor. Biosens. Bioelectron. **25**, 1548–1552 (2010)
166. S. Zhu, F.F. Pang, T.Y. Wang, Single-mode tapered optical fiber for temperature sensor based on multimode interference. Opt. Sens. Biophotonics III **8311**, 83112B (2011)

Chapter 7
Planar Optofluidics for On-Chip Particle Manipulation

Hong Cai, Kaelyn D. Leake, and Holger Schmidt

Abstract Optical forces can be used to mechanically manipulate small particles on the micrometre and sub-micrometre scales. We review the recent development of planar optofluidic devices that harness optical forces to control particle movement on a chip. Complementary implementations using solid-core and liquid-core optical waveguides are described, along with potential applications in biosensing and other fields.

7.1 Introduction

Optical forces on massive particles have been known for a long time and operate on a truly cosmic scale. Indeed, Kepler postulated as early as 1619 that the sun exerts a repulsive force on the tail of a comet, and Lebedev, in 1901, demonstrated the existence and basic properties of such light-induced forces [1]. It was, however, not until the development of the laser as a convenient, powerful light source—and the seminal experiments by Ashkin in 1970 [2, 3]—that optical forces were systematically used to manipulate the motion of small particles. Over the course of a few years, two main directions have developed: one being the control of atoms, which has culminated in the demonstration of a new type of matter in the form of a Bose–Einstein condensate [4]. The other is the application to biological particles of micrometre and sub-micrometre size. Ashkin developed both dual-beam and single-beam ("optical tweezer") traps that allow for holding and affecting individual cells and bioparticles [5]. Over the years, a number of scientific breakthroughs have been achieved with

H. Cai · K. D. Leake · H. Schmidt (✉)
School of Engineering, University of California, Santa Cruz, 1156 High St., Santa Cruz, CA 95064, USA
e-mail: hschmidt@soe.ucrc.edu

Present Address:
K. D. Leake
Margaret Jones Wyllie '45 Engineering Program, Sweet Briar College, Sweet Briar, VA 24595, USA

R. De La Rue et al. (eds.), *Biomedical Optical Sensors*,
Biological and Medical Physics, Biomedical Engineering,
https://doi.org/10.1007/978-3-030-48387-6_7

these optical traps, for example the observation of single viruses [6], and the observation of the movement of messenger RNA molecules through a ribosome during protein synthesis [7]. Due to the experimental complexity and relatively elaborate components and setups, optical particle manipulation has largely been confined to specialized research labs, while mainstream commercial applications have remained virtually non-existent.

It could be argued that we are on the cusp of yet another quantum leap in the history of optical particle manipulation. The recent emergence of the new field of optofluidics has led to a surge in activities aimed at harnessing light-matter interactions in the context of integrated, lab-on-a-chip devices [8–12]. It has already been demonstrated that not only can we replicate some of the previously known bulk optics techniques on an integrated chip, but also the use of planar, waveguide-based structures has already led to the introduction of new particle manipulation concepts and devices that could be the basis of stand-alone instruments, or assist in optimizing other functions, in particular bioparticle detection, sensing, and analysis. This progress in research suggests that in the near future optical particle manipulation will have transcended from providing a celestial spectacle to being an integral part of powerful and easy-to-use instruments for application in laboratory and clinical diagnostics.

In this chapter, we review the current state-of-the-art of optical particle manipulation using planar, optofluidic approaches. We describe two canonical approaches in detail, give an overview of other promising approaches, and discuss possible practical applications.

To this end, we need to classify the types of optical force that are being utilized as well as the types of optical waveguide that form the basis of the photonic integration. In the photon picture, each photon in a laser beam carries a linear momentum of $p = h/\lambda$ where h is Planck's constant and λ is the wavelength. As a light beam interacts with a particle via reflection, refraction, or absorption, the photon momentum is changed. In order to conserve the total momentum, an optical force $F_{\mathrm{opt}} = \mathrm{d}p_{\mathrm{tot}}/\mathrm{d}t$ has to be imparted on the particle. This force can be used to control the motion of the particle. We distinguish two components of the optical force as shown in Fig. 7.1,

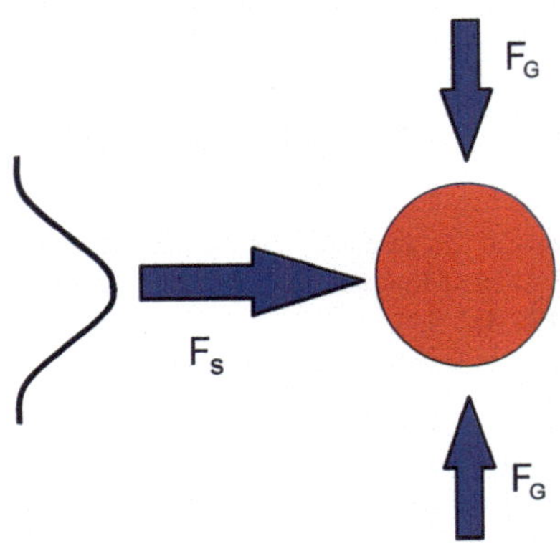

Fig. 7.1 Optical forces on dielectric particle

which shows a dielectric particle subjected to a propagating light beam with an inhomogeneous (here Gaussian) intensity profile.

The *scattering force* F_S acts along the propagation direction of the light beam, and its magnitude is proportional to the amount of power (intensity) that hits the particle. The *gradient force* F_G occurs in beams with spatially varying intensity profiles—and its magnitude is proportional to the local power (intensity) gradient. For most practical cases (refractive index of particle larger than index of surrounding medium), the gradient force pulls the object towards higher intensity regions of the beam. Both types of force are present in integrated waveguide structures, and one of the attractive motivations for using waveguides for particle manipulation is the possibility to precisely control the spatial dependence of these forces with proper fabrication. As we shall see, one can take advantage of either or both of these forces to manipulate particles in many different ways and configurations. Moreover, the optical forces can be readily combined with other forces on a chip such as electrokinetic forces, flow-based forces or forces from physical walls, providing even more opportunities for device design. The fact that the gradient force F_G typically pulls an object towards regions of higher light intensity implies that a concentrated (e.g. focused) light beam acts like a potential well, in which the object on which the light is acting is trapped in the higher-intensity region—with a stiffness that increases with light intensity.

The second fundamental design criterion for on-chip particle manipulation is the choice of optical waveguide. There are two canonical approaches to creating the necessary interaction between light and particles as illustrated in Fig. 7.2, dictated by the fact that liquids and gases typically have lower refractive indices than all solid materials used in standard micro-device fabrication.

The first approach is to use conventional solid-core optical waveguides in which light is guided by total internal reflection in a high index core. Because the light will not be guided efficiently in the liquid, interaction is facilitated either near the core surface via the evanescent tail of the guided mode (Fig. 7.2a) or over a short channel segment in which the light is not guided, but not too much power is lost. These approaches allow for more straightforward integration of optical waveguides and microfluidic channel devices. However, proper mechanisms to ensure particle

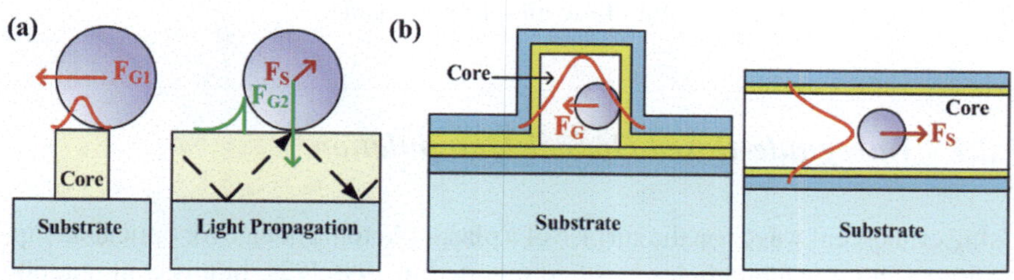

Fig. 7.2 Strategies for waveguide-based optical particle manipulation. **a** Evanescent field of solid-core waveguide; **b** guided mode field of liquid-core waveguide. Both approaches are shown in cross-sectional and side view, respectively.

transport to the surface, in order to experience an optical force, have to be built into the device design. The second option is to guide both light and particles through the same channel of a hollow-core or liquid-core waveguide (Fig. 7.2b). This has the advantage that particles experience the full intensity of the optical field over extended distances. Here, the main challenge lies in creating efficient optical confinement in a low-index non-solid core and developing an interface to deliver light and liquid to the interaction region. A number of liquid-core waveguide types including leaky waveguides [13], porous cladding waveguides [14], slot waveguides [15], liquid-liquid [16] and liquid-air waveguides [17], hollow-core photonic crystal fibers (HC-PCF) [18], and liquid-core antiresonant reflecting optical waveguides (LC-ARROWs) [19–21] have been developed for this purpose.

The remainder of this chapter is dedicated to providing a more in-depth understanding of the fundamental challenges and possibilities for on-chip particle manipulation. To this end, Sect. 7.2 is devoted to implementations using solid-core waveguides, while Sect. 7.3 discusses liquid-core waveguide-based approaches. In each case, we first provide an overview of several key demonstrations to illustrate the full breadth of possibilities, followed by a more thorough discussion of a specific system that has been used to implement a number of particle manipulation techniques. Section 7.4 summarizes the chapter and provides an outlook on future developments and applications.

7.2 Particle Manipulation Using Solid-Core Waveguides

There are two types of planar photonic devices for surface wave optical manipulation: waveguide-based and microresonator-based devices. They both have the key advantages of overcoming the diffraction limit to manipulate nanoparticles and parallel manipulation of multiple particles compared to conventional optical tweezers. Furthermore, waveguide-based devices offer the advantages of relatively simple device design, high fabrication tolerance and wide operation wavelength band. Microresonator-based devices enable wavelength-tunability, optical manipulation force enhancement, and two-dimensional manipulation, with the compromise of more challenging design and fabrication requirements.

7.2.1 Waveguide-Based Optical Manipulation

Using evanescent waves on the surface of a planar photonic device for optical manipulation of micro/nanoparticles was first reported in 1996 [22]. In this configuration (Fig. 7.3a), light is coupled into a channeled glass waveguide and propagates inside the glass core. The evanescent field near the waveguide surface provides enough gradient force to confine the micro-objects in both the lateral and vertical directions. Due to the propagation of light along the waveguide, the scattering force propels the

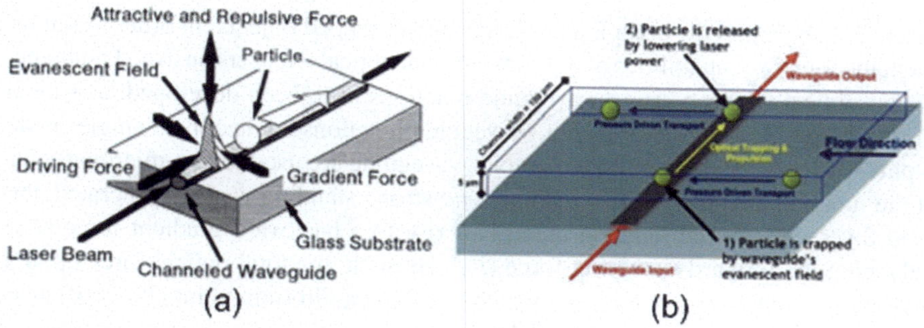

Fig. 7.3 Schematic of optical manipulation **a** on a channeled glass waveguide (reproduced from [22] with permission), and **b** on an optofluidic chip with integration of SU8 rib-waveguides and PDMS microfluidic channels (reproduced from [23] with permission)

trapped particles (dielectric and metallic micro-sized particles) along the waveguides with a maximum recorded velocity of ~14 μm/s.

High refractive index (RI) contrast between the waveguides and the surrounding medium is needed to obtain a sufficiently large evanescent field/optical force. This requirement is more critical for biological applications, which require a water-based environment and low RI contrast between the environment and the biological cells/systems. For example, silicon nitride (SiN) with an RI contrast to water of ~2/1.33, enhances optical forces on the microparticles by up to ~20 times compared to a glass waveguide with an RI contrast of ~1.45/1.33 [24]. Optical manipulation of biological cells using evanescent fields [24] is very promising, given the compact size and possible mass-production capability leverage from the traditional CMOS fabrication process [25], and could complement conventional optical tweezers. Evanescent fields can even be used for optical manipulation of particles in the nanometre range [26, 27], in which the conventional optical tweezers are limited by the diffraction limit.

Thanks to soft lithography technology, planar waveguides on a photonic chip can be immediately integrated with microfluidic channels, which serve as an efficient sample delivery method, as shown in Fig. 7.3b. Using SU8 (RI of ~1.57) waveguides to optically manipulate microparticles delivered by a microfluidic channel on top was demonstrated in 2007 [23].

7.2.2 *Functional Waveguide Devices*

7.2.2.1 Waveguide Junctions

One main feature of conventional optical tweezers is their ability to trap micro-sized objects in a fixed location for further analysis. The simple straight waveguides discussed above, on the other hand, always move trapped objects. In order to implement the trapping capabilities of 3D conventional optical tweezers on to a 2D scheme, planar optical tweezers using waveguide junctions have been developed, as shown in Fig. 7.4a [28, 29]. In this design, waveguide junctions connecting a single-mode input-waveguide and a multimode output-waveguide are used to trap microparticles at the junction area (Fig. 7.4b). Although no strong standing wave is generated, the field divergence in the junction area gives rise to a backward gradient force (F_g) balancing the forward scattering force (F_s). As such, the total optical force upon a particle is given as $F_{0z} = F_g + F_s$ which finds an equilibrium point ($F_{0z} = 0$) near the junction area, as shown in Fig. 7.4c, d.

Figure 7.5a shows SEM pictures of one of the waveguide junctions on a SiN platform. The input single-mode waveguide is ~0.5 μm wide and the output multimode waveguide is ~4 μm wide. Experimental results show that a single 1 μm-sized polystyrene particle can be trapped at the junction each time—until it is replaced by another incoming particle, as shown in Fig. 7.5b–e. The dependence of this trapping capability of the waveguide junction on the input waveguide width has also been studied [28].

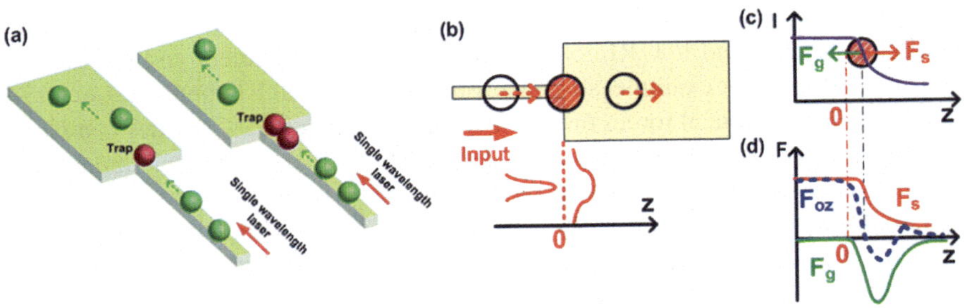

Fig. 7.4 **a** Schematic of optical manipulation using single-mode to multi-mode waveguide junctions. **b** Top-view schematic of particle trapping on a waveguide junction. Curved profiles: the mode-field profiles in the vicinity of the junction at $z = 0$. Hollow circles: travelling particles. Shaded circle: trapped particle. **c** Schematic mode-field intensity distribution along the z-direction. **d** Schematic of the optical forces exerted on a Rayleigh particle along the z-direction (reproduced from [28, 29] with permission)

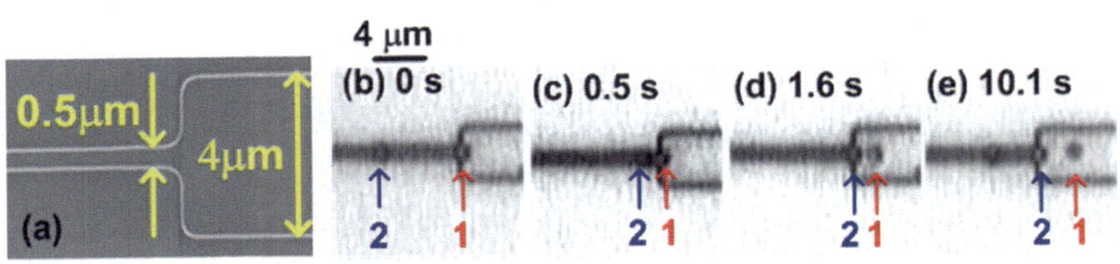

Fig. 7.5 **a** SEM pictures of the fabricated waveguide junctions: input width ~0.5 μm and output width ~4 μm. **b–e** Optical micrographs of the waveguide junctions with particles at various times. (reproduced from [28, 29] with permission)

Like conventional optical tweezers the focal volume of which can be adjusted by changing focusing objective lens, the trapping volume of this kind of planar optical tweezers can also be adjusted by adding a shallow taper in between the input and output waveguides (not visible in Fig. 7.4a). Up to two 1 μm-sized polystyrene particles can be trapped each time [28]. This kind of waveguide-junction promises a key feature of holding a certain number of particles with successive substitutions by the incoming particles delivered from the input waveguide.

7.2.2.2 MMI Power Splitters

Another natural extension of a single waveguide manipulation device is to use multiple waveguides to form structures with various output ports. For example, one can split a waveguide into two output ports to form a Y–junction. The input optical power is also split into the output waveguides. By simply adjusting the input-coupling fiber positions to differentiate the power between the output ports, dielectric particles can be trapped by the input waveguide and routed to the preferred output port with higher guided power [30].

Another way to structure this kind of 1 × 2 power splitter is to use multimode-interference (MMI)-based power splitters. A symmetric 1 × 2 MMI-based power splitter consists of one single-mode input waveguide, one multi-mode waveguide and two output waveguides. The multimode waveguide supports multiple modes with various phase velocities. These can be excited simultaneously with the single-mode input waveguide. If the phase difference between any of the two excited modes is 2π or an integer multiple of 2π, a constructive interference pattern is formed at the output [31, 32]. By proper design of the output-port location, one can route the particles to the two output ports.

Figure 7.6a shows a microscope picture of a fabricated 1 × 2 MMI-based power splitter on a SiN platform. The multimode waveguide is designed to be ~4 μm in width

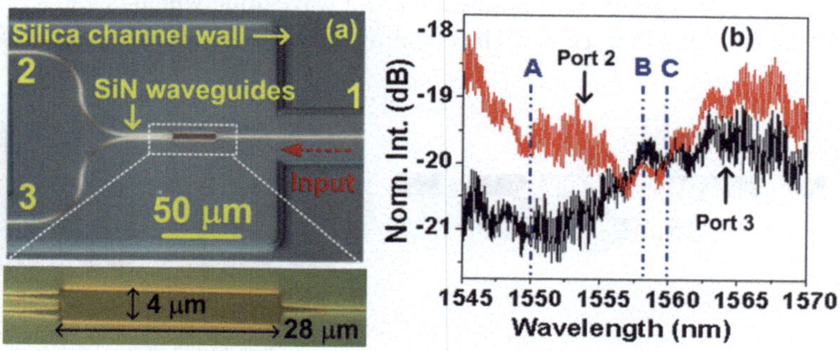

Fig. 7.6 **a** Optical micrograph of the optofluidic device with air cladding. Inset: zoomed-in image of the 1 × 2 MMI power splitter. **b** Measured TM-polarized optical transmission spectra of the MMI from output-ports 2 and 3. The dashed-dotted lines label the wavelengths A, B and C used in the particle manipulation experiment. (reproduced from [33] with permission)

and ~28 μm in length. The input waveguide and output waveguides are ~0.4 μm wide and are coupled to the multimode waveguide by tapers of 6 μm in length (tapering angle θ of ~1.91°). The device is designed for a 50/50 power split between the two output ports at 1550 nm wavelength. Figure 7.6b shows the measured TM-polarized optical transmission spectra from output ports 2 and 3 with 1 μm-sized polystyrene particles colloidal solution. The output power levels from ports 2 and 3 vary with the input wavelength. Instead of adjusting the input coupling fiber position mechanically, the output power ratio between the two output ports can be tuned by wavelength. Various particle branching ratios were observed upon tuning the output power ratios from the two output ports at various wavelength A, B and C (Fig. 7.6b) [33]. Similar results are reported later on an SU8 platform with 10 μm-sized particles [34].

7.2.2.3 Directional Couplers

If two waveguides are placed in close proximity, optical power is evanescently coupled back and forth between the input waveguide and the coupled waveguide. By proper design of the coupling gap and coupling length, one can achieve different output power ratios. Such devices are called directional couplers. Figure 7.7a schematically shows microparticle manipulation on a waveguide-based directional coupler. The microparticles are initially propelled by the surface wave of the input waveguide and can be transferred laterally by gradient force to the coupled waveguide through small gap spacing along a certain interaction length. Because the coupling length also depends on the input laser wavelength, the coupling transfer efficiency can be controlled by tuning the input laser wavelength. For example, microparticles can be routed to the left output port using laser wavelength λ_1 and to the right output-port using laser wavelength λ_2, as shown in Fig. 7.7b, c [34].

By introducing a structural perturbation near the junction of the coupling region, particle sorting on directional couplers was demonstrated [35]. Figure 7.8a shows an SEM picture of a directional coupler on an SOI platform. The input-coupled waveguide is ~420 nm wide and coupled to a slot waveguide which is 450 nm wide, with a 50 nm wide slot in its center. The coupling region is designed to achieve a 3 dB

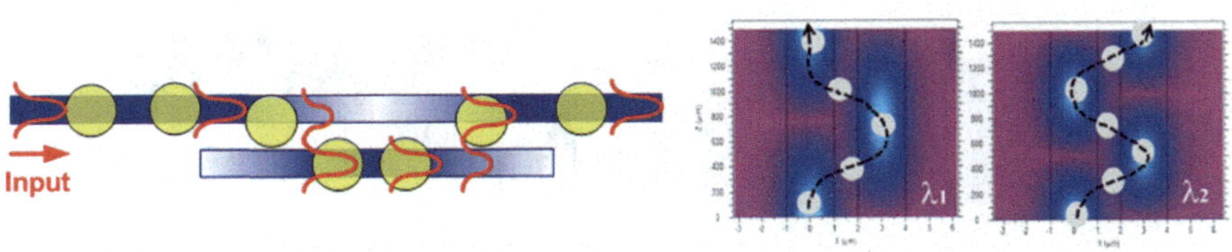

(a) (b) (c)

Fig. 7.7 **a** Schematic of microparticle manipulation on an evanescently coupled waveguide-based directional coupler. **b, c** Schematic of particle trajectories on the directional coupler for different wavelengths. (reproduced from [34] with permission)

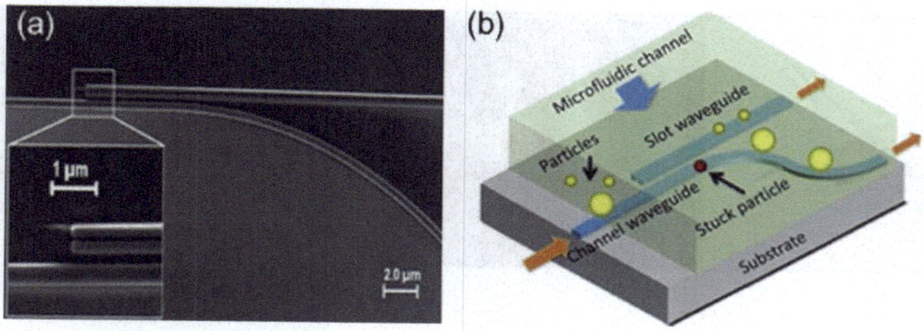

Fig. 7.8 **a** SEM picture of a directional coupler with a rib-waveguide coupled to a slot waveguide. **b** Schematic of particle sorting with presence of a stuck particle. (reproduced from [35] with permission)

splitting of the TM mode with a coupling gap of ~200 nm over a coupling length of 4.6 μm.

The two waveguides provide two potential energy wells for the smaller particles, but only one broad potential well for the larger particles. By introducing a stuck particle (red) near the output junction of the coupling region, as shown in Fig. 7.8b, the smaller particles were bounced to the second potential well associated with the slot waveguide, while the larger particles were brought to the region between the two waveguides and eventually followed the channel waveguide, which is associated with a deeper potential well. Therefore, small particles with a diameter of ~320 nm and larger sized-particles with a diameter of ~2 μm were separated.

7.2.3 *Resonator-Based Optical Manipulation*

Because the optical forces scale with the guided optical power, a stronger waveguide field is always desirable. One way of increasing the optical forces is to increase the laser power. Another viable solution is to use microresonators that can route light and store strong cavity fields. The field enhancement inside the resonator can enable efficient optical manipulation while keeping the input laser power relatively low. Microresonators can also greatly expand the functionalities of waveguide-based micro/nanoparticle routing and sorting.

7.2.3.1 Photonic Crystal Resonators

The first planar resonator-based optical manipulation has been reported by S. Mandal et al. using a photonic crystal cavity, in 2010 [36]. Photonic crystal cavities are formed by introducing well-defined defects into waveguiding slabs of which the refractive index is modulated by a group of submicron-scale periodic air holes. The defects in the photonic crystal act as optical resonators or cavities by creating a pass-band in the

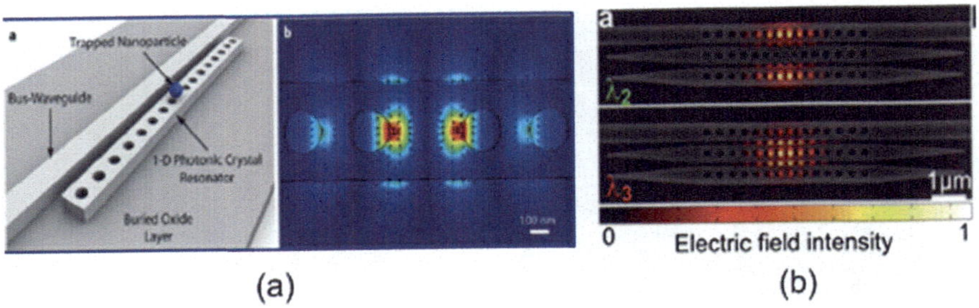

Fig. 7.9 **a** Schematic of optical manipulation of particles on a one-dimensional photonic-crystal resonator (left) and 3D FEM simulation of optical mode field inside the resonator (right) (reproduced from [36] with permission); **b** 3D FDTD simulation of coupled resonance modes on 2D resonators upon different laser wavelengths. (reproduced from [38] with permission)

photonic band gap. Small modal volume with large optical field enhancement can be achieved inside the cavities. For example, a one-dimensional silicon photonic crystal resonator can be constructed by taking out one hole in the center of a row of periodic air holes, as shown in Fig. 7.9a. The photonic resonator can be evanescently coupled to a single-mode waveguide. When the photonic crystal cavity is on resonance, a strong standing wave is formed inside the cavity, which provides a three-dimensional stationary potential well to confine micro/nano particles, as shown in Fig. 7.9a. For such a design with a measured quality factor (Q-factor) of ~2500, dielectric nanoparticles with sizes from ~48 nm to ~500 nm can be trapped, and an order-of-magnitude trapping stiffness enhancement and several orders stronger confinement than from straight waveguides was reported. Large bacteria such as B. subtilis and E. coli were successfully trapped using a different kind of one-dimensional silicon photonic crystal cavity for trapping [37].

One can also greatly enhance the trapping capability through reduction of the thermal background by exploring laser wavelengths in the visible to near-IR range where water has lower optical absorption than at 1.55 μm. An alternative optical platform, e.g. one based on SiN or glass, with high transmission in that range is required. For example, with a similar design in a SiN platform using a 1064 nm laser wavelength, 22 nm-sized polymer particles could be trapped stably [39].

Through expansion of the one-dimensional photonic crystal cavity into two-dimensional form using coupled resonators, one can shape the resonance mode field pattern by changing the laser wavelength, as shown in Fig. 7.9b. Microparticles that were trapped by the field can then be rotated by switching the laser wavelength [38].

7.2.3.2 Microring Resonators

Among all the different kinds of microresonators, circular-shaped microring [40, 41] and microdisk [42–44] resonators have been the two most investigated structures for silicon photonics. For integrated photonic devices, light is usually evanescently coupled into the microresonator through planar input waveguides and confined by total-internal-reflection (TIR) at the microresonator sidewalls. When the wavefront of the cavity lightwave is matched with the input-coupled lightwave upon traveling each round trip, an enhanced cavity field builds up due to constructive interference over multiple round-trips. If a drop-waveguide is placed near the microresonator, part of the on-resonance cavity field can be coupled out to the drop-waveguide. Only a few selective resonance wavelengths can travel in round-trips and be coupled out to the drop-port. The spectral selectivity and field enhancement inside the resonators enable microresonators to be a versatile and powerful building block for tunable optical manipulation devices/circuits.

Microring-based notch filters usually consist of one input-coupling waveguide and one ring-shaped resonator. At an off-resonance wavelength (λ_{off}), most of the light from the input waveguide is transmitted to the through-put waveguide and therefore any particles trapped on the input waveguide will be directed to the throughput port. At an on-resonance wavelength (λ_{on}), particles from the input waveguide are transferred to the enhanced field above the microring, where stronger trapping/drive forces are present, as shown in Fig. 7.10a. For example, 250% enhancement of optical intensities and particle velocity compared to that of the input waveguide was reported in a SU8 solid-core microring resonator [45]. In a later demonstration of such trapping devices on an SOI platform, the on-resonance microring provided the same trapping capability as the input waveguide, but required an order of magnitude lower input laser power. In this case, the optically created potential energy profile over the entire microring circumference reached a depth of 25 k_BT on resonance. The microparticles traveled around the microring at hundreds of micrometres per second, producing periodic revolutions at a few Hertz, with highly accurate positioning [46].

In order to facilitate a three-port particle switch, one can add a drop waveguide to the microring-notch filter to form a microring add-drop filter [47, 48], or add another microring resonator that is coupled to the input waveguide [49].

For the microring add-drop filter, as shown in Fig. 7.10b, at the microring off-resonance wavelength (λ_{off}), optically driven microparticles are transported to the

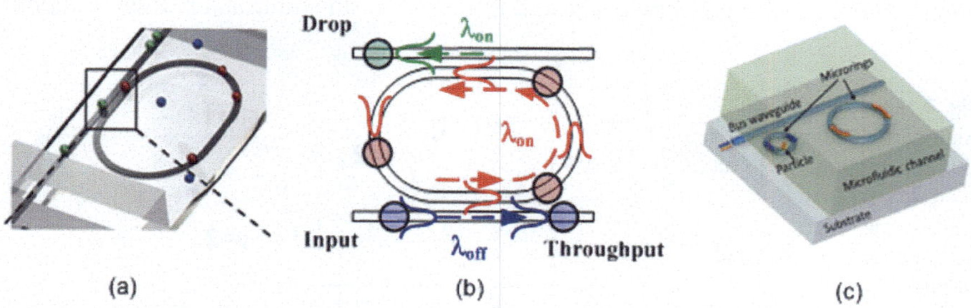

Fig. 7.10 **a** Schematic of optical manipulation on a micro-racetrack notch filter: particles are transported to throughput (green) or buffered on the resonator (red) (reproduced from [45] with permission). **b** Schematic of optical manipulation on a micro-racetrack add-drop filter: particle throughput (blue), buffering (red) and dropping (green) (reproduced from [47, 48] with permission). **c** Schematic of particle trapping on double-ring notch filters. (reproduced from [49] with permission)

throughput port as in the case of the microring notch filter. At the microring on-resonant wavelengths (λ_{on} and λ'_{on}), both the guided field in the microring and the drop-waveguide interfere constructively. For a sufficiently enhanced microring field (namely, one larger than the drop-waveguide field), the particles can be trapped on the microring in round trips, as shown in Fig. 7.10b. For a smaller microring field (namely, comparable to the drop-waveguide field), the particles can be output-coupled to the drop waveguide, as shown in Fig. 7.10b. By tuning the input laser wavelength and choosing proper microring Q-factors, three-port optical manipulation (throughput, buffering and dropping) was demonstrated and design rules for implementing these functions were derived [47, 48, 50].

For the double-ring notch filters, the two microring resonators are designed to have different sizes in order to have different resonance wavelengths. Microparticles are routed to the throughput port at off-resonance wavelengths for both microresonators. By switching the input laser wavelength to different microring resonances, microparticles can be routed and stored on different microrings [49].

7.2.3.3 Microdisk Resonators

The preceding discussion illustrates the point that single microring-based devices for trapping particles along the microring waveguides only provide a single trapping channel and limited trapping area. It is natural to seek for wider trapping area and possible parallel processing in optical manipulation. Microdisk resonators in which the surface comprises the entire disk solve this problem and form another important building block for on-chip particle manipulation. Unlike microring resonators, which are single-waveguide-mode devices, microdisk resonators usually support multiple whispering gallery modes (WGM), as illustrated in Fig. 7.11 [51]. Inside the microdisk resonators, the high-order WGMs exhibit multiple mode-field maxima (MFM) along the radial direction, resulting in multiple tracks for particle trapping. The extended mode-field distribution inside the microresonator of the high-order mode extends the particle trapping range inward from the disk edge. The evanescent field outside the microdisk sidewall also enables an additional particle trapping

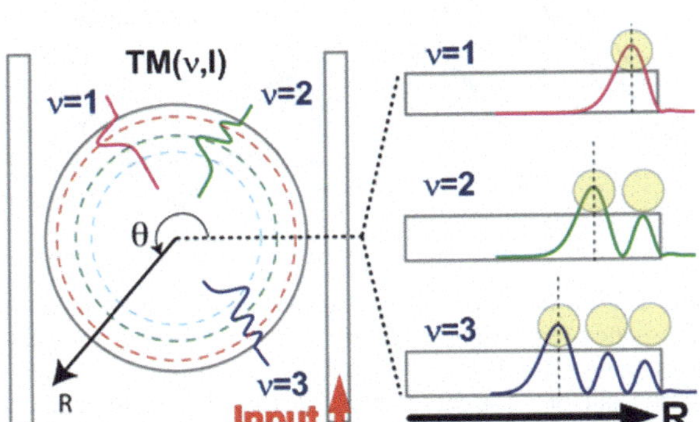

Fig. 7.11 Schematic of microparticle trapping by multiple trapping tracks on a microdisk resonator. υ: radial mode order. (reproduced from [51] with permission)

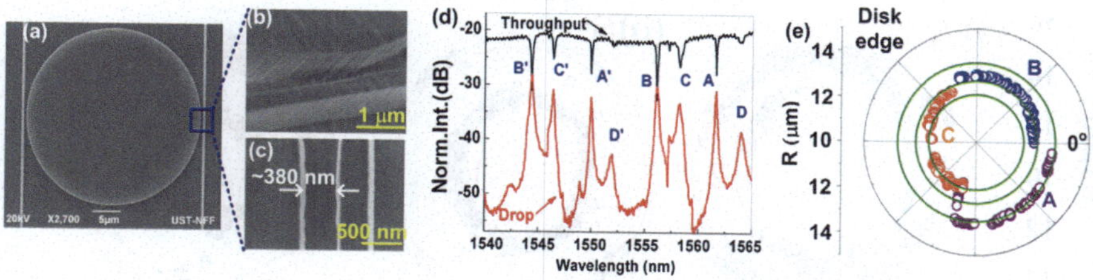

Fig. 7.12 **a** SEM pictures of a SiN microdisk resonator. Zoomed-in SEM pictures of the **b** coupling region and **c** coupling gap spacing. **d** Measured throughput- and drop-spectra of the microdisk add-drop device. **e** Measured particle trajectories upon laser wavelength on B-C-A in an R versus θ plot. (reproduced from [51] with permission)

track. Particles trapped on these tracks are driven by the optical force of these resonant traveling waves, as in the microring resonator case.

Figure 7.12a shows SEM images of a fabricated microdisk add-drop device on a SiN platform. The resonator is 30 μm in diameter, and 0.7 μm-thick, on top of a 1.8 μm-thick silica under-cladding layer on a silicon wafer. Figure 7.12b, c show the coupling region between the input waveguide and the microdisk. The sidewall of the microdisk and the waveguide appear relatively rough due to the dry etching process used. The coupling gap is ~380 nm.

Figure 7.12d shows the measured throughput- and drop-spectra of the waveguide-coupled microdisk add-drop device with colloidal solution cladding. The spectra show two free spectral ranges containing multiple WGMs with Q factors from ~3000 to ~6000. Each WGM has a different major trapping track from each other and it is, therefore, possible to switch the trapped particle trajectories on the microresonator by tuning the input laser wavelength. Figure 7.12e shows a particle trajectory on the microdisk resonator in a polar plot with various trapping tracks indicated by the lines when the laser wavelength is tuned away from the resonances B-A-C.

On each WGM, particles can be trapped in multiple tracks within a single resonance wavelength. For example, Fig. 7.13a–f show the radial position, R, and the angular position, θ, of the accumulated particle trajectories upon resonances A′, B′, C′, A, B and C in Fig. 7.12d. The particles travel in multiple tracks on the disk (tracks 1 and 2) with $R < 15$ μm and along the microresonator sidewall edge (track 3) with $R > 15$ μm. The experimental data show that the particles cannot be trapped stably between the trapping tracks. The experimental results are consistent with the simulated modal field distribution (for 2nd, 3rd and 4th order TM modes) using the finite-element method, as shown in Fig. 7.13g–i.

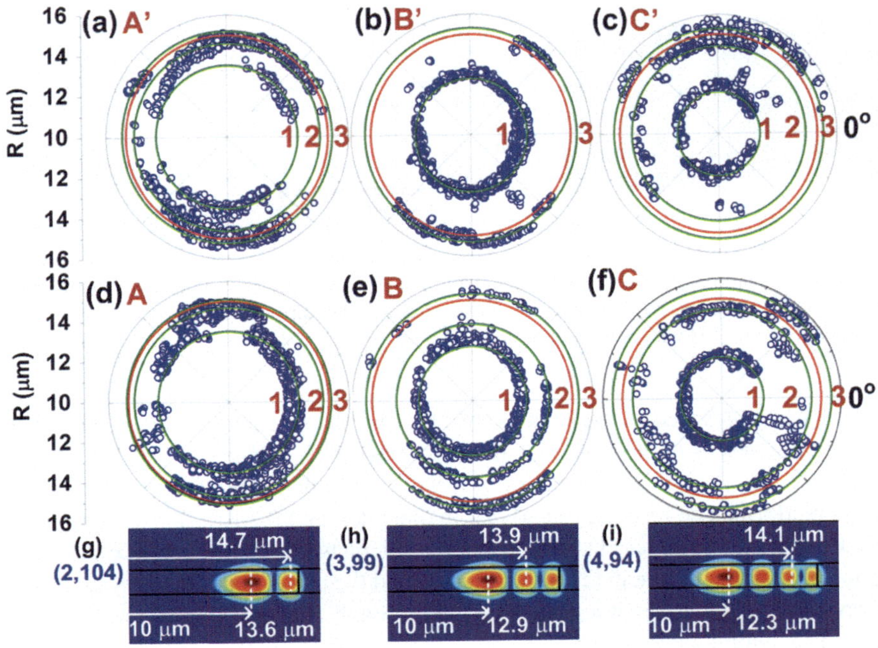

Fig. 7.13 **a–f** Measured R versus θ plots for the resonator-trapped particles upon resonances. **a** A′, **b** B′, **c** C′, **d** A, **e** B and **f** C. **g–i** Simulated cross-section of the mode-field distributions of the second to fourth order radial modes (reproduced from [51] with permission)

7.3 Particle Manipulation Using Liquid-Core Waveguides

7.3.1 Particle Pushing in Hollow-Core Fibers

The earliest demonstration of moving a particle *inside* a waveguide was reported in 1995 by Renn et al., who used a laser guided through a hollow core fiber to move rubidium (Rb) atoms along the interior of the fiber [52]. The propulsion along the fiber was due to the scattering force while the atoms were drawn to the center of the waveguide by the gradient force generated by the guided optical mode. By changing the wavelength of a titanium-sapphire laser around the Rb D2 resonance lines, the magnitude of the forces and thus the number of atoms pushed through the fiber per second could be tuned. With a 40 μm core diameter, a 45 mW laser source, and a detuning of −6 GHz from the D2 resonance, a potential depth of 71 mK was achieved with a transverse capture velocity of 3.7 m/s [52]. The maximum atomic flux was found to be about 2400 s^{-1} at a detuning of −3 GHz. The effects of fiber bending and varying laser intensities on the atomic flux were also investigated. At lower intensities the atom flux scales as a power law, but saturates at higher intensities. As the fiber is bent it takes larger intensities to get the same amount of flux. A critical

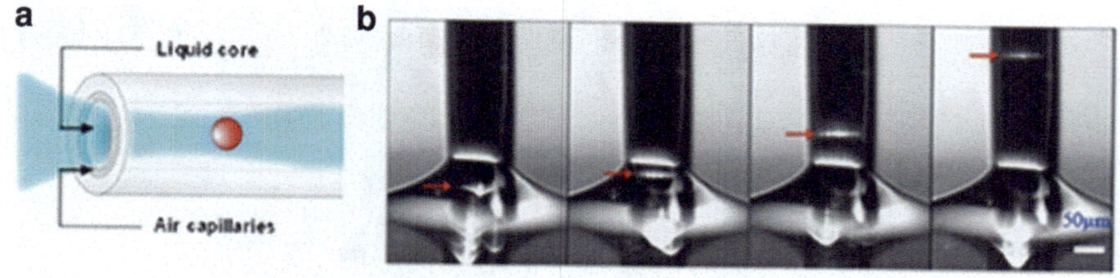

Fig. 7.14 **a** Schematic of particle in PCF fiber particle would be pushed to the right. **b** Near the fiber inlet a 3 μm polystyrene bead is pushed upwards (red arrow shows particle movement) (reproduced from [53] with permission)

radius of about $r = 6$ cm was observed, beyond which the movement of the atoms was no longer possible since the guiding properties of the fiber were too weak.

In 2007, Mandal et al. pushed 3 μm polystyrene beads in a liquid-core photonic crystal fiber (PCF) with a core diameter of 20 μm. By shining a 488 nm argon laser upwards particles could be stopped by balancing the scattering force with the gradient force due to fiber loss and with gravity (see Fig. 7.14) [53].

By equating these forces with the Stokes viscous drag force, the velocity of the particle can be determined using [53]

$$v_{\mathrm{p}} = \frac{64\pi^2 a^5 n_{\mathrm{m}}}{9\mu_{\mathrm{m}}\lambda^4 c}\left(\frac{(n_{\mathrm{p}}/n_{\mathrm{m}})^2 - 1}{(n_{\mathrm{p}}/n_{\mathrm{m}})^2 + 2}\right)^2 I(r,\theta,z) + \frac{a^5 n_{\mathrm{m}}}{3\mu_{\mathrm{m}} c}\left(\frac{(n_{\mathrm{p}}/n_{\mathrm{m}})^2 - 1}{(n_{\mathrm{p}}/n_{\mathrm{m}})^2 + 2}\right)^2 \frac{\partial I(r,\theta,z)}{\partial z} - \frac{m_{\mathrm{p}} g}{6\pi\mu_{\mathrm{m}} a} \tag{7.1}$$

where a is the particle radius, and n_{m} and n_{p} are the indices of refraction for the medium and the particle, respectively; λ is the wavelength of light, c is the speed of light, μ_{m} is the viscosity of the medium, I is the intensity of light incident on the particle, m_{p} is the mass of the particle and g is the acceleration of gravity. A maximum particle movement of 2 cm was found at a laser power level of 210 mW before coupling. At a laser power of 120 mW, the measured velocity of the particles varied normally over a large range from 30 to 100 μm/s with its median at 70 μm/s. This large variation was most likely due to the movement of the particles between modes in this large-core multi-mode fiber.

7.3.2 On-Chip Particle Trapping

In 2006 Cran-McGreehin et al. reported an integrated microfluidic chip with built-in semiconductor lasers for particle trapping in a fluidic channel [54]. Their design features two pairs of opposing lasers that can be independently switched on and tuned (see Fig. 7.15a). This geometry corresponds to an on-chip implementation of a dual-beam trap in which opposing scattering forces from diverging beams balance at the trapping point. The device consists of a GaAs/AlGaAs heterostructure with InAs quantum dots that emit at 1290 nm. Each laser produces up to 20 mW of optical

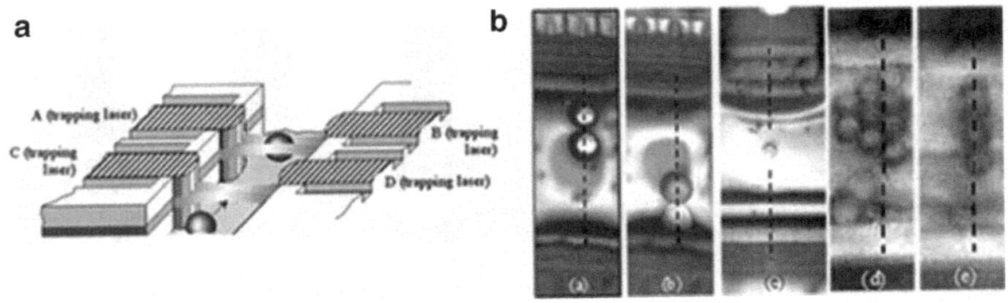

Fig. 7.15 **a** Schematic of device with four built in lasers, one particle is shown trapped. **b** Multiple trapped particles; fluid channel is shown left to right, dashed line represents beam axis. **a, b** Shows 20 μm particles pushed back and forth across channel. **c** Image of trapped SA2 cell. **d, e** Trapped crystal in divergent beams. (reproduced from [54] with permission)

power, which is sufficient for trapping of biological and colloidal microparticles. The movement of a 5 μm particle was detected by shining a 411 nm laser diode from above and then detecting the fluorescence of the dyed spheres. By tuning the relative intensities of opposing lasers, the particle position was controlled (See Fig. 7.15b). By switching the lasers off, the particles could be released and transport along the fluidic channel could resume.

Depending on the flow directions, particles can be switched back and forth between the two sets of opposing lasers, i.e. the two trapping points. Cran-McGreehin et al. determined that the trapping efficiency, the efficiency of photons in transfering their momentum to the particle, increases with particle size as well as with the distance from the output facet. This behavior is due to the lower intensity, since the farther the particle is from the facet, the more divergent the light. The trap stiffness, defined as the effective spring constant that relates optical force and spatial displacement, was determined to be 3.2×10^{-8} Nm^{-1}.

In 2009 Yang et al. manipulated particles with a sub-wavelength-scale slot waveguide [55]. They were able to capture, trap, and transport polystyrene nanoparticles and λ-DNA. Figure 7.16a shows that particles are drawn into the slot via the gradient force and propelled along the nano-channel by the scattering force.

Slots with widths of 100 nm and 120 nm were used for 75 nm and 100 nm diameter nanoparticles, respectively. The optical power levels were as high as 300 mW at 1550 nm, in TE polarization. Flow in a microfluidic channel defined above the slot waveguides was directed perpendicularly to the slot waveguides (See Fig. 7.16). With flow speeds of 80 μm/s, less than 25% of the particles were successfully captured. Once a 100 nm particle was trapped, optical transport at 1.5 μm/s at a laser power of 250 mW was possible. λ-DNA is difficult to trap when uncoiled since most optical traps work at a focused location. Yang et al. were able to draw partially uncoiled λ-DNA into the slot waveguide because of the extended distribution of the gradient force in the guided mode along a line rather than at a single point. The trap had a stronger stiffness for particles in the slot than for those outside, and, due to the higher force levels there, the majority of the trapped particles was found inside the slot. A trap stiffness of 0.2 pN/nm was estimated for a 100 nm nanoparticle being trapped

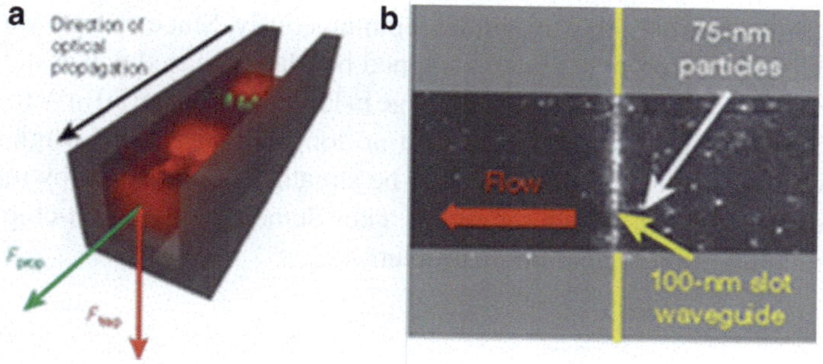

Fig. 7.16 **a** Schematic of slot waveguide showing force directions. **b** 75 nm particles are shown trapped in 100 nm slot waveguide, LC is shown left to right, flow moves right to left. (reproduced from [55] with permission)

by 1 W of light. Due to the lower stiffness of the trapping outside the slot, particles there had lower stability and were therefore released from the trap after less time. It was also shown that the stability of the trapping of DNA was lower due to the extended state of the coils.

In 2010 Bellini et al. fabricated an optofluidic fused silica glass-based chip using a femtosecond laser that was designed to trap and stretch a single red blood cell (See Fig. 7.17) [56].

The chip featured two optical waveguides that intersect with a microfluidic channel, both fabricated with the same femtosecond laser, but with different laser repetition rates. The trapping and stretching was carried out with a 1070 nm laser coupled into a fiber and then split between the two waveguides. The power on each side could be independently tuned by adjusting variable optical attenuators, thereby allowing the particle position to be changed. The cells were flown through the microchannel at 10–50 μm/s and could be trapped with 20 mW of laser power from each waveguide. Once a particle was stably trapped it could be stretched by

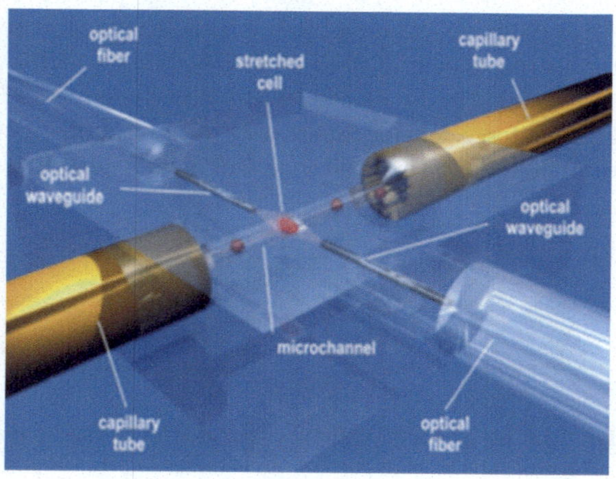

Fig. 7.17 Schematic of stretching chip. Red blood cell shown being stretched at the intersection of optical and fluid channels. (reproduced from [56] with permission)

increasing the power of both waveguides simultaneously. Since the two waveguides are still balanced, the particle remains trapped but the increased power deforms the cell. A 25% elongation was seen with a change in laser power from 20 mW to 300 mW, which stretches the cell to the point that it no longer returns to its original shape. Lower power levels allowed for the cell to be stretched and released by turning the power level up and down, which represents a key demonstration of on-chip trapping and manipulation of biological microstructures.

7.3.3 Particle Manipulation in Liquid-Core ARROW Waveguides

Liquid-core antiresonant reflecting optical waveguides (LC-ARROWs) are a specific type of hollow-core optical waveguide in which both light and liquid are guided through the same microchannel [57]. Optical confinement in a low-index liquid is created using dielectric layers of alternating materials that are antiresonant, i.e. highly reflective, for the transverse wavevector component of a propagating mode [57, 58]. The hollow cores are fabricated with either a sacrificial core process [59] or by wafer bonding [58]. The hollow channels are optically interfaced with solid-core ARROW (SC-ARROW) waveguides that ensure light delivery to and from the edges of the chip. LC-ARROW devices have successfully been used for particle detection and sensing using fluorescence or Raman scattering all the way down to single DNA molecules [60–62]. They have also proven to be an ideal platform for a number of optical and electro-optical particle manipulation techniques, and we will review these capabilities in more detail here.

7.3.3.1 Single Beam Particle Pushing and Confinement

In 2008 it was shown by Measor et al. that particles could be pushed along an LC-ARROW by optical forces [63]. The ARROW chip used has an LC-ARROW which runs perpendicular to an SC-ARROW (See Fig. 7.18).

A particle in the channel experiences three types of forces. The scattering force, gradient force, and the Stokes drag force resulting from pressure-based or electrokinetic fluid flow. The main source of the forward movement of the particle is due to

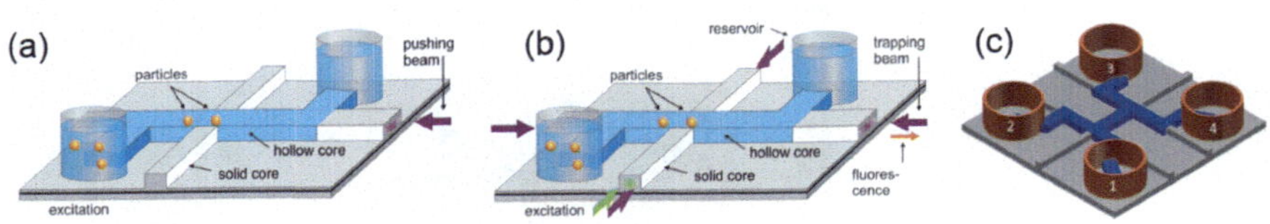

Fig. 7.18 Schematic of ARROW chip **a** used for pushing, **b** used for trapping, and detection [64] and **c** used for sorting (reproduced from [65] with permission)

the scattering force that is acting along the direction of light propagation, z. In addition, transverse gradient forces exist that pull the particle towards the high-intensity regions of the mode. The scattering force is highly dependent on the amount of power acting on the particle. Due to the propagation loss, α, along a waveguide the power entering the waveguide P_0 decays exponentially with distance [63]. This leads to a scattering force equation that varies with propagation distance:

$$F_S(z) = Q\frac{n}{c}P_0\exp(-\alpha z) \tag{7.2}$$

where Q is a dimensionless factor, n is the refractive index of the medium, and c is the speed of light. In the steady state, the Stokes drag force must counteract and balance the scattering force:

$$F_{\text{Stokes}}(z) = -6\pi\eta' r\frac{\mathrm{d}z}{\mathrm{d}t} = -F_s(z) \tag{7.3}$$

where r is the radius of the particle and η' is the average dynamic viscosity of the fluid. Therefore, the particle trajectory along the channel can be calculated as:

$$z(t) = \frac{1}{\alpha}\ln\left[v_0\alpha t + \exp(\alpha z_0)\right] \tag{7.4}$$

where z_0 is the initial particle position, and $v_0 = QP_0n/(6\pi\eta' rc)$.

The gradient force acts to center the particle in the channel since the direction of the gradient force, F_G, is towards the highest intensity. The motion of the particle diffusion perpendicular to the propagation direction is defined by the corresponding potential $U(x)$:

$$U(x) = -kT\ln p(x) = -\frac{\gamma}{c\varepsilon}I(x) = -\int F_g(x)\mathrm{d}x \tag{7.5}$$

where $p(x)$ is the probability of finding the particle at a given location, k is the Boltzmann constant, T is the temperature, ε is permittivity, γ is the effective polarizability of the particle, and $I(x)$ is the intensity profile of the guided mode.

The aforementioned dependence of the scattering force on channel position (7.2) enables a new way for characterizing the loss and mode profile of a hollow-core waveguide, as shown by Measor et al. [63]. The particle trajectory along the channel $z(t)$ is recorded using a CCD camera in top-down view. Equation (7.4) can then be fitted to the measured trajectory to extract the waveguide loss, as shown in Fig. 7.19a.

Similarly, the transverse location of the particle in the channel $x(z, t)$ gives information on how the intensity of light changes along that axis. Since the gradient force pulls particles into high intensity regions, the likelihood of finding a particle in such a region is higher.

A histogram of lateral position distributions was created from the movie of the particle moving along the channel under the influence of a single-beam optical force

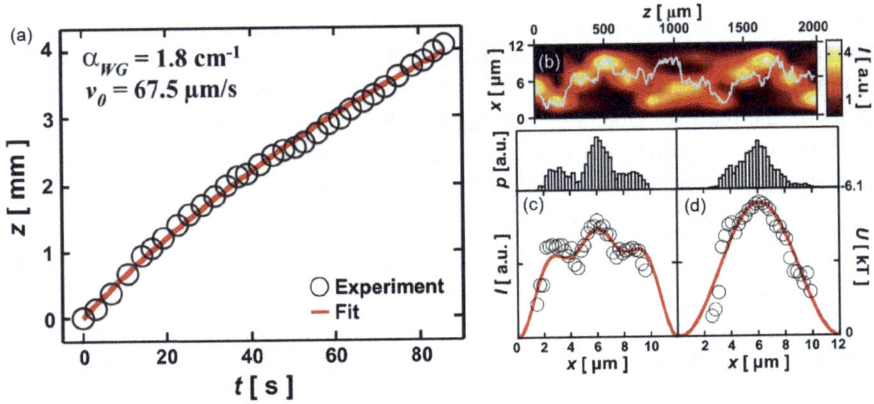

Fig. 7.19 **a** Plot of experimental data, fit used to calculate loss and initial velocity. **b** Calculated mode and measured particle trajectory of multimode LC. **c** Multi-mode LC histogram (inset), calculated mode (line) and averaged intensity potential (circles). **d** Single-mode LC histogram (inset), calculated mode (line) and averaged intensity potential (circles). (reproduced from [63] with permission)

(Fig. 7.19c, d). The figure shows excellent agreement between the shape of these histograms and the waveguide mode profiles calculated using commercial photonic simulation software.

Measor et al. introduced another use for a single optical beam in 2009 [66]. One of the main issues with particle sensing in liquid-core waveguides is the fact that particles can assume lateral positions in the channel that result in poor coupling of any generated fluorescence into the liquid-core waveguide modes. In the typical fluorescence experiment shown in Fig. 7.20b, the excitation light is coupled into a SC-ARROW using a single-mode fiber, exciting particles in the location where the SC-ARROW intersects with the LC-ARROW channel. Fluorescence is then collected along the flow channel and detected using a sensitive photodetector. By adding a near

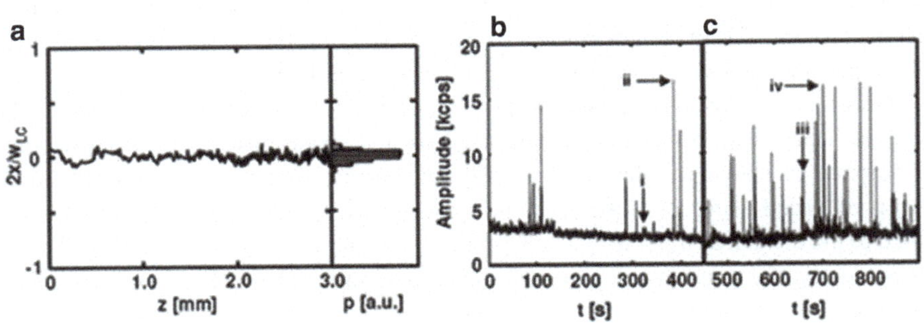

Fig. 7.20 **a** Measured particle trajectory in LC and histogram of particle's transverse (x) position, showing single mode guiding. Fluorescence collected from LC **b** without and **c** with guiding beam. (reproduced from [66] with permission)

infrared guiding beam along the liquid channel, particles are pulled into the center of the channel by the gradient force. This "optical focusing" results in higher and more uniform coupling of fluorescence into the LC-ARROW modes, and therefore in a large improvement in collection efficiency, by up to 85% (See Fig. 7.20).

A practical application of exerting optical forces on small particles with a single beam is optical particle sorting on an optofluidic chip. This capability was demonstrated by Leake et al. [65], who used a combination of fluid flow and a single optical beam to implement novel size-based sorting methods for micron and sub-micron sized particles. Sorting requires an extended fluid channel layout—which was implemented using an "H-shaped" channel configuration, as shown in Fig. 7.18c. Two geometries for the relative orientations of particle movement, flow, and optical beam propagation were introduced. They offer size tunability and up to 100% sorting efficiency, and good agreement between experimental results and calculated trajectories was found.

As before, both the scattering force and the Stokes drag force act upon a particle, and it is their relative magnitude that enables size selectivity and therefore sorting. The scattering force can be recast as:

$$F_{\mathrm{LASER}} = Q\pi \frac{r^2}{c} I. \tag{7.6}$$

Notice that F_{LASER} scales quadratically with particle size while F_{Stokes} (7.3) scales linearly. By adjusting the relative flow velocity and laser power, smaller particles can be made to follow the flow direction, while larger particles will move in the direction of the propagating optical mode. Experiments were carried out using mixtures of sulfate latex beads of varying size ($r = 0.25$, 0.5, 1 and 1.5 μm). In the first configuration, flow acted perpendicularly to the optical propagation direction, while the second orientation had the flow in the opposite direction. To determine the position of the particle in the channel, similar methods were used to those for the propagation loss measurements. The measured trajectories were compared with dynamic simulations for which the optical forces were calculated at each location using formulae for loosely focused and collimated beams [67]—and known mode profiles for LC-ARROWs. Similarly, position-dependent drag forces were determined using 3D COMSOL simulations. The instantaneous velocity at a particular location is therefore the vector sum of the velocities in all three directions.

The first orientation that used the perpendicular beam orientation has the bulk of the flow, which contains the bead mixture, moving through the LC-ARROW bottom to the top of the channel (Fig. 7.21a). The intersecting LC has a small amount of flow that escapes and departs from the intersection to the right. The sorting beam is incident from the left through an SC-ARROW, pushing the beads to the right. Large particles will move to the right and are considered "sorted"—and small particles will continue upwards with the bulk of the flow (See Fig. 7.21a). Depending on the ratio of the flow speed to the power output of the laser, all the larger beads can be removed—or alternatively a well-defined percentage of each bead size.

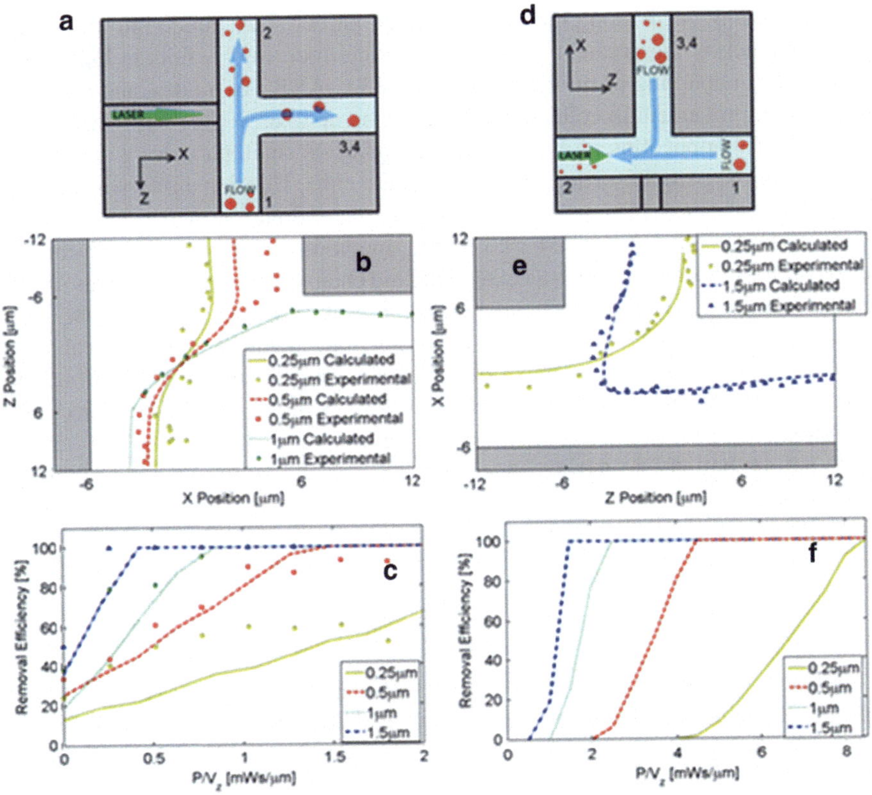

Fig. 7.21 **a** Schematic of first orientation with perpendicular flow and laser. **b** Comparison of experimental and calculated trajectories for perpendicular orientation. **c** Removal efficiency plot for perpendicular orientation showing calculated and experimental values. **d** Schematic of second orientation with optical beam counter-propagating to flow. **e** Comparison of experimental and calculated trajectories for counterpropagating orientation. **f** Removal efficiency plot for counterpropagating orientation. (reproduced from [65] with permission)

Since particles can start at any location in the channel and therefore have both different amounts of lateral movement required to "sort" and have different flow speeds and optical intensities, this method is very position dependent. Some small particles along the right-hand edge of the channel are pushed out of the flow despite their small size. Therefore, some small beads are removed from the flow, when the scattering force is not strong enough to remove all the large beads (See Fig. 7.21b), which leads to removal efficiencies (ratio of the number of particles removed to the total number of particles) greater than zero for all particle sizes (See Fig. 7.21c). Depending on the ratio of the flow speed to the laser output power, the removal efficiencies of each size can be tuned.

In the second orientation (counter-propagating), beads flow into the T-junction from the top in a mixture—and a second flow originates from the right. The flows meet to the left of the intersection where the laser is coupled in. Since the beads come from the top, they either turn towards the laser or they move in the direction of light propagation, depending on the balance between scattering and drag forces (See Fig. 7.21d, e). In this case, the particles are considered "sorted" if they move to the right with the optical beam. This configuration exhibits only a very weak dependence on the lateral starting position of each particle, allowing for achievement of removal efficiencies of zero for small particles, in this orientation (See Fig. 7.21f). This enables the complete removal of particles from the flow, above a tunable cutoff size.

7.3.3.2 Dual Beam Particle Trapping

While a single beam manipulating particles inside a liquid-core waveguide clearly adds a diverse range of capabilities to a planar optofluidic platform, it is the use of multiple beams that enables additional particle manipulation options, in particular particle trapping.

In 2009 Kühn et al. introduced a new type of dual-beam trap based on counter-propagating beams [64]. Unlike Ashkin's original dual-beam trap, which relied on spatial variations of the beam area, this method used the propagation loss in confined LC-ARROW modes to create beam-dependent force profiles and to trap particles (See Fig. 7.22a). This loss-based (LB) trap is a long-range trap in which the particles can be held at any point along a several millimetres long channel by adjusting the relative power of the two beams. In this case, the scattering force takes on a similar form to (7.2),

$$F_{r/l}^{\mathrm{Sca}} = \mp Q\frac{c}{n}P_{r/l}^{0}\mathrm{Exp}\left(-\alpha\left(\frac{L}{2}\pm z\right)\right), \tag{7.7}$$

where L is the length of the LC-ARROW, P_i^0 is the input power at each end of the liquid core waveguide, and z is the particle position relative to the center of the liquid core in the lateral direction. The trapping point occurs where the scattering forces are balanced:

$$z_1 = \frac{1}{2\alpha}\mathrm{Log}\left(\frac{P_r^0}{P_l^0}\right) \tag{7.8}$$

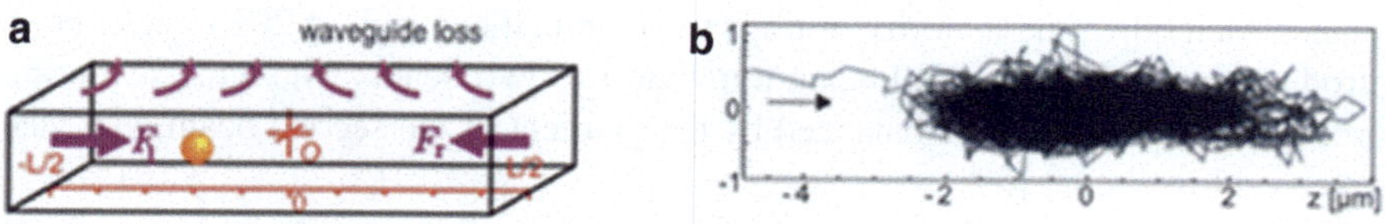

Fig. 7.22 **a** Schematic of forces and loss in liquid-core waveguide. **b** Trajectory of particle in trap. Particle enters from the left. (reproduced from [64] with permission)

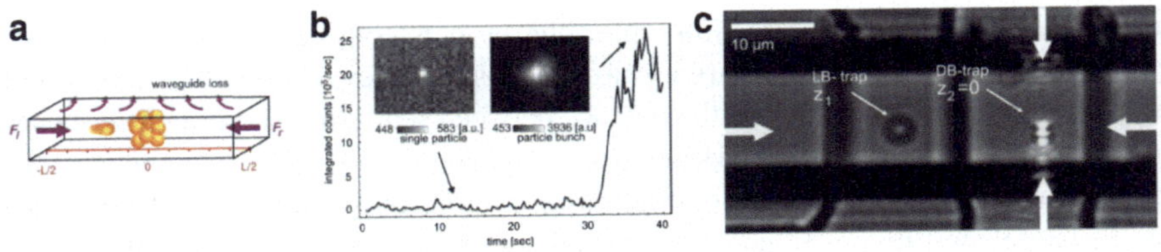

Fig. 7.23 **a** Schematic of forces and loss in liquid-core waveguide. **b** Photon counts of trapped particles, initially one particle is trapped, particles concentrate to create larger signal [66]. **c** Image showing two simultaneously trapped particles: one particle is trapped with a loss based trap (horizontal) and the second is trapped with a divergence-based dual beam trap (vertical). (reproduced from [64] with permission)

For a waveguide with $\alpha = 5.2\,\text{cm}^{-1}$ the forces were found to be about 0.14 pN at the beginning of the LC and the trap was found to have a stiffness of about 50 nN/m (See Fig. 7.22b).

With this method, Kühn et al. were able to trap single silica beads as well as silica beads tagged with Escherichia coli (E. coli) bacteria the DNA of which was stained with Acridine Orange intercalating dye. This LB trapping method can be extended to multiple particles to function as a particle concentrator for increased detection efficiencies [68]. When trying to detect particles at low concentrations, near the single particle level, it can be difficult to detect the fluorescence. Here, particles from either end of the channel are drawn to the trapping point under the influence of the two trapping beams (see Fig. 7.23a), which can increase the concentration of the particles by more than two orders of magnitude (see Fig. 7.23b). Once trapped, the group of particles can easily be moved by adjusting the relative input powers.

Moreover, a traditional divergence based (DB) trap can be created by matching beam powers via two SC-ARROWS, similarly to [54]. Once the beam exits the solid core it diverges as it propagates across the liquid core. Due to the chip architecture, this DB trap can be used simultaneously with the LB trap—as shown in Fig. 7.22 [64].

It was shown that this "dual dual-beam trap" can be used to control the relative spacing of two particles trapped in each beam pair. A separation of 7.4 μm from each other was achieved without losing the two separate traps. Such a configuration could be used to facilitate controlled interactions and binding between two molecules tethered to trapped microbeads.

Due to the simplicity of counter-propagating beam traps where scattering forces balance each other, the majority of traps work on that concept. In 2013 Leake et al. introduced an orthogonal dual beam trap that uses two beams where the scattering force from the first beam is balanced by the gradient of the second beam, and vice versa (see Fig. 7.24a) [69].

Unlike free-space dual-beam or single-beam traps that need to be focused, the orthogonal trap works with collimated beams—making it ideal for on-chip applications. If the intensity distributions of the two intersecting beams take on the form of:

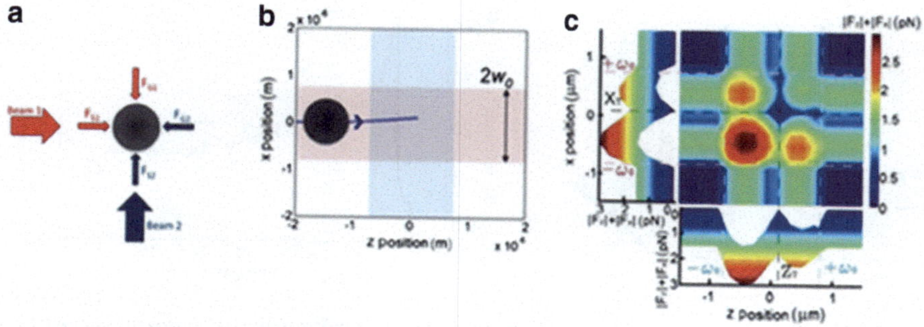

Fig. 7.24 **a** Schematic of beams and forces in orthogonal trap. **b** Simulated trajectory of particles in trap. **c** Total force at intersection. The trapping point is shown in black, beam 1 (red) and beam 2 (blue) have beam waists shown by dotted lines. (reproduced from [64] with permission)

$$I_1(x) = I_0 \cdot e^{-\gamma x^2} \; ; \; I_2(z) = I_0 \cdot e^{-\gamma z^2} \tag{7.9}$$

where $\gamma = 2/w_0^2$, and w_0 is the beam waist, then the scattering and gradient forces become:

$$F_{S1}(x) = k_S \cdot I_1(x) = Q_S \cdot \frac{A}{c} \cdot I_1(x) \tag{7.10}$$

$$F_{G1}(x) = k_G \cdot \left| \vec{\nabla} I_1(x) \right| = 2\gamma k_G \cdot x \cdot I_1(x) = Q_G(x) \cdot \frac{A}{c} \cdot I_1(x). \tag{7.11}$$

In this case, k_S and k_G are constant coefficients and Q_S and Q_G are the scattering and gradient force specific pre-factors [2]. It can be found, by setting the gradient force for one beam equal to the scattering force of the other, that the trapping point is located at:

$$x_T = z_T = \frac{k_S}{k_G}\frac{1}{2\gamma} = \frac{k_S}{k_G}\frac{w_0^2}{4}. \tag{7.12}$$

This trapping method is stable when:

$$w_0 \le 2\frac{k_G}{k_S}. \tag{7.13}$$

The stability and depth of the trap depends on the beam power, wavelength, and w_0. This trap is self-loading, which makes it ideal for on-chip applications (See Fig. 7.25b). According to simulations performed in the same manner as the sorting method mentioned earlier, for beams with $w_0 = 0.75$ μm, $\lambda = 532$ nm, and $P =$ 60mW, the trapping point is located at $(x_T, z_T) = (0.1$ μm, 0.1 μm$)$ and the trapping potential has a depth of $\Delta W = 1.9 \times 10^{-18}$ J $= 12$ eV (See Fig. 7.25c).

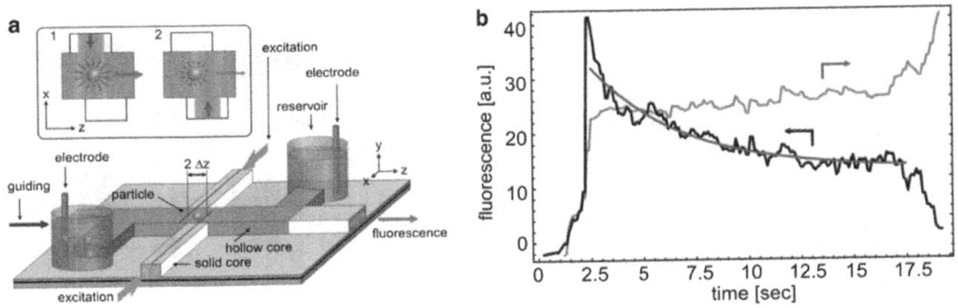

Fig. 7.25 **a** Schematic of on-chip implementation of Anti-Brownian Electrokinetic (ABEL) trap. **b** Time-dependent fluorescence signal and z-position showing photobleaching of trapped E. coli, fluorescence signal fitted with a monoexponential decay curve (reproduced from [70] with permission)

The major drawback of all-optical traps, including the chip implementations described thus far, is the relatively high laser power that is needed to generate a sufficiently deep trapping potential. One method of reducing the amount of power is to use electro-optical traps that combine an ultra-low power optical source with electrical feedback. The optical source can be used for either generation of the feedback signal required for trapping and fluorescence or scattering analysis of the trapped particle. Such traps are termed Anti-Brownian electrokinetic (ABEL) traps and exist in various bulk optical implementations [71, 72]. The ABEL trap scales favorably with reduced particle size, and objects on the nanometer scale down to single dye molecules have been trapped in bulk versions [73]. Kühn et al. demonstrated an optofluidically integrated version of an ABEL trap, as shown in Fig. 7.25 [70]. This trap has two solid core waveguides that are offset from each other so that they overlap by half their modal beam-waist. These waveguides are alternately excited, creating a modulated fluorescence or scattering signal from a particle in the intersection. By correlating the fluorescence signal with the excitation sequence, the location of the particle can be determined—and an electrical force can be applied to push the particle towards the center of the intersection, which compensates for Brownian motion along the channel direction and was proven to be sufficient to trap micro- and nano-scale beads with ultralow optical power levels as low as 10 μW, approximately five orders of magnitude lower than the all-optical version on a similar chip [70]. Moreover, single, fluorescently stained E. coli bacteria were successfully trapped—and the photobleaching process of their stained DNA was observed. In addition to the low power requirement, another unique characteristic of ABEL traps is that they are true single-particle traps, since random Brownian motion can be corrected only for one particle.

7.4 Summary

Planar integration of optofluidics has seen a lot of progress over the past few years. We have shown that optical particle manipulation on a chip has been a particularly fruitful area in this regard. Numerous inventive approaches to affecting the location and velocity of a particle in a fluidic microchannel with light have been implemented on a number of platforms. These were broadly categorized into surface-based and channel-based arrangements in which the light is guided in solid- and liquid-core waveguides, respectively. While the initial emphasis has been on exploring the possibilities of creating sufficiently strong forces and exploring the operational parameters and limitations of each scheme, the research focus is gradually shifting towards developing applications that incorporate particle manipulation concepts. In particular, the combination with ultrasensitive particle detection and analysis on a chip for applications in biosensing, single molecule spectroscopy, and molecular biology promises a bright future indeed.

Acknowledgements Financial support for this work by the National Institutes of Health (grants R21EB003430, R01EB006097, and 1R21AI100229), the National Science Foundation (grants ECS-0528730, CBET-1159453, CBET-1159423, CBET-1402848, and CBET-1402880), the Rogers Family Foundation, the Defense Advanced Research Projects Agency (DARPA) under Contract No. HR 0011-10-1-0075, and the W.M. Keck Foundation through the W.M. Keck Center for Nanoscale Optofluidics at UC Santa Cruz, is gratefully acknowledged.

References

1. P.N. Lebedev, Experimental examination of light pressure. Ann. Der Phys. **6**, 433 (1901)
2. A. Ashkin, Acceleration and trapping of particles by radiation pressure. Phys. Rev. Lett. **24**, 156 (1970)
3. A. Ashkin, J.M. Dziedzic, Optical levitation by radiation pressure. Appl. Phys. Lett. **19**, 283 (1971)
4. M.H. Anderson, J.R. Ensher, M.R. Matthews et al., Observation of Bose-Einstein condensation in a dilute atomic vapor. Science **269**, 198–201 (1995)
5. A. Ashkin, J.M. Dziedzicz, J.E. Bjorkholm, S. Chu, Observation of a single-beam gradient force optical trap for dielectric particles. Opt. Lett. **11**, 288 (1986)
6. A. Ashkin, J.M. Dziedzicz, Optical trapping and manipulation of viruses and bacteria. Science **235**, 1517–1520 (1987)
7. J.D. Wen, L. Lancaster, C. Hodges, A.C. Zeri, S.H. Yoshimura, H.F. Noller, C. Bustamante, I. Tinoco, Following translation by single ribosomes one codon at a time. Nature **452**, 598–603 (2008)
8. C. Monat, P. Domachuk, B.J. Eggleton, Integrated optofluidics: a new river of light. Nat. Photonics **1**, 106 (2007)
9. D. Psaltis, S.R. Quake, C. Yang, Developing optofluidic technology through the fusion of microfluidics and optics. Nature **442**, 381 (2006)
10. H. Schmidt, A.R. Hawkins, The photonic integration of non-solid media using optofluidics. Nat. Photonics **5**, 598–604 (2011)
11. X. Fan, I.M. White, Optofluidic microsystems for chemical and biological analysis. Nat. Photonics **5**, 591–597 (2011)

12. A.R. Hawkins, H. Schmidt (eds.), *Handbook of Optofluidics* (CRC Press, 2010). ISBN 9781420093544
13. B. Schelle, P. Dreß, H. Franke, K.F. Klein, J. Slupek, Physical characterization of lightguide capillary cells. J. Phys. D Appl. Phys. **32**, 3157–3163 (1999)
14. W.P. Risk, H.C. Kim, R.D. Miller, H. Temkin, S. Gangopadhyay, Optical waveguides with an aqueous core and a low-index nanoporous cladding. Opt. Exp. **12**, 6446–6455 (2004)
15. V.R. Almeida, Q. Xu, C.A. Barrios, M. Lipson, Guiding and confining light in void nanostructures. Opt. Lett. **29**, 1209–1211 (2004)
16. D.B. Wolfe, R.S. Conroy, P. Garstecki, B.T. Mayers, M.A. Fischbach, K.E. Paul, M. Prentiss, G.M. Whitesides, Dynamic control of liquid-core/liquid-cladding optical waveguides. PNAS **101**, 12434–12438 (2004)
17. J.-M. Lim, S.-H. Kim, J.-H. Choi, S.-M. Yang, Fluorescent liquid-core/air-cladding waveguides towards integrated optofluidic light sources. Lab Chip **8**, 1580–1585 (2008)
18. G. Antonopoulos, F. Benabid, T.A. Birks, D.M. Bird, J.C. Knight, P.St.J. Russell, Experimental demonstration of the frequency shift of bandgaps in photonic crystal fibers due to refractive index scaling. Opt. Express **14**, 3000–3006 (2006)
19. R. Bernini, E. DeNuccio, A. Minardo, L. Zeni, P.M. Sarro, Liquid-core/liquid-cladding integrated silicon ARROW waveguides. Opt. Commun. **281**, 2062–2066 (2008)
20. H. Schmidt, A.R. Hawkins, Optofluidic waveguides: I. Concepts and implementations. Microfluid. Nanofluid. **4**, 3–16 (2008)
21. A.R. Hawkins, H. Schmidt, Optofluidic waveguides: II. Fabrication and structures. Microfluid. Nanofluid. **4**, 17–32 (2008)
22. S. Kawata, T. Tani, Optically driven Mie particles in an evanescent field along a channeled waveguide. Opt. Lett. **21**, 1768–1770 (1996)
23. B.S. Schmidt, A.H.J. Yang, D. Erickson, M. Lipson, Optofluidic trapping and transport on solid core waveguides within a microfluidic device. Opt. Express **15**, 14322–14334 (2007)
24. S. Gaugiran, S. Getin, J.M. Fedeli, G. Colas, A. Fuchs, F. Chatelain, J. Derouard, Optical manipulation of microparticles and cells on silicon nitride waveguides. Opt. Express **13**, 6956–6963 (2005)
25. H. Cai, A. Poon, Optical manipulation of microbeads in an integrated optofluidic device, in *Proceedings of Conference on Lasers and Electro-Optics*, San Jose, CA, USA, 16–21 May 2010
26. S. Gaugiran, S. Getin, J.M. Fedeli, J. Derouard, Polarization and particle size dependence of radiative forces on small metallic particles in evanescent optical fields. Evidences for either repulsive or attractive gradient forces. Opt. Express **15**, 8146–8156 (2007)
27. D. Neel, S. Getin, P. Ferret, M. Rosina, J.M. Fedeli, O.G. Helleso, Optical transport of semiconductor nanowires on silicon nitride waveguides. Appl. Phys. Lett. **94**, 253115 (2009)
28. H. Cai, A.W. Poon, Optical trapping of microparticles using silicon nitride waveguide junctions and tapered-waveguide junctions on an optofluidic chip. Lab Chip **12**, 3803–3809 (2012)
29. H. Cai, A.W. Poon, Planar optical tweezers using tapered-waveguide junctions. Opt. Lett. **37**, 3000–3002 (2012)
30. K. Grujic, O.G. Helleso, J.P. Hole, J.S. Wilkinson, Sorting of polystyrene microspheres using a Y-branched optical waveguide. Opt. Express **13**, 1–7 (2005)
31. R. Ulrich, T. Kamiya, Resolution of self-images in planar optical-waveguides. J. Opt. Soc. Am. **68**, 583–592 (1978)
32. L.B. Soldano, E.C.M. Pennings, Optical multi-mode interference devices based on self-imaging: principles and applications. IEEE J. Lightwave Technol. **13**, 615–627 (1995)
33. H. Cai, A.W. Poon, Optofluidic microparticle splitters using multimode-interference-based power splitters, in *Conference on Lasers and Electro-Optics*, San Jose, CA, USA, paper CTu2L.5, 6–11 May 2012
34. L.-C. Hsu, T.-C. Chen, Y.-T. Yang, C.-Y. Huang, D.-W. Shen, Y.-T. Chen, M.-C.M. Lee, Manipulation of micro-particles through optical interference patterns generated by integrated photonic devices. Lab Chip **13**, 1151–1155 (2013)

35. S.Y. Lin, K.B. Crozier, An integrated microparticle sorting system based on near-field optical forces and a structural perturbation. Opt. Express **20**, 3367–3374 (2012)
36. S. Mandal, X. Serey, D. Erickson, Nanomanipulation using silicon photonic crystal resonators. Nano Lett. **10**, 99–104 (2010)
37. T. van Leest, J. Caro, Cavity-enhanced optical trapping of bacteria using a silicon photonic crystal. Lab Chip **13**, 4358–4365 (2013)
38. C. Renaut, B. Cluzel, J. Dellinger, L. Lalouat, E. Picard, D. Peyrade, E. Hadji, F. De Fornel, On chip shapeable optical tweezers. Sci. Rep. **3**, 2290 (2013)
39. Y.F. Chen, X. Serey, R. Sarkar, P. Chen, D. Erickson, Controlled photonic manipulation of proteins and other nanomaterials. Nano Lett. **12**, 1633–1637 (2012)
40. B.E. Little, J.S. Foresi, G. Steinmeyer, E.R. Thoen, S.T. Chu, H.A. Haus, E.P. Ippen, L.C. Kimerling, W. Greene, Ultra-compact Si-SiO2 microring resonator optical channel dropping filters. IEEE Photonics Technol. Lett. **10**, 549–551 (1998)
41. S.T. Chu, B.E. Little, W.G. Pan, T. Kaneko, Y. Kokubun, Second-order filter response from parallel coupled glass microring resonators. IEEE Photonics Technol. Lett. **11**, 1426–1428 (1999)
42. D.K. Armani, T.J. Kippenberg, S.M. Spillane, K.J. Vahala, Ultra-high-Q toroid microcavity on a chip. Nature **421**, 925–928 (2003)
43. T.J. Kippenberg, S.M. Spillane, D.K. Armani, K.J. Vahala, Fabrication and coupling to planar high-Q silica disk microcavities. Appl. Phys. Lett. **83**, 797–799 (2003)
44. P. Koonath, T. Indukuri, B. Jalali, Add-drop filters utilizing vertically coupled microdisk resonators in silicon. Appl. Phys. Lett. **86**, 091102 (2005)
45. A.H.J. Yang, D. Erickson, Optofluidic ring resonator switch for optical particle transport. Lab Chip **10**, 769–774 (2010)
46. S.Y. Lin, E. Schonbrun, K. Crozier, Optical manipulation with planar silicon microring resonators. Nano Lett. **10**, 2408–2411 (2010)
47. H. Cai, A.W. Poon, Optical manipulation and transport of microparticles on a silicon nitride microracetrack resonator add-drop device, in *IEEE 7th International Conference on Group IV Photonics*, Beijing, China, 1–3 Sept 2010
48. H. Cai, A.W. Poon, Optical manipulation and transport of microparticles on silicon nitride microring-resonator-based add-drop devices. Opt. Lett. **35**, 2855–2857 (2010)
49. S.Y. Lin, K.B. Crozier, Planar silicon microrings as wavelength-multiplexed optical traps for storing and sensing particles. Lab Chip **11**, 4047–4051 (2011)
50. J. Wang, A.W. Poon, Unfolding a design rule for microparticle buffering and dropping in microring-resonator-based add-drop devices. Lab Chip (2014). https://doi.org/10.1039/c3lc51186c
51. H. Cai, A.W. Poon, Optical manipulation of microparticles using whispering-gallery modes in a silicon nitride microdisk resonator. Opt. Lett. **36**, 4257–4259 (2011)
52. M.J. Renn, D. Montgomery, O. Vdovin, D.Z. Anderson, C.E. Wieman, E.A. Cornell, Laser-guided atoms in hollow-core optical fibers. Phys. Rev. Lett. **75**, 3253 (1995)
53. S. Mandal, D. Erickson, Optofluidic transport in liquid core waveguiding structures. Appl. Phys. Lett. **90**, 184103 (2007)
54. S. Cran-McGreehin, T.F. Krauss, K. Dholakia, Integrated monolithic optical manipulation. Lab Chip **6**, 1122 (2006)
55. A.H.J. Yang, S.D. Moore, B.S. Schmidt, M. Klug, M. Lipson, D. Erickson, Optical manipulation of nanoparticles and biomolecules in sub-wavelength slot waveguides. Nature **457**, 71–75 (2009)
56. N. Bellini, R. Osellame, K.C. Vishnubhatla, F. Bragheri, L. Ferrara, P. Minzioni, R. Ramponi, I. Cristiani, Femtosecond laser fabricated monolithic chip for optical trapping and stretching of single cells. Opt. Express **18**, 4679–4688 (2010)
57. M.A. Duguay, Y. Kokubun, T.L. Koch, L. Pfeiffer, Antiresonant reflecting optical waveguides in SiO_2-Si multilayer structures. Appl. Phys. Lett. **49**, 13 (1986)
58. R. Bernini, S. Campopiano, L. Zeni, Silicon micromachined hollow optical waveguides for sensing applications. IEEE J. Sel. Top. Quantum Electron. **8**, 106–110 (2002)

59. D. Yin, H. Schmidt, J.P. Barber, A.R. Hawkins, Integrated ARROW waveguides with hollow cores. Opt. Express **12**, 2710 (2004)
60. D. Yin, J.P. Barber, D.W. Deamer, A.R. Hawkins, H. Schmidt, Single-molecule detection sensitivity using planar integrated optics on a chip. Opt. Lett. **31**, 2136 (2006)
61. P. Measor, L. Seballos, D. Yin, J.Z. Zhang, E.J. Lunt, A.R. Hawkins, H. Schmidt, On-chip surface-enhanced Raman scattering detection using integrated liquid-core waveguides. Appl. Phys. Lett. **90**, 211107 (2007)
62. D. Yin, E.J. Lunt, M.I. Rudenko, D.W. Deamer, A.R. Hawkins, H. Schmidt, Planar optofluidic chip for single particle detection, manipulation, and analysis. Lab Chip **7**, 1171 (2007)
63. P. Measor, S. Kühn, E.J. Lunt, B.S. Phillips, A.R. Hawkins, H. Schmidt, Hollow-core waveguide characterization by optically induced particle transport. Opt. Lett. **7**, 672–674 (2008)
64. S. Kühn, P. Measor, E.J. Lunt, B.S. Phillips, D.W. Deamer, A.R. Hawkins, H. Schmidt, Loss-based optical trap for on-chip particle analysis. Lab Chip **9**, 2212 (2009)
65. K.D. Leake, B.S. Phillips, T.D. Yuzvinsky, A.R. Hawkins, H. Schmidt, Optical particle sorting on an optofluidic chip. Opt. Express **21**, 32605 (2013)
66. P. Measor, S. Kühn, E.J. Lunt, B.S. Phillips, A.R. Hawkins, H. Schmidt, Multi-mode mitigation in an optofluidic chip for particle manipulation and sensing. Opt. Express **17**, 24342–24348 (2009)
67. S.B. Kim, S.S. Kim, Radiation forces on spheres in loosely focused Gaussian beam: ray-optics regime. JOSA B **23**, 897–903 (2006)
68. S. Kühn, E.J. Lunt, B.S. Phillips, A.R. Hawkins, H. Schmidt, Optofluidic particle concentration by a long-range dual-beam trap. Opt. Lett. **34**, 2306–2308 (2009)
69. K.D. Leake, A.R. Hawkins, H. Schmidt, All-optical particle trap using orthogonally intersecting beams. Photonics Res. **1**, 47 (2013)
70. S. Kühn, B.S. Phillips, E.J. Lunt, A.R. Hawkins, H. Schmidt, Ultralow power trapping and fluorescence detection of single particles on an optofluidic chip. Lab Chip **10**, 189–194 (2010)
71. J. Enderlein, Trapping of fluorescent molecules diffusing within membranes. Appl. Phys. B **71**, 773–777 (2000)
72. A.E. Cohen, W.E. Moerner, Controlling Brownian motion of single protein molecules and single fluorophores in aqueous buffer. Opt. Express **16**, 6941 (2008)
73. A.P. Fields, A.E. Cohen, Electrokinetic trapping at the one nanometer limit. Proc. Natl. Acad. Sci. **108**, 8937–8942 (2011)

Chapter 8
New Directions in Sensing Using Raman Analysis on Paper and Microfluidic Platforms

Derek Craig and Kishan Dholakia

Abstract The vibration and rotational dynamics of molecules are very specific and offer an interesting route for characterisation, analysis and sensing. Spectroscopic probing of the bond dynamics, notably their energies, particularly using an optical source has emerged as a powerful route to ascertaining the chemical composition of a sample. This type of information can be crucial for a detailed molecular analysis and indeed can complement other information within a sample, for example morphological information. Light can be very selective and probe specific transitions in atoms and molecules at will. In terms of sensing, optical approaches can be attractive—due to the fact that they can be highly specific, sensitive and often can be engineered into compact systems that are relatively inexpensive.

8.1 Basic Principles of Raman Spectroscopy

The vibration and rotational dynamics of molecules are very specific and offer an interesting route for characterisation, analysis and sensing. Spectroscopic probing of the bond dynamics, notably their energies, particularly using an optical source has emerged as a powerful route to ascertaining the chemical composition of a sample. This type of information can be crucial for a detailed molecular analysis and indeed can complement other information within a sample, for example morphological information. Light can be very selective and probe specific transitions in atoms and molecules at will. In terms of sensing, optical approaches can be attractive—due to the fact that they can be highly specific, sensitive and often can be engineered into compact systems that are relatively inexpensive.

Raman spectroscopy has emerged as a powerful technique that has generated major interest and importance for the sensing community over the last few decades.

D. Craig · K. Dholakia (✉)
SUPA, School of Physics and Astronomy, University of St Andrews, North Haugh, St Andrews, Fife KY16 9SS, UK
e-mail: kd1@st-andrews.ac.uk

R. De La Rue et al. (eds.), *Biomedical Optical Sensors*,
Biological and Medical Physics, Biomedical Engineering,
https://doi.org/10.1007/978-3-030-48387-6_8

This method relies on the inelastic scattering of light from a sample, which is indicative of its molecular composition. Importantly, the method can be label-free, implying minimal intrusion or interference with the sample in question. Indeed, though it has been around for several decades, its increasing applicability, the enhancement through processes such as surface plasmonics and the potential in healthcare mean that now is a very timely juncture in this burgeoning field.

In this chapter we describe the basics of the Raman spectroscopic approach, with an emphasis on some of the recent work for sensing that use microfluidic approaches. We concentrate upon native and 'linear' Raman approaches that rely neither on nonlinearity nor on surface enhanced approaches. We remark that surface enhanced Raman scattering (SERS) typically refers to the interaction of the sample with a metal surface that exhibits roughness, typically at the nanoscale. The enhancement can occur when molecules are adsorbed onto this surface, which may dramatically increase the Raman signal and potentially offer single molecule detection [1]. Such enhancements can be crucial where native Raman measurement is too weak (for example, in trace material analysis). Nonlinear approaches to acquisition of molecular signatures can include CARS and SRS. Though such approaches offer enhancements in terms of the speed of data acquisition as well as increased sensitivity, they also often involve complex fabrication and surface preparation—or laser apparatus that may offset any advantages gained.

In contrast, native Raman spectroscopy can be relatively inexpensive and straightforward to implement, label-free and is the main focus of this chapter, though we do mention some key SERS experiments for both microfluidic and paper platforms. We direct the reader to other excellent reviews for further information on SERS and nonlinear Raman approaches [2].

The virtual transitions that lead to the Raman process can be visualised by looking at Fig. 8.1.

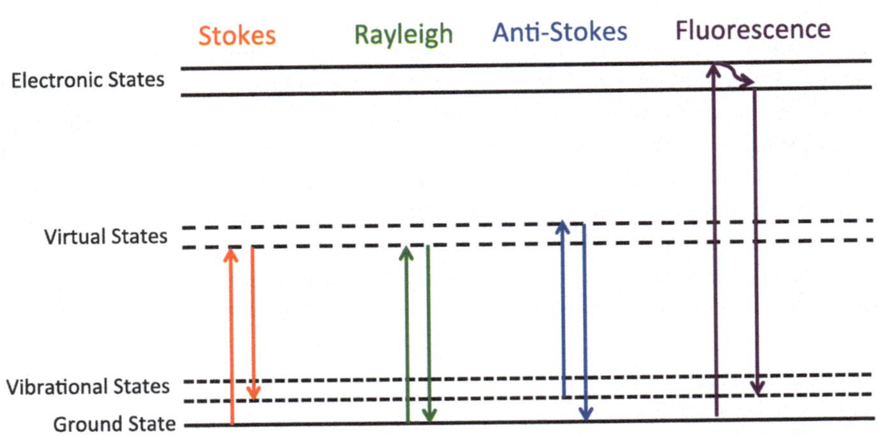

Fig. 8.1 Diagram showing the energy transitions that occur during Raman scattering (both Stokes and Anti-Stokes lines), Rayleigh scattering and Fluorescence. It is typically the Stokes line that is observed in Raman analysis

Here we see that Raman transitions denote a virtual process that involves a ground state and a virtual excited state. These can be amplified by exploring transitions close to an actual real transition, a process referred to as resonance Raman excitation. The fact that this is a virtual process means that Raman emission is very fast, in contrast to processes such a fluorescence or phosphorescence. Raman is often accompanied by fluorescence and there are numerous approaches in the literature that can be used to obviate its influence. The most popular approaches include numerical approaches where polynomial fits to data are used to subtract the background, though care has to be taken as artifacts can arise in such an approach [3]. An emergent sub-topic in recent years has been more physics-based approaches that extract the Raman signature in the presence of a background. This includes use of the fact that the Raman signal is generated through the interaction with virtual states and so is emitted before fluorescence. Thus gating the signal can lead to directly recording the Raman signal. Alternative approaches include modulating the laser wavelength, which results in a direct modulation of the Raman signal, but if the range of modulation is small the background fluorescence remains essentially constant. In turn this means that it is possible to extract the Raman signal as a derivative-like feature [4].

8.2 Soft Lithography Geometries for Microfluidic Raman Analysis

Microfluidic geometries have rapidly come to the fore over the last few decades, since they offer a route to the analysis of very small volumes of analyte (typically mL or less) in very compact, often disposable, systems. Various forms of microfluidic platform have emerged. A popular route—due to its ease of fabrication, cost and implementation—has been the use of soft lithography approaches based on elastomers such as poly(dimethylsiloxane) (PDMS) for the fabrication of microfluidic systems [5].

PDMS has been at the heart of numerous microfluidic systems. It is an optically transparent elastomer. The stiffness of PDMS can be varied: at the soft end of its range, it is readily deformed by hand. Importantly, fabrication with PDMS is straightforward and lends itself to reproducibility and mass production. From a research perspective, this means that rapid prototyping of devices to realise new modalities at the microfluidic scale is feasible. If we compare PDMS to other materials such as glass or silicon, PDMS can be used to make a reversible Van der Waals force contact between two surfaces (but one that is dependent upon smoothness). In turn, this makes microfluidic PDMS chips demountable and resealable, which has obvious advantages for reusability. Liquid confinement in very small channels embedded in PDMS leads to interesting physics, due to the typically low Reynolds number regime, and strategies have emerged for enhanced mixing and directing of fluid in an environment where inertia and turbulence play an insignificant role.

Various approaches have emerged to realise Raman spectroscopy on-chip. Many embodiments use a standard microscope platform into which the chip of interest is inserted. In several instances, this platform has been combined with optical trapping. The latter modality allows one to isolate the sample from the surrounding environment and literally suspend the cell or sample in a suitable medium—which can lead to reduced background signals from interfering cover slips, for example. Optical forces are relatively weak and—in the ray-optics picture—originate from the reflection or refraction of light at a refractive index boundary. The forces are gentle and in the piconewton range—and so they are not typically suitable for holding more than a single cell. The laser wavelength for trapping has to be chosen carefully in Raman trapping systems—to ensure that there is no interference between modalities. Geometries that have used so-called single-beam optical tweezers and also dual-beam trap geometries, with Raman analysis performed in a direction orthogonal to the trapping fields, have appeared. A key feature of trapping and Raman scattering in a microfluidic environment is the prospect of a system for higher throughput, akin to a flow cytometer—though based on Raman analysis. Novel geometries have enabled realization for studies of key cell types [6].

The research described above relies on Raman apparatus that requires bulk free-space optics for its realisation. The size of a typical Raman microfluidic system makes it impractical for portable applications [7], which is typically due to the requirement of appropriate laser sources—and analysis via a bulk spectrometer. Separately, fiber Raman probes have been developed with clinical aims in mind and offer ease of alignment. The Raman approach described here has advanced to the stage of achieving both portable spectrometers and compact lasers for Raman spectroscopy [8]. However it is to be noted that fiber based probes may not be the best choice to record the Raman spectrum from a small amount of analyte. This problem can arise for a number of reasons, including low numerical aperture and reduced collection efficiency. One way that fiber optics can add value is by turning to the popular Lab-on-a-Chip (LoC) techniques, which aim to use the methods of soft lithography for chamber construction and then use this combination for spectroscopic analysis of small amounts of sample—on the order of micro-litres or even nano-litres [9–11].

We now turn to a description of some of the key approaches in this area. A recent study has worked towards assembling an all-fiber based Raman analysis on-chip. The aim of this approach is to reduce the physical footprint of the device by the use of fibers on a chip platform. A key attribute of this study is the absence of lenses in either the illumination or the collection paths. Close proximity of the fibers used to the sample and the sample geometry design used permitted useful excitation and collection of the Raman signal. This methodology, developed at the University of St Andrews was termed waveguide confined Raman spectroscopy (WCRS) and has been realised as a fiber-based microfluidic chip, where the Raman and fluorescence signatures are delivered and recorded using optical fibers [9]. The microfluidic chip used embedded multimode optical fibers to optimise the collection of the Raman signal, while simultaneously minimising any background signal. The device used is shown in Fig. 8.2.

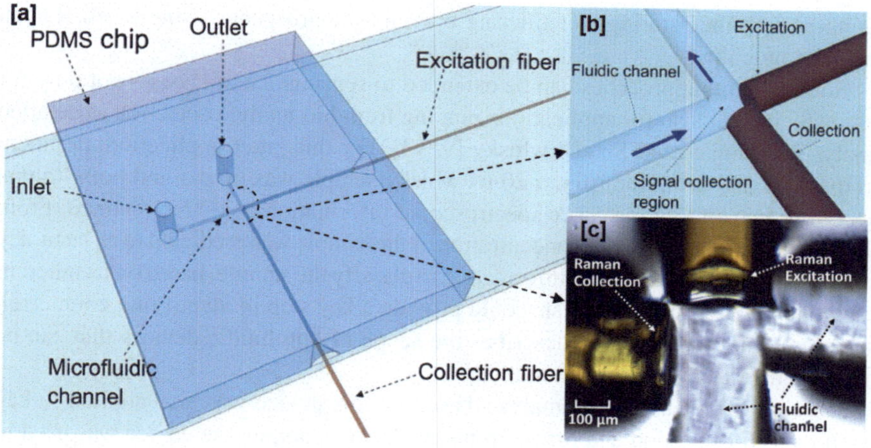

Fig. 8.2 On-chip, all-fiber based Raman analysis based on a Polydimethylsiloxane (PDMS) chip fabricated using soft-lithography. Two optical fibers were used. One to deliver the Raman excitation beam and the other to record both the Raman and fluorescence signal. **a** A diagram to show the geometry of the chip-based Raman approach. **b** A diagram to illustrate the collection region where confined collection of Raman signal was achieved. **c** Top view of the Raman signal collection region in the PDMS microfluidic chip [9]. Reproduced from [9] with permission from the Royal Society of Chemistry

This device was used in a number of studies to prove the versatility of such an approach. The on-chip approach was used for process monitoring. Specifically, the all-fibre WCRS chip was placed within a micro-reactor to explore a binary reaction on-chip, with sulfuric acid being used as the catalyst. In this instance, the reaction was acid-catalysed esterification of ethanol with acetic anhydride—and resulted in ethyl acetate. Further work on the same all-fiber WCRS platform looked at the detection of micro-droplets. These droplets were ethanol based and generated in silicone oil. The major application of this device is to perform Raman spectroscopy of analytes. As envisaged, the WCRS device records spectra that contain both Raman information and fluorescence information from the sample. A study of the urea concentration showed detection down to around 80 mM, with reasonable excitation power and relatively low acquisition time [9, 11, 12]. More extensive work was performed on urea detection using the same CRS architecture to look at modulated Raman spectroscopy, which is an approach that removes the background fluorescence and may yield a higher signal-to-noise ratio [4].

In 2013 a group in Jena, Germany, developed another form of quartz-plate based microfluidic chip, with the aim of achieving on-chip Raman analysis. A key innovation in this approach was the use of multi-core single-mode fibres, with integrated fiber Bragg-gratings being used for detection. Raman detection limits of 7.5 mM for urea and 2.5 mM for nicotine respectively were recorded with this set-up. The fibre Bragg-grating assisted with the removal of unwanted signals in the system. Typically scattered light is removed in Raman apparatus using low-pass filters or notch filters.

Long-period fiber gratings for filtering open up the prospect of single-cell Raman analysis on-chip.

Such on-chip approaches can be extended to sensing and analysis—not only for biomedicine but also for applications ranging from bio-analyte detection to detailed analysis of single malt Scotch whisky [9, 13]. For this latter application, pertinent to the food and drink industry, a 20 μl whisky sample was used—and both Raman spectroscopic and fluorescence spectroscopic information could be obtained. From these signatures, the alcohol concentration of the sample was predicted to an accuracy of 1%—along with additional information to classify the sample, based on its inherent aromatic features. Such analysis could provide a key step in identifying counterfeit samples. This article briefly describes the scope of optofluidic devices that can be implemented using the WCRS technique.

On-chip compact Raman analysis based on this device was also applied to key challenges for the food and drinks industry. As an example, the device was used to record spectra from various single-malt Scotch whiskies, with the aim of enabling discrimination and their detailed characterization [9, 13]. Whisky spectra had Raman bands from both ethanol and a broad fluorescence background. Using the partial least-square (PLS) method, the alcohol concentration of each whisky investigated was determined. Furthermore, using the multivariate statistical analysis approach of principal component analysis (PCA) based clustering, a correlation between the broad fluorescence background and the aromatic features of the Scotch whisky was revealed—allowing a clear delineation of whisky type. This is shown in Fig. 8.3.

Beyond being an analyte detection tool, a WCRS device may also be combined with other microfluidic functionalities. It has already been demonstrated that it is possible to monitor micro-reaction and probe micro-droplets using WCRS [11]. Hence a WCRS based device may have a variety of applications ranging from chemical fingerprinting of a low volume of analyte to on-line process monitoring. The completely alignment-free nature makes it possible for a non-optics expert to use this device.

Another key feature of the device is its fabrication cost. Since the chip can be fabricated by soft-lithography or injection moulding and does not require any millimeter size filters at the tip of the probe, the cost of the device can be at least two orders of magnitude lower than that of a commercially available fiber Raman probe. Since the optical signal is delivered and collected through optical fibers, with appropriate light-proof shielding of the chip, it is even possible to use the device for measuring the Raman spectrum of an analyte in broad daylight, which would make it more amenable for various field applications.

"Lab-on-a-chip" technologies envisage the use of miniscule sample volumes for a range of analyses. In optofluidic devices where optical techniques are combined with microfluidics, it is important to ensure a smaller form-factor and an alignment-free nature—thereby making it appealing for point-of-care applications. WCRS is an example where these criteria are satisfied, with the added advantage of the low fabrication cost of this optofluidic chip. These devices can have applications in various fields—such as point-of-care diagnostics, environmental sensing, food and drink sensing and on-line process monitoring.

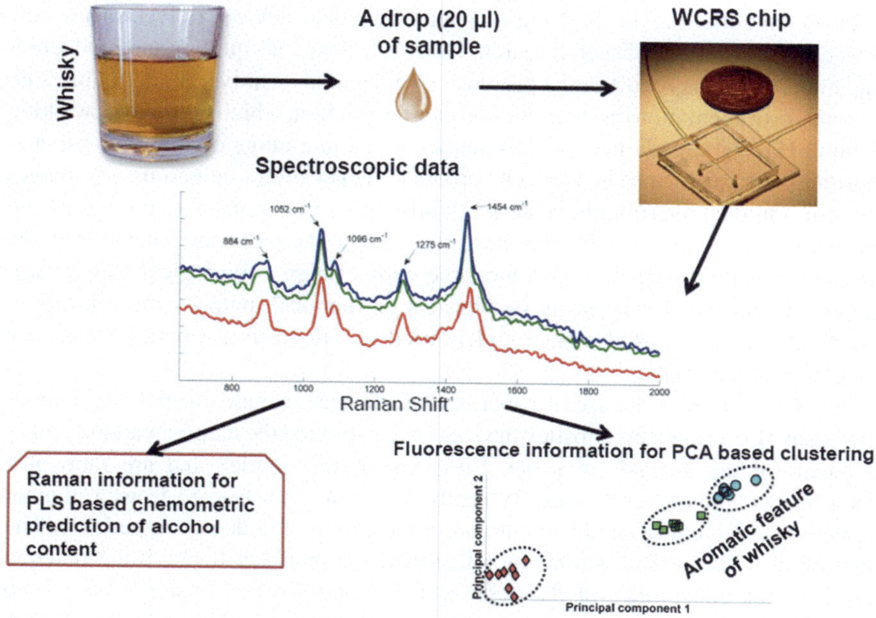

Fig. 8.3 Whisky analysis on the WCRS chip [13]. The Raman analysis delivers the ethanol content whereas the fluorescence spectra can be used to cluster based on each type of whisky. Reprinted with permission from [13], published by OSA

8.3 Paper: An Alternative Platform for Microfluidic Raman Analysis

In a biomedical sensing environment microfluidic devices have been shown to significantly reduce the consumption of samples and reagents, as well as reducing the complexity of operating procedures. The overall cost of assaying time is reduced, without any significant disadvantage in terms of specificity or sensitivity. However, the majority of microfluidic devices are constructed from glass, silicon or thermo-elastic polymers such as poly(dimethylsulfoxide) (PDMS) or poly(methylmethacrylate) (PMMA). These devices all require complex fabrication processes and a range of external instrumentation, which discourages their implementation in point-of-care (POC) testing. In low resource settings such as the developing world, commonly available POC tests are regarded as being too expensive. These nations lack basic infrastructure, which is a key requirement for common POC testing—such as appropriate laboratory conditions that include power and refrigeration, as well as a lack of trained personnel to undertake and understand the analyses. The World Health Organization maintains that diagnostic devices for use in low cost resource settings should follow the ASSURED proverb, i.e. they should be: affordable, sensitive, specific, user-friendly, rapid and robust, equipment-free and deliverable to end users.

Paper microfluidics in the form of paper analytical devices (μPADs) has now moved to the forefront of such research. Paper is a cheap, ubiquitous and biodegradable material that is robust in the presence of most organic solvents and requires no external instrumentation due to its natural capillary action, which enables the wicking of fluid. Having already been widely implemented in a range of analyses, paper is a strong platform for producing POC products. Paper offers numerous advantages over conventional microfluidic materials whilst being free of many of the significant disadvantages also associated with these other materials. As paper is made from the meshing of cellulose fibers it is compatible with a range of biological materials. It can also be modified very easily by physical or chemical methods for tailoring to a particular need. It has also been widely used as a platform for both filtering and separation of materials.

Initial examples of the use of paper in POC devices include dipstick and lateral-flow assay (LFA) devices. Dipstick devices have exploited the natural wicking ability of paper for the analysis of urine. LFAs are more complex and are fabricated using nitrocellulose membranes. Typically, LFAs are constructed from a number of sections—including an absorbent pad, a wicking pad, a detection pad, etc. The nitrocellulose has a dual action since it provides a platform for both the reaction and detection capabilities of the devices. LFAs and dipstick analyses have been widely implemented for applications such as home pregnancy tests, blood and DNA testing, respectively. As such these devices are outwith the remit of this discussion. In what follows, the discussion will focus on the emergence of both two-dimensional and three-dimensional μPADs—and their implementation in the field of biomedical analysis.

8.4 Device Fabrication for Paper Based Analysis

Whitesides et al. are widely acknowledged as the pioneers of the field of μPADs [14]. Their initial research conveyed means by which paper could be modified in a cheap and simple manner for low-cost resource settings. Typically paper can be modified in one of two ways; (i) physical modification, such as cutting and shaping—or by: (ii) chemical modification of the paper surface. Chemical modification involves the formation of hydrophobic barriers on the paper surface, in order to create micro-channels. The dimensions of these micro-channels, as well as the porosity and chemical structure of the paper used and the ambient conditions, will control the wicking rate of any fluid through the device. The patterning of the micro-channels on the paper surface can be achieved by a variety of methods. The most common techniques used are photolithography, ink-jet etching, plasma etching and wax patterning. The various advantages and disadvantages of each technique are set out in Table 8.1.

Wax printing is currently the most commonly used technique for the production of paper microfluidic devices. Using a cheap and readily available solid wax printer it is possible to deposit hydrophobic wax barriers of controllable thickness and resolution onto the surface of the paper. Heating of the wax results in its re-distribution, both

Table 8.1 Various approaches to the patterning of microchannels, along with their advantages and disadvantages

Patterning technique	Advantage(s)	Disadvantage(s)
Photolithography	Can pattern a variety of paper	Hydrophilic areas are exposed to polymers and solvents
Ink-jet etching	Reagents can be inkjet printed into the test zones using the printer	Requires a customized inkjet printer. Hydrophilic areas are exposed to polymers and solvents
Plasma etching	Useful in laboratories with a plasma cleaner which wish to make many replicates of the same pattern	Hydrophilic areas are exposed to polymers and solvents. Individual metal masks are required for each pattern
Wax printing	Rapid. Requires only a commercial wax printer and a hotplate. Hydrophilic areas are not exposed to polymers and solvents	Consideration must be given, when designing the pattern, to the spreading of wax, which will occur upon heating

laterally and vertically, through the paper to create 2D hydrophobic barriers. Various microfluidic devices have been designed, due to the ease with which multiple wax designs can be conceived and printed onto the paper surface. As a result, wax printing is currently the most routinely used technique for the fabrication of paper microfluidic devices.

Paper has been shown to be a hospitable environment for many biological molecules and chemical substrates, which can be bound onto the paper surface by either physical deposition or chemical attachment. Physical deposition of the analyte into the reagent zone can be achieved by either hand dispensing or by printing the reagents into these zones, using techniques such as ink-jet printing. Chemical or biological molecules can also be immobilized onto the paper surface by physical absorption, chemical coupling and by carrier mediated deposition. When the reagents have dried upon the substrate, the analyses can be performed. If the desired application of the device requires multiple or complex functions to occur, such as pre-mixing of the sample or filtering of sample components, then the use of two dimensional μPADs in this area is very limited. To overcome these issues and to perform more complex analyses 3D paper analytical devices (3D μPADs) are favoured.

3D μPADs possess many advantages in comparison with their two-dimensional counterparts. The combination of multiple layers of μPADs on a single device allows for multiple assays to be conducted on the same area as used previously by a 2D μPAD, due to it enabling the fluid to move through multiple layers of paper. As a consequence, 3D μPADs are able to perform multiple-step assays within a single device and the sample volume typically lost due to paper swelling is vastly reduced, which in turn lowers the initial sample volume required. There currently exist two main types of 3D μPAD: (i) those formed by affixing layers of paper on top of one another, using an ordered layer-by-layer approach, and: (ii) a 3D stack of layers being assembled and then compressed together, using either a clamp or an adhesive to bring the layers into contact with one another.

Whitesides et al. proposed the original concept for 3D μPADs, which were constructed using double-sided sticky tape to fix layers of paper together to form overlapping channels [15]. This approach allowed for the alignment of hydrophilic channels through the device and a reduction in the sample volume required. However, constructing these 3D μPADs in this manner is a tedious and time-consuming process. As a result other approaches have been sought. One approach that has been considered is the gluing of adjacent sheets together using an adhesive, which results in the rapid construction of the 3D stacks and enables the layering of many paper devices to be achieved. A 'cold-lamination' approach has also been used, which involves stacking layers of devices together prior to affixing them in place by sealing. Typically a metal framework such as a clamp has been used to seal the layers together. Recently a new approach has emerged for the construction of 3D μPADs—which is inspired by origami, the Japanese art of folding, where the layers are folded into place prior to being fixed together using a clamp. The origami approach has the advantage of permitting the opening of the device and the alteration of sheet orientation, which is not possible when adhesives are used (see Fig. 8.4). Liu and Crooks were the first to propose the origami approach and their initial research highlighted the capabilities of 3D μPADs to perform parallel reactions on the same device [16]. The application of origami paper analytical devices (oPADs) has since been extended to many studies—including the aptamer sensing of adenosine and immunosensing, for serum-based cardiac biomarkers [17].

Functional devices such as switches, valves and fluidic timers can also be incorporated into μPADs. Shen and co-workers produced a reconfigurable paper switch to connect and disconnect microfluidic channels and control fluid wicking [18]. Noh

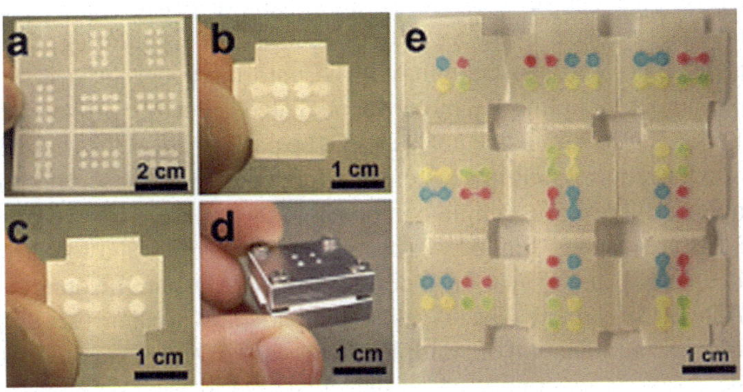

Fig. 8.4 Paper origami device: **a** Chromatography paper with photolithography defined channels. All channels were 900 μm wide and the reservoirs were 2.5 mm in diameter. **b** Folded device revealing the four inlet reservoirs. **c** Bottom of the paper device showing the four detection reservoirs. The four corners of the paper device were cut so that it could be housed in the aluminum housing (**d**). **e** The unfolded device, following injection of four-coloured devices through the four injection ports of the aluminum housing [16]. Reprinted (adapted) with permission from J. Am. Chem. Soc. 2011, 133, 17564–17566. Copyright 2011 American Chemical Society

and Phillips chose a different approach by modulating the wettability of the microfluidic channels to control the flow of the fluid to individual channels and Martinez et al. created single use buttons that could be activated to connect two paper strips together and activate the wicking of fluid by using a ballpoint pen or stylus [19]. Fluidic timers have been implemented for tracking of the end-point of a time-based assay on paper. Phillips et al. have integrated a fluidic timer in a paper device to signify the end-point of a glucose analysis test. The fluidic timer constructed consists of three components; (i) hydrophilic channels of paper which link the timer to the assay, (ii) a graduated area that modulates the flow rate of the fluid (in this case made of paraffin wax) and a signaling component which unequivocally communicates to the user that the assay could be read. Sugar bridges have also been used to regulate fluid flow. By saturating the paper strip to different lengths in sugar, the movement of fluid can be both retarded and accelerated in comparison with surrounding paper channels. This technique has been employed in an antigen-antibody immunoassay by Yager and co-workers [20]. By employing fluidic timers, the paper device can be programmed to carry out multi-step reactions in sequence—without the need for an external user to intervene.

The ability to pattern the paper surface for biological analyses has resulted in the emergence of an abundance of paper microfluidic devices for a variety of applications. In parallel with the emergence of these devices, a variety of detection methods has been implemented for the translation of the information they hold to biologically relevant data. In what follows, a non-exhaustive summary will be provided of a number of biomedical applications in which paper microfluidic devices have been successfully implemented, together with a discussion of the detection methodologies that have been interfaced with these devices to make the analysis possible.

8.5 Sensing Approaches Using Paper Based Analysis

8.5.1 Colorimetric Detection

Label-free detection methodology lends itself to paper microfluidic assays, due to the avoidance of any expensive instrumental costs for analysis. The most commonly used detection method that has been interfaced with paper microfluidics has been colorimetric detection. Colorimetric detection is based upon the production of a colour change that accompanies the interaction of components embedded in the paper substrate. This technique is typically exploited in devices where a yes or no answer can be achieved by visual inspection of the sample, either by eye or by using inexpensive imaging devices. Such imaging devices include digital cameras, mobile phone cameras and desktop scanners. The relative simplicity and inexpensive nature of these materials make them ideal for use in low-cost resource settings. As colorimetric detection is widely implemented in the field of paper microfluidics, it can be sub-classified into three main areas: enzymatic reactions, immunoassays and subsequent applications.

Enzymatic reactions are the most commonly observed in paper microfluidics—and have been used for the detection of important analytes such as glucose. There exist numerous devices in the literature that have been employed for this type of analysis. Methods of analysis typically focus on enzymatic reactions through the reaction of glucose with glucose oxidase—or the combination reaction of glucose with glucose oxidase and other enzymes, such as horseradish peroxidase (HRP). The enzymatic detection of proteins such as serum albumin is also a common application of μPAD technology. One such example was one of the eminent publications in this field by Whitesides et al. A quantitative 2D μPAD for the detection of both glucose and bovine serum albumin (BSA) in urine was developed. The assay was based upon well-known chemical and enzymatic reactions that are commonly used in dipstick analysis. The benefits of this device were that multiple assays could be completed on the one device, simple colorimetric detection was possible—with the colour developing as a function of the analyte concentration present—and very small volumes of fluid ($\approx$5 μl) were required [14]. Some data from this work is seen in Fig. 8.5.

Tian et al. have also uniquely incorporated text reporting into their assay designed to test a patient's blood type [21]. Using a bioactive paper device that has been impregnated with specific blood group antibodies and water-soluble inks, the testing of all eight blood types of the ABO Rh system was shown to be possible and highly accurate. However, the use of only colorimetric detection restricts the use of these devices to mainly enzymatic coloured reactions. A number of disadvantages are also attributed to colorimetric detection, such as non-homogenous distribution of colour and the negative influence of the background colour of the paper.

8.5.2 Fluorescence Spectroscopy

Optical detection methods such as fluorescence spectroscopy are widely used for diagnostic assays, due to their being highly sensitive and producing low limits of detection (LODs). The combination of fluorescence spectroscopy and paper microfluidics has been retarded due to the inherent fluorescent background of the paper substrate (arising from whitening additives added to the paper) and the inability to yield the low LODs required. However, studies have recently begun to emerge that have overcome these limitations and have established LODs on the scale required for the implementation of these devices in diagnostic applications. Examples of such studies have led to the implementation of fluorescence-based devices for the detection of bacteria, proteins and cancer biomarkers [22].

Potential methods to circumvent the inherent fluorescent background of paper and couple fluorescence analysis to this technology have been proposed. One such successful method has been the embedding of an up-conversion (UC) fluorescence assay within a paper device. Using near-Infrared light (NIR) to excite up-conversion

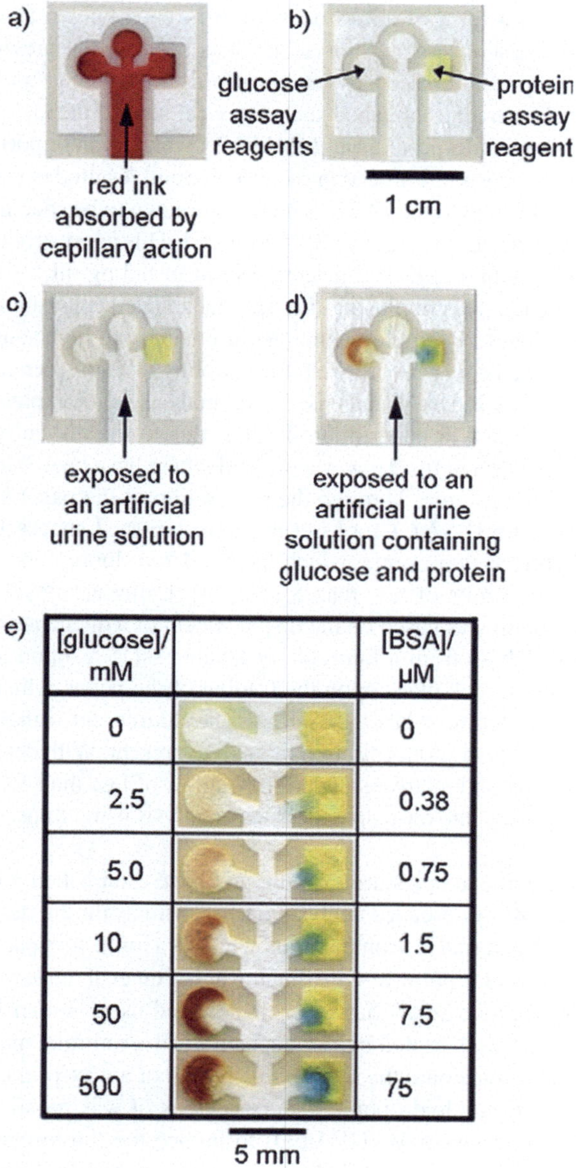

Fig. 8.5 2D µPAD for analysis of glucose and BSA. **a** Photoresist patterned chromatography paper following absorption of red ink. **b** Paper device prepared for analytical testing following the spotting of the protein and glucose reagents into the appropriate reservoirs. **c** Paper device following exposure to the reactant solution and **d** the completed assay following exposure to the solution containing glucose and BSA. **e** Comparative analysis of the colorimetric difference between devices exposed to different concentrations of glucose and BSA. Figure from [14]. Figure reproduced with permission from Wiley publishers.

phosphors results in a form of anti-Stokes luminescence known as up-conversion fluorescence. This up-conversion fluorescence has been shown previously to overcome background auto-fluorescence—and it affords a promising tool for point-of-care diagnostics. Liu et al. established such a device, termed the UC-μPAD, for the detection of matrix metallo-proteinase-2 (MMP-2), which is an important biomarker in blood [23]. MMP-2 was introduced into the device and resulted in the cleavage of a peptide substrate attached to the UCPs, which was known to be specific for MMP-2. Following this peptide cleavage the UCPs fluoresced. This assay produced clinically relevant sensitivity with an MMP-2 detection limit of 8.3 pg mL^{-1}. Therefore, the merits of fluorescence detection using this approach have been established through its ease of fabrication, low sample volume requirements and high sensitivity.

Another area which could benefit from the coupling of fluorescence spectroscopy to paper microfluidics is DNA analysis. DNA analysis is an expanding field that has largely been avoided in paper microfluidics, due to the difficulty of achieving sub-nanomolar detection limits. However, recently there have been interesting developments reported in this area. Through the passivation of the paper substrate using bovine serum albumin (BSA), Crooks et al., have reported successful analysis of on-chip DNA hybridization events, which have utilized fluorescence spectroscopy to achieve detection limits of less than 5 nM [24]. Using an oPAD approach, the DNA detection system was based on the displacement of a quencher-labeled single-strand of DNA (ssDNA) from a fluorophore labeled ssDNA, upon addition of the analyte. This reaction took place upon the folding of the paper, which brought into contact the different types of labeled ssDNA. The fluorescent signal achieved was directly proportional to the concentration of analyte present. With detection limits in the nanomolar region and relative standard deviations of less than 3% reported, this marks an exciting foray into on-chip paper DNA analysis using fluorescent detection [24].

While these recent examples are exciting and have established a paradigm shift in the capabilities of fluorescence analysis on paper microfluidic devices, they still require the need for external instrumentation, which is a limiting capability for implementing these devices for point-of-care diagnostics. Yun et al. have reported a paper microfluidic fluorescence assay that was self-powered using a microfluidic battery [25]. A galvanic cell was created by pre-depositing silver nitrate, magnesium chloride and barium chromate onto the hydrophilic region of a wax-printed oPAD. After affixing electrodes to the hydrophilic battery regions, it was possible to power an ultra-violet light emitting diode (UV-LED) to induce the fluorescence of alkaline phosphatases on the paper device. Alkaline phosphatases are relevant analytes as they are indicators of a variety of diseases such as liver cancer and hepatitis, etc. Following analysis of the fluorescent responses for the presence of different concentrations of alkaline phosphatases, it was possible to determine a calibration curve dependent on the concentrations observed. Phillips and co-workers published a similar study that used a paper device self-powered by a galvanic battery for the quantitative detection of the enzyme β-D-galactosidase at 700 pM levels [26]. Again this approach emphasises the point that, through the use of an LED and smartphone technology combination, fluorescent assays can be conducted with high sensitivity—and, as such, this approach can expand the application of this technology for a variety of analyses. By avoiding the use of expensive equipment such as excitation light sources and spectrometers, it is possible to produce low-cost, reproducible devices that can utilise fluorescence detection in this manner.

8.5.3 Raman and Surface-Enhanced Raman Spectroscopy (SERS) Detection

Raman scattering is an inherently weak process, with only 1 in 10^6 photons being inelastically scattered. Due to this weakness, the unique fingerprint spectrum afforded by Raman spectroscopy can be masked by auto-fluorescence from the sample or substrate being analysed. Typically, in order to overcome these failings, surface-enhanced Raman spectroscopy has been deployed [27]. SERS overcomes these failings by using a roughened metal surface to provide an increased sensitivity, which negates the problems with fluorescence, encountered using non-surface enhanced mechanisms. Typically roughened metal surfaces such as densely-packed arrays of metallic nanoparticles made from coinage materials, such as gold and silver, are exploited for this technique. In order to utilise SERS with paper microfluidics, metal nanoparticles are typically deposited onto the paper surface. Methods for depositing the nanoparticles onto the surface range from the simplistic approach of dip-coating the paper device in a concentrated solution of nanoparticles to using a low-cost inkjet printer to deposit 'nanoparticle ink' into specific reaction zones.

SERS analysis has mainly been used with μPAD technology for the trace detection of substances of abuse. Narcotics such as cocaine and heroin were both detected using paper based SERS devices, as reported by Yu and White et al. Through creation of a nanoparticle ink, silver nanoparticles were deposited onto a reaction zone configured on a paper dipstick [28]. The dipstick was then used to swab different concentrations of the narcotic components prior to being allowed to dry and then analysed using a 785 nm excitation wavelength. Following analysis, this nanoparticle-dipstick test was shown to permit the detection of cocaine and heroin in the ng mL^{-1} range. This limit of detection range is comparable to the current detection ranges for such analytes through the use of more complex and expensive techniques. Ngo et al. have utilised SERS detection with paper microfluidics to build a bio-diagnostic platform [29]. Gold nanoparticles were absorbed onto the paper to create a metal enhanced substrate for analysis. Following sequential functionalisation steps, involving the addition of functionalising ligands and antibody receptors to the particles, an immunoassay platform was created. This preliminary immunoassay platform was used to detect the interactions between specific antibodies and antigens such as rabbit IgG antigen using SERS. The detection of spectral bands related to specific amino acid residues was used to confirm that binding between the antibody and antigen had occurred on the paper device.

Chen et al. have also reported the fabrication of an immunoassay detection platform on a paper device. Passivation of the paper surface using poly-vinyl pyrrolidone (PVP) was first completed followed by the drop-wise addition of silver nanoparticles which had been pre-concentrated and then aggregated to ensure that an increased SERS enhancement was achieved. A magnetic-bead based immunoassay was conducted 'off-chip', involving the formation of a sandwich complex of mouse IgG antibodies and antigens. The anti-mouse IgG antibodies were labelled with alkaline phosphatase, which underwent a catalytic reaction when the aromatic compound BCIP was added. This produced a dimeric BCIP substrate, which showed an enhanced surface-enhanced *resonant* Raman spectroscopy (SERRS) response. This solution was added to the enhanced paper substrate and the SERRS spectra recorded. The concentration of dimeric BCIP present was in direct correlation with the concentration of mouse IgG antibody present in the immunoassay solution. The immunoassay solution was then spotted onto the enhanced paper substrate and the SERRS spectra were recorded. A linear range of detection of mouse IgG of 1–500 ng mL^{-1} was achieved. This assay was also extended to other important biomarkers such as α-fetoprotein (AFP) and carcinoembryonic antigen (CEA), with detection limits of sub 20 and 5 ng mL^{-1} levels being achieved, respectively.

The limiting factors that have prevented the wider implementation of SERS for paper microfluidic analysis are the loss of key functionalities such as on-device pre-concentration and separation of analytes. These key processes are inhibited due to the covering of the paper substrate with an enhancing material. However, Abbas et al. have produced a SERS-based paper microfluidic device that encompasses both of these unique functionalities, whilst affording unprecedented sensitivities in the attomolar range. Faster drying of the solvent occurs at the tips, in comparison with the bulk of the paper substrate. As a result of this drying process, rapid flow is induced from the wet area to the dry area, resulting in the concentration of analytes into the test regions. Further to this process, concentration and separation of the analytes present was achieved by exposing the tips of the paper substrate to a polyelectrolyte solution—which forms a concentration gradient of polyelectrolyte charge along the tips of the paper platform and permits the subsequent separation of the analytes, due to electrostatic interactions with charged substrates.

Recent studies by Craig et al. have returned the focus to implementing Raman spectroscopy as a detection technology for paper microfluidics [30]. To circumvent the auto-fluorescence that plagues the coupling of standard Raman spectroscopy with paper microfluidics, this study implemented wavelength-modulated Raman spectroscopy (WMRS). WMRS involves the recording of a series of spectra whilst switching the excitation wavelength by less than 1 nm, which results in a significant increase in the signal to noise ratio of the spectra and, while the Raman-active bands shift with the excitation wavelength, the fluorescence remains insensitive to small wavelength changes and, as a result, it can be easily removed. Through the use of WMRS the Raman spectral features of the paper substrate become unmasked and, in turn, the spectral features of the analytes swabbed onto the paper surface are also revealed. Craig et al. capitalised on the benefits of using WMRS in this study to identify pharmaceutical compounds at nano-molar concentrations (see Fig. 8.6) [30].

The emergence of SERS analysis in this field could yield significant advances in the sensitivity of current paper microfluidic devices. However, until miniaturization of expensive SERS instrumentation can be achieved at an affordable cost it will remain on the periphery of techniques deployed for point-of-care diagnostic analyses.

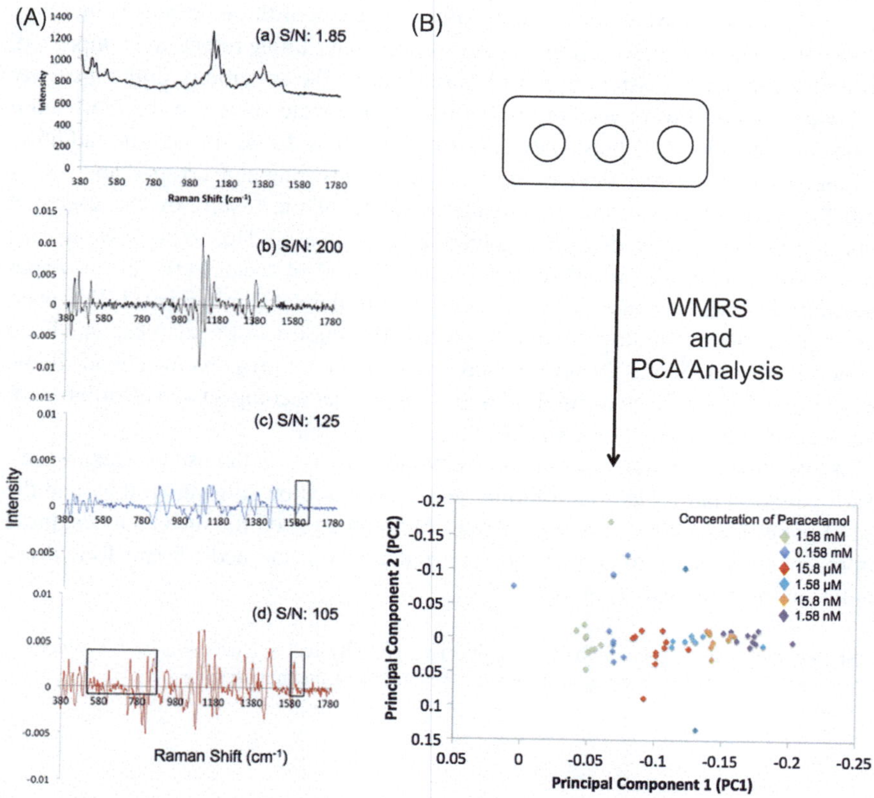

Fig. 8.6 Wavelength modulated Raman spectroscopy (WMRS) was employed for the quantitative detection of paracetamol solutions of different concentration. **a** Displays the spectra obtained from (a) Standard Raman spectroscopy (785 nm) of the microfluidic paper device, (b) WMRS spectra of the paper device and (c), (d) represent the spectra obtained with paracetamol (c) and ibuprofen (d) on the paper substrate respectively. **b** The type of paper device used in the experiment and following swabbing of the paracetamol and WMRS analysis it was possible to obtain a PCA plot of the detected concentrations of paracetamol. Figure adapted and reproduced from Craig et al., PLoS ONE 2015, 10(5): e0123334; https://doi.org/10.1371/journal.pone.0123334

8.6 Challenges and Future Directions

Although there are many advantages in using paper microfluidics, this technology remains in the early stages of development. As such, this technology will continue to evolve, as will the detection technologies that each device will incorporate. In the meantime there are several limitations that paper microfluidics must overcome. The evaporation of the sample from microfluidic channels has a destructive affect on sample retention. Since one of the main advantages of these devices is that they require very small sample volumes, it is essential that sample retention is improved. Currently problems exist in terms of variations in the specificity and sensitivity of devices. These variations can result from discrepancies between the fabrication of batches of individual paper devices, user error introduced through the addition of reagents to the paper device due to common instrumental errors—and issues with the device read-out due to misinterpretation of the results by the user and poor interfacing with the detection technologies being used. Due to these variations, false positives and false negatives can occur. In an ideal scenario the use of paper microfluidic devices would be become common in the developing world. However, current detection technologies being explored still require vast resources, which are economically not viable. Therefore, there exists scope for progress to be made in the development of detection technologies that can be made compact and affordable for use in developing areas of the world.

In this chapter we have shown some emergent themes in the use of Raman analysis—using, in particular, microfluidic platforms based on soft lithography and the use of paper-based platforms. As we see a burgeoning need for such analysis, these approaches are likely to stay at the forefront for both biomedical and food/drink studies in the foreseeable future.

Acknowledgements We thank the UK Engineering and Physical Sciences Research Council for funding and Graham Bruce for assistance with preparation of the manuscript.

References

1. M. Fleischmann, P.J. Hendra, Aj. McQuilla, Chem. Phys. Lett. **26**, 163–166 (1974); D.L. Jeanmaire, R.P. Van Duyne, J. Electroanal. Chem. **84**, 1–20 (1977); M.G. Albrecht, J.A. Creighton, J. Am. Chem. Soc. **99**, 5215–5217 (1977)
2. S. Schlücker, Angew. Chem. Inte. Ed. **53**, 4756–4795 (2014); J. Kneipp, H. Kneipp, K. Kneipp, Chem. Soc. Rev. **37**, 1052–1060 (2008); J.P.R. Day, K.F. Domke, G. Rago, H. Kano, H.-O. Hamaguchi, E.M. Vartiainen, M. Bonn, J. Phys. Chem. B **115**, 7713–7725 (2011)
3. J. Zhao, H. Lui, D.I. McLean, H. Zeng, Appl. Spectrosc. **61**, 1225–1232 (2007); M.G. Shim, B.C. Wilson, J. Raman Spectrosc. **28**, 131–142 (1997); Z. Huang, H. Lui, X.K. Chen, A. Alajlan, D.I. McLean, H. Zeng, J. Biomed. Opt. **9**, 1198–1205 (2004)
4. P.C. Ashok, A.C. De Luca, M. Mazilu, K. Dholakia, J. Biophotonics **4**, 514–518 (2011); A.C. De Luca, M. Mazilu, A. Riches, C.S. Herrington, K. Dholakia, Anal. Chem. **82**, 738–745 (2010); M. Mazilu, A.C. De Luca, A. Riches, C.S. Herrington, K. Dholakia, Opt. Express **18**, 11382–11395 (2010)

5. S.K.Y. Tang, G.M. Whitesides, in *Optofluidics: Fundamentals, Devices, and Applications*, ed. by Y. Falnman, pp. 0-07-160156-2
6. P.R.T. Jess, V. Garces-Chavez, D. Smith, M. Mazilu, L. Paterson, A. Riches, C.S. Herrington, W. Sibbett, K. Dholakia, Opt. Express **14**, 5779–5791 (2006); S. Rao, S. Raj, S. Balint, C.B. Fons, S. Campoy, M. Llagostera, D. Petrov, Appl. Phys. Lett. **96** (2010); S. Dochow, C. Krafft, U. Neugebauer, T. Bocklitz, T. Henkel, G. Mayer, J. Albert, J. Popp, Lab Chip **11**, 1484–1490 (2011)
7. P.J. Viskari, J.P. Landers, Electrophoresis **27**, 1797–1810 (2006)
8. J.T. Motz, M. Hunter, L.H. Galindo, J.A. Gardecki, J.R. Kramer, R.R. Dasari, M.S. Feld, Appl. Opt. **43**, 542–554 (2004)
9. P.C. Ashok, G.P. Singh, H.A. Rendall, T.F. Krauss, K. Dholakia, Lab Chip **11**, 1262–1270 (2011)
10. S. Mozharov, A. Nordon, J.M. Girkin, D. Littlejohn, Lab Chip **10**, 2101–2107 (2010)
11. P. Ashok, Dholakia K, in *Optical Nano- and Microsystems for Bioanalytics*, ed. by W. Fritzsche, J. Popp (Springer, Berlin, Heidelberg, 2012), pp. 247–268
12. P.C. Ashok, G.P. Singh, K.M. Tan, K. Dholakia, Opt. Express **18**, 7642–7649 (2010)
13. P.C. Ashok, B.B. Praveen, K. Dholakia, Opt. Express **19**, 22982–22992 (2011)
14. A.W. Martinez, S.T. Phillips, M.J. Butte, G.M. Whitesides, Angew. Chem. Int. Ed. **46**, 1318–1320 (2007)
15. A.W. Martinez, S.T. Phillips, G.M. Whitesides, Proc. Natl. Acad. Sci. U.S.A. **105**, 19606–19611 (2008)
16. H. Liu, R.M. Crooks, J. Am. Chem. Soc. **133**, 17564–17566 (2011)
17. H. Liu, Y. Xiang, Y. Lu, R.M. Crooks, Angew. Chem. Int. Ed. **51**, 6925–6928 (2012)
18. X. Li, J.F. Tian, T. Nguyen, W. Shen, Anal. Chem. **80**, 9131–9134 (2008)
19. H. Noh, S.T. Phillips, Anal. Chem. **82**, 8071–8078 (2010); A.W. Martinez, S.T. Phillips, Z.H. Nie, C.M. Cheng, E. Carrilho, B.J. Wiley, G.M. Whitesides, Lab Chip **10**, 2499–2504 (2010)
20. B. Lutz, T. Liang, E. Fu, S. Ramachandran, P. Kauffman, P. Yager, Lab Chip **13**, 2840–2847 (2013)
21. M. Li, J. Tian, M. Al-Tamimi, W. Shen, Angew. Chem. Int. Ed. **51**, 5497–5501 (2012)
22. T.S. Park, W.Y. Li, K.E. McCracken, J.Y. Yoon, Lab Chip **13**, 4832–4840 (2013); G. Wu, J. Srivastava, M.H. Zaman, Anal. Biochem. **449**, 147–154 (2014); A.D. Warren, G.A. Kwong, D.K. Wood, K.Y. Lin, S.N. Bhatia, Proc. Natl. Acad. Sci. U.S.A. **111**, 3671–3676 (2014)
23. M.Y. He, Z.H. Liu, Anal. Chem. **85**, 11691–11694 (2013)
24. K. Scida, B.L. Li, A.D. Ellington, R.M. Crooks, Anal. Chem. **85**, 9713–9720 (2013)
25. Y. Koo, J. Sankar, Y. Yun, Biomicrofluidics **8** (2014)
26. N.K. Thom, G.G. Lewis, K. Yeung, S.T. Phillips, RSC Adv. **4**, 1334–1340 (2014)
27. J.F. Betz, W.W. Yu, Y. Cheng, I.M. White, G.W. Rubloff, Phys. Chem. Chem. Phys. **16**, 2224–2239 (2014)
28. W.W. Yu, I.M. White, Analyst **138**, 1020–1025 (2013)
29. N. Ying Hui, T. Whui Lyn, W. Shen, G. Garnier, J. Colloid Interface Sci. **409**, 59–65 (2013)
30. D. Craig, M. Mazilu, K. Dholakia, PLoS One **10**, e0123334 (2015)

Index

R. De La Rue et al. (eds.), *Biomedical Optical Sensors*,
Biological and Medical Physics, Biomedical Engineering,
https://doi.org/10.1007/978-3-030-48387-6

The manufacturer's authorised representative in the EU is Springer Nature Customer Service Centre GmbH, Europaplatz 3, 69115 Heidelberg, Germany. If you have any concerns regarding our products, please contact ProductSafety@springernature.com

Printed and bound by CPI Group (UK) Ltd, Croydon, CR0 4YY
22/09/2026
02220948-0014